ORGANIC SYNTHESES

ORGANIC SYNTHESES

AN ANNUAL PUBLICATION OF SATISFACTORY
METHODS FOR THE PREPARATION OF
ORGANIC CHEMICALS
VOLUME 90
2013

DAVID L. HUGHES
VOLUME EDITOR

ORGANIC SYNTHESES

VOLUME	VOLUME EDITOR	PAGES
I*	† ROGER ADAMS	84
II*	† JAMES BRYANT CONANT	100
III*	† HANS THACHER CLARK	105
IV*	† OLIVER KAMM	89
V*	† CARL SHIPP MARVEL	110
VI*	† HENRY GILMAN	120
VII*	† FRANK C. WHITMORE	105
VIII*	† ROGER ADAMS	139
IX*	† JAMES BRYANT CONANT	108
Collective Vol. I	A revised edition of Annual Volumes I-IX	580
	†HENRY GILMAN, *Editor-in-Chief*	
	2nd Edition revised by † A. H. BLATT	
X*	† HANS THACHER CLARKE	119
XI*	† CARL SHIPP MARVEL	106
XII*	† FRANK C. WHITMORE	96
XIII*	† WALLACE H. CAROTHERS	119
XIV	† WILLIAM W. HARTMAN	100
XV*	† CARL R. NOLLER	104
XVI*	† JOHN R. JOHNSON	104
XVII*	† L. F. FIESER	112
XVIII*	† REYNOLD C. FUSON	103
XIX*	† JOHN R. JOHNSON	105
Collective Vol. II	A revised edition of Annual volumes X-XIX	654
	† A. H. BLATT, *Editor-in-Chief*	
20*	† CHARLES F. H. ALLEN	113
21*	† NATHAN L. DRAKE	120
22*	† LEE IRVIN SMITH	114
23*	† LEE IRVIN SMITH	124
24*	† NATHAN L. DRAKE	119
25 *	† WERNER E. BACHMANN	120
26*	† HOMER ADKINS	124
27*	† R. L. SHRINER	121
28*	† H. R. SNYDER	121
29*	† CLIFF S. HAMILTON	119
Collective Vol. III	A revised edition of Annual Volumes 20–29	890
	† E. C. HORNING, *Editor-in-Chief*	
30*	† ARTHUR C. COPE	115
31*	† R. S. SCHREIBER	122
32*	† RICHARD ARNOLD	119
33*	† CHARLES PRICE	115
34*	† WILLIAM S. JOHNSON	121
35*	† T. L. CAIRNS	122

Out of print.
†*Deceased.*

Out of print.
†*Deceased.*

*Out of print.
†Deceased.

NOTICE

Beginning with Volume 84, the Editors of *Organic Syntheses* initiated a new publication protocol, which is intended to shorten the time between submission of a procedure and its appearance as a publication. Immediately upon completion of the successful checking process, procedures are assigned volume and page numbers and are then posted on the Organic Syntheses website (www.orgsyn.org). The accumulated procedures from a single volume are assembled once a year and submitted for publication. The annual volume is published by John Wiley and Sons, Inc., and includes an index. The hard cover edition is available for purchase through the publisher. Incorporation of graphical abstracts into the Table of Contents began with Volume 77. Annual volumes 70–74, 75–79 and 80–84 have been incorporated into five-year versions of the collective volumes of *Organic Syntheses*. Collective Volumes IX, X and XI are available for purchase in the traditional hard cover format from the publishers.

Beginning with Volume 88, a new type of article, referred to as Discussion Addenda, appeared. In these articles submitters are provided the opportunity to include updated discussion sections in which new understanding, further development, and additional application of the original method are described. Organic Syntheses intends for Discussion Addenda to become a regular feature of future volumes.

Organic Syntheses, Inc., joined the age of electronic publication in 2001 with the release of its free web site (www.orgsyn.org). The site is accessible through internet browsers using Macintosh and Windows operating systems, and the database can be searched by key words and sub-structure. John Wiley & Sons, Inc., and Accelrys, Inc., partnered with Organic Syntheses, Inc., to develop a database (www.mrw.interscience.wiley.com/osdb) that is available for license with internet solutions from John Wiley & Sons, Inc. and intranet solutions from Accelrys, Inc.

Both the commercial database and the free website contain all annual and collective volumes and indices of *Organic Syntheses*. Chemists can draw structural queries and combine structural or reaction transformation queries with full-text and bibliographic search terms, such as chemical name, reagents, molecular formula, apparatus, or even hazard warnings or phrases. The preparations are categorized into reaction types, allowing search by category. The contents of individual or collective volumes can be browsed by lists of titles, submitters' names, and volume and page references, with or without structures.

The commercial database at www.mrw.interscience.wiley.com/osdb also enables the user to choose his/her preferred chemical drawing package, or to utilize several freely available plug-ins for entering queries. The user is also able to cut and paste existing structures and reactions directly into the structure search query or their preferred chemistry editor, streamlining workflow. Additionally, this database contains links to the full text of primary literature references via CrossRef, ChemPort, Medline, and ISI Web of Science. Links to local holdings for institutions using open url technology can also be enabled. The database user can limit his/her search to, or order the search results by, such factors as reaction type, percentage yield, temperature, and publication date, and can create a customized table of reactions for comparison. Connections to other Wiley references are currently made via text search, with cross-product structure and reaction searching to be added in the near future. Incorporations of new preparations will occur as new material becomes available.

INFORMATION FOR AUTHORS OF PROCEDURES

Organic Syntheses welcomes and encourages submissions of experimental procedures that lead to compounds of wide interest or that illustrate important new developments in methodology. Proposals for *Organic Syntheses* procedures will be considered by the Editorial Board upon receipt of an outline proposal as described below. A full procedure will then be invited for those proposals determined to be of sufficient interest. These full procedures will be evaluated by the Editorial Board, and if approved, assigned to a member of the Board for checking. In order for a procedure to be accepted for publication, each reaction must be successfully repeated in the laboratory of a member of the Editorial Board at least twice, with similar yields (generally ±5%) and selectivity to that reported by the submitters.

Organic Syntheses Proposals

A cover sheet should be included providing full contact information for the principal author and including a scheme outlining the proposed reactions (an *Organic Syntheses* Proposal Cover Sheet can be downloaded at orgsyn.org). Attach an outline proposal describing the utility of the methodology and/or the usefulness of the product. Identify and reference the best current alternatives. For each step, indicate the proposed scale, yield, method of isolation and purification, and how the purity of the product is determined. Describe any unusual apparatus or techniques required, and any special hazards associated with the procedure. Identify the source of starting materials. Enclose copies of relevant publications (attach pdf files if an electronic submission is used).

Submit proposals by mail or as e-mail attachments to:

Professor Charles K. Zercher
Associate Editor, *Organic Syntheses*
Department of Chemistry
University of New Hampshire
23 Academic Way, Parsons Hall
Durham, NH 03824

For electronic submissions: *org.syn@unh.edu*

Submission of Procedures

Authors invited by the Editorial Board to submit full procedures should prepare their manuscripts in accord with the Instructions to Authors which are described below or may be downloaded at orgsyn.org. Submitters are also encouraged to consult this volume of *Organic Syntheses* for models with regard to style, format, and the level of experimental detail expected in *Organic Syntheses* procedures. Manuscripts should be submitted to the Associate Editor. Electronic submissions are encouraged; procedures will be accepted as e-mail attachments in the form of Microsoft Word files with all schemes and graphics also sent separately as ChemDraw files.

Procedures that do not conform to the Instructions to Authors with regard to experimental style and detail will be returned to authors for correction. Authors will be notified when their manuscript is approved for checking by the Editorial Board, and it is the goal of the Board to complete the checking of procedures within a period of no more than six months.

Additions, corrections, and improvements to the preparations previously published are welcomed; these should be directed to the Associate Editor. However, checking of such improvements will only be undertaken when new methodology is involved.

NOMENCLATURE

Both common and systematic names of compounds are used throughout this volume, depending on which the Volume Editor felt was more appropriate. The Chemical Abstracts indexing name for each title compound, if it differs from the title name, is given as a subtitle. Systematic Chemical Abstracts nomenclature, used in the Collective Indexes for the title compound and a selection of other compounds mentioned in the procedure, is provided in an appendix at the end of each preparation. Chemical Abstracts Registry numbers, which are useful in computer searching and identification, are also provided in these appendices.

ACKNOWLEDGMENT

Organic Syntheses wishes to acknowledge the contributions of Amgen, Inc. and Merck & Co. to the success of this enterprise through their support, in the form of time and expenses, of members of the Board of Editors.

INSTRUCTIONS FOR AUTHORS

All organic chemists have experienced frustration at one time or another when attempting to repeat reactions based on experimental procedures found in journal articles. To ensure reproducibility, *Organic Syntheses* requires experimental procedures written with considerably more detail as compared to the typical procedures found in other journals and in the "Supporting Information" sections of papers. In addition, each *Organic Syntheses* procedure is carefully "checked" for reproducibility in the laboratory of a member of the Board of Editors.

Even with these more detailed procedures, the experience of *Organic Syntheses* editors is that difficulties often arise in obtaining the results and yields reported by the submitters of procedures. To expedite the checking process and ensure success, we have prepared the following "Instructions for Authors" as well as a **Checklist for Authors** and **Characterization Checklist** to assist you in confirming that your procedure conforms to these requirements. Please include a completed Checklist together with your procedure at the time of submission. Procedures submitted to *Organic Syntheses* will be carefully reviewed upon receipt and procedures lacking any of the required information will be returned to the submitters for revision.

Scale and Optimization

The appropriate scale for procedures will vary widely depending on the nature of the chemistry and the compounds synthesized in the procedure. However, some general guidelines are possible. For procedures in which the principal goal is to illustrate a synthetic method or strategy, it is expected, in general, that the procedure should result in **at least 5 g** and no more than 50 g of the final product. In cases where the point of the procedure is to provide an efficient method for the preparation of a useful reagent or synthetic building block, the appropriate scale may be larger, but in general should not exceed 100 g of final product. Exceptions to these guidelines may be granted in special circumstances. For example, procedures describing the preparation of reagents employed as catalysts will often be acceptable on a scale of less than 5 g.

In considering the scale for an *Organic Syntheses* procedure, authors should also take into account the cost of reagents and starting materials. In general, the Editors will not accept procedures for checking in which the cost of any one of the reactants exceeds **$500** for a single full-scale run. Authors are requested to identify the most expensive reagent or starting material on the procedure submission checklist and to estimate its cost per run of the procedure.

It is expected that all aspects of the procedure will have been optimized by the authors prior to submission, and it is required that each reaction will have been carried out at least twice on exactly the scale described in the procedure, and with the results reported in the manuscript.

It is appropriate to report the weight, yield, and purity of the product of each step in the procedure as a range. In any case where a reagent is employed in significant excess, a Note should be included explaining why an excess of that reagent is necessary. If possible, the Note should indicate the effect of using amounts of reagent less than that specified in the procedure.

The Checking Process

A unique feature of papers published in *Organic Syntheses* is that each procedure and all characterization data is carefully checked for reproducibility in the laboratory of a member of the Board of Editors. In the event that an editor finds it necessary to make any modifications in an experimental procedure, then the published article incorporates the modified procedure, with an explanation and mention of the original protocol often included in a Note. The yields reported in the published article are always those obtained by the checkers. In general, the characterization data in the published article also is that of the checkers, unless there are significant differences with the data obtained by the authors, in which case the author's data will also be reported in a Note.

Reaction Apparatus

Describe the size and type of flask (number of necks) and indicate how *every* neck is equipped.

"A 500-mL, three-necked, round-bottomed flask equipped with an 3-cm Teflon-coated magnetic stirbar, a 250-mL pressure-equalizing addition funnel fitted with an argon inlet, and a rubber septum is charged with ... ."

Indicate how the reaction apparatus is dried and whether the reaction is conducted under an inert atmosphere. Note that balloons are not acceptable as a means of maintaining an inert atmosphere. The description of the reaction apparatus can be incorporated in the text of the procedure or included in a Note.

"The apparatus is flame-dried and maintained under an atmosphere of argon during the course of the reaction."

In the case of procedures involving unusual glassware or especially complicated reaction setups, authors are encouraged to include a photograph or drawing of the apparatus in the text or in a Note (for examples, see *Org. Syn.*, Vol. 82, 99 and Coll. Vol. X, pp 2, 3, 136, 201, 208, and 669).

Use of Gloveboxes

When a glovebox is employed in a procedure, justification must be provided in a Note and the consequences of carrying out the operation without using a glovebox should be discussed.

Reagents and Starting Materials

All chemicals employed in the procedure must be commercially available or described in an earlier *Organic Syntheses* or *Inorganic Syntheses* procedure. For other compounds, a procedure should be included either as one or more steps in the text or, in the case of relatively straightforward preparations of reagents, as a Note. In the latter case, all requirements with regard to characterization, style, and detail also apply. Authors are encouraged to consult with the Associate Editor if they have any question as to whether to include such steps as part of the text or as a Note.

In one or more Notes, indicate the purity or grade of each reagent, solvent, etc. It is highly desirable to also indicate the source (company the chemical was purchased from), particularly in the case of chemicals where it is suspected that the composition (trace impurities, etc.) may vary from one supplier to another. In cases where reagents are purified, dried, "activated" (e.g., Zn dust), etc., a detailed description of the procedure used should be included in a Note. In other cases, indicate that the chemical was "used as received".

> "Diisopropylamine (99.5%) was obtained from Aldrich Chemical Co., Inc. and distilled under argon from calcium hydride before use. THF (99+%) was obtained from Mallinckrodt, Inc. and distilled from sodium benzophenone ketyl. Diethyl ether (99.9%) was purchased from Aldrich Chemical Co., Inc. and purified by pressure filtration under argon through activated alumina. Methyl iodide (99%) was obtained from Aldrich Chemical Co., Inc. and used as received."

The amount of each reactant must be provided in parentheses in the order mL, g, mmol, and equivalents with careful consideration to the correct number of **significant figures**. Avoid indicating amounts of reactants with more significant figures than makes sense. For example, "437 mL of THF" implies that the amount of solvent must be measured with a level of precision that is unlikely to affect the outcome of the reaction. Likewise, "5.00 equiv" implies that an amount of excess reagent must be controlled to a precision of 0.01 equiv.

The concentration of solutions should be expressed in terms of molarity or normality, and not percent (e.g., 1 N HCl, 6 M NaOH, not "10% HCl").

Reaction Procedure

Describe every aspect of the procedure clearly and explicitly. Indicate the order of addition and time for addition of all reagents and how each is added (via syringe, addition funnel, etc.).

Indicate the temperature of the reaction mixture (preferably internal temperature). Describe the type of cooling (e.g., "dry ice-acetone bath") and heating (e.g., oil bath, heating mantle) methods employed. Be careful to describe clearly all cooling and warming cycles, including initial and final temperatures and the time interval involved.

Describe the appearance of the reaction mixture (color, homogeneous or not, etc.) and describe all significant changes in appearance during the course of the reaction (color changes, gas evolution, appearance of solids, exotherms, etc.).

Indicate how the reaction can be monitored to determine the extent of conversion of reactants to products. In the case of reactions monitored by TLC, provide details in a Note, including eluent, R_f values, and method of visualization. For reactions followed by GC, HPLC, or NMR analysis, provide details on analysis conditions and relevant diagnostic peaks.

"The progress of the reaction was followed by TLC analysis on silica gel with 20% EtOAc-hexane as eluent and visualization with *p*-anisaldehyde. The ketone starting material has $R_f = 0.40$ (green) and the alcohol product has $R_f = 0.25$ (blue)."

Reaction Workup

Details should be provided for reactions in which a "quenching" process is involved. Describe the composition and volume of quenching agent, and time and temperature for addition. In cases where reaction mixtures are added to a quenching solution, be sure to also describe the setup employed.

"The resulting mixture was stirred at room temperature for 15 h, and then carefully poured over 10 min into a rapidly stirred, ice-cold aqueous solution of 1 N HCl in a 500-mL Erlenmeyer flask equipped with a magnetic stirbar."

For extractions, the number of washes and the volume of each should be indicated as well as the size of the separatory funnel.

For concentration of solutions after workup, indicate the method and pressure and temperature used.

"The reaction mixture is diluted with 200 mL of water and transferred to a 500-mL separatory funnel, and the aqueous phase is separated and extracted with three 100-mL portions of ether. The combined organic layers are washed with 75 mL of water and 75 mL of saturated NaCl solution, dried over 25 g of $MgSO_4$, filtered through a 250-mL medium porosity sintered glass funnel, and concentrated by rotary evaporation (25 °C, 20 mmHg) to afford 3.25 g of a yellow oil."

"The solution is transferred to a 250-mL, round-bottomed flask equipped with a magnetic stirbar and a 15-cm Vigreux column fitted with a short path distillation head, and then concentrated by careful distillation at 50 mmHg (bath temperature gradually increased from 25 to 75 °C)."

In cases where solid products are filtered, describe the type of filter funnel used and the amount and composition of solvents used for washes.

" … and the resulting pale yellow solid is collected by filtration on a Büchner funnel and washed with 100 mL of cold (0 °C) hexane."

When solid or liquid compounds are dried under vacuum, indicate the pressure employed (rather than stating "reduced pressure" or "dried *in vacuo*").

"… and concentrated at room temperature by rotary evaporation (20 mmHg) and then at 0.01 mmHg to provide … ."

"The resulting colorless crystals are transferred to a 50-mL, round-bottomed flask and dried overnight in a 100 °C oil bath at 0.01 mmHg."

Purification: Distillation

Describe distillation apparatus including the size and type of distillation column. Indicate temperature (and pressure) at which all significant fractions are collected.

"…. and transferred to a 100-mL, round-bottomed flask equipped with a magnetic stirbar. The product is distilled under vacuum through a 12-cm, vacuum-jacketed column of glass helices (Note 16) topped with a Perkin triangle. A forerun (ca. 2 mL) is collected and discarded, and the desired product is then obtained, distilling at 50–55 °C (0.04-0.07 mmHg) … ."

Purification: Column Chromatography

Provide information on TLC analysis in a Note, including eluent, R_f values, and method of visualization.

Provide dimensions of column and amount of silica gel used; in a Note indicate source and type of silica gel.

Provide details on eluents used, and number and size of fractions.

"The product is charged on a column (5 x 10 cm) of 200 g of silica gel (Note 15) and eluted with 250 mL of hexane. At that point, fraction collection (25-mL fractions) is begun, and elution is continued with 300 mL of 2% EtOAc-hexane (49:1 hexanes:EtOAc) and then 500 mL of 5% EtOAc-hexane (19:1 hexanes:EtOAc). The desired product is obtained in fractions 24–30, which are concentrated by rotary evaporation (25 °C, 15 mmHg) … ."

Purification: Recrystallization

Describe procedure in detail. Indicate solvents used (and ratio of mixed solvent systems), amount of recrystallization solvents, and temperature protocol. Describe how crystals are isolated and what they are washed with. A photograph of the crystalline product is often valuable to indicate the form and color of the crystals.

"The solid is dissolved in 100 mL of hot diethyl ether (30 °C) and filtered through a Büchner funnel. The filtrate is allowed to cool to room temperature, and 20 mL of hexane is added. The solution is cooled at –20 °C overnight and the resulting crystals are collected by suction filtration on a Büchner funnel, washed with 50 mL of ice-cold hexane, and then transferred to a 50-mL, round-bottomed flask and dried overnight at 0.01 mmHg to provide … ."

Characterization

Physical properties of the product such as color, appearance, crystal forms, melting point, etc. should be included in the text of the procedure. Comments on the stability of the product to storage, etc. should be provided in a Note.

In a Note, provide data establishing the identity of the product. This will generally include IR, MS, ^{1}H NMR, and ^{13}C NMR data, and in some cases UV data. Copies of the proton and carbon NMR spectra for the products of each step in the procedure should be submitted showing integration for all resonances. Submission of copies of the NMR spectra for other nuclei are encouraged as appropriate.

In the same Note, provide analytical data establishing the purity of the product. **Note that this data should be obtained for the material on which the yield of the reaction is based**, not for a sample that has been subjected to further purification by chromatography, distillation, or crystallization. Elemental analysis for carbon and hydrogen (and nitrogen if present) agreeing with calculated values within 0.4% is preferred. However, GC data (for distilled or vacuum-transferred samples) and/or HPLC data (for material isolated by column chromatography) may be acceptable in some cases. Provide details on equipment and conditions for GC and HPLC analyses.

In procedures involving non-racemic, enantiomerically enriched products, optical rotations should generally be provided, but **enantiomeric purity must be determined by another method** such as chiral HPLC or GC analysis.

In cases where the product of one step is used without purification in the next step, a Note should be included describing how a sample of the product can be purified and providing characterization data for the pure material. Copies of the proton NMR spectra of both the product both *before* and *after* purification should be submitted.

Hazard Warnings

Any significant hazards should be indicated in a statement at the beginning of the procedure in italicized type. Reference to standard operating procedures for work with organic chemicals such as working in a hood, avoiding skin contact, etc. need not be highlighted with a special caution note. Efforts should be made to avoid the use of toxic and hazardous solvents and reagents when less hazardous alternatives are available.

Discussion Section

The style and content of the discussion section will depend on the nature of the procedure.

For procedures that provide an improved method for the preparation of an important reagent or synthetic building block, the discussion should focus on the advantages of the new approach and should describe and reference all of the earlier methods used to prepare the title compound.

In the case of procedures that illustrate an important synthetic method or strategy, the discussion section should provide a mini-review on the new methodology. The scope and limitations of the method should be discussed, and it is generally desirable to include a table of examples. Please be sure each table is numbered and has a title. Competing methods for accomplishing the same overall transformation should be described and referenced. A brief discussion of mechanism may be included if this is useful for understanding the scope and limitations of the method.

Titles of Articles

In cases where the main thrust of the article is the illustration of a synthetic method of general utility, the title of the article should incorporate reference to that method. Inclusion of the name of the final product is acceptable but not required. In the case of articles where the objective is the preparation of a specific compound of importance (such as a chiral ligand), then the name of that compound should be part of the title.

Examples

Title without name of product:

"Stereoselective Synthesis of 3-Arylacrylates by Copper-Catalyzed Syn Hydroarylation" (*Org. Synth.* **2010**, *87*, 53).

Title including name of final product (note name of product is not required):

"Catalytic Enantioselective Borane Reduction of Benzyl Oximes: Preparation of (S)-1-Pyridin-3-yl-ethylamine Bis Hydrochloride" (*Org. Synth.* **2010**, *87*, 36).

Title where preparation of specific compound is the subject:

"Preparation of (S)-3,3'-Bis-Morpholinomethyl-5,5',6,6',7,7',8,8'-octahydro-1,1'-bi-2-naphthol" (*Org. Synth.* **2010**, *87*, 59).

Style and Format for Text

Articles should follow the style guidelines used for organic chemistry articles published in the ACS journals such as *J. Am. Chem. Soc.*, *J. Org. Chem.*, *Org. Lett.*, etc. as described in the ACS Style Guide (3rd Ed.). The text of the procedure should be created using the Word template available on the *Organic Syntheses* website. Specific instructions with regard to the manuscript format (font, spacing, margins) is available on the website in the "Instructions for Article Template" and embedded within the Article Template itself.

Style and Format for Tables and Schemes

Chemical structures and schemes should be drawn using the standard ACS drawing parameters (in ChemDraw, the parameters are found in the "ACS Document 1996" option) with a maximum full size width of 15 cm (5.9 inches). The graphics files should then be pasted into the Word document at the correct location and the size reduced to 75% using "Format Picture" (Mac) or "Size and Position" (Windows). Graphics files must also be submitted separately. All Tables that include structures should be entirely prepared in the graphics (ChemDraw) program and inserted into the word processing file at the appropriate location. Tables that include multiple, separate graphics files prepared in the word processing program will require modification.

Tables and schemes should be numbered and should have titles. The title for a Table should be included within the ChemDraw graphic and placed immediately above the table. The title for a scheme should be included within the ChemDraw graphic and placed immediately below the scheme. Use 12 point Palatino Bold font in the ChemDraw file for all titles. For footnotes in Tables use Helvetica (or Arial) 9 point font and place these immediately below the Table.

Acknowledgments and Author's Contact Information

Contact information (institution where the work was carried out and mailing address for the principal author) should be included as footnote 1. This footnote should also include the email address for the principal author. Acknowledgment of financial support should be included in footnote 1.

Biographies and Photographs of Authors

Photographs and 100-word biographies of all authors should be submitted as separate files at the time of the submission of the procedure. The format of the biographies should be similar to those in the Volume 84 procedures found at the orgsyn.org website. Photographs can be accepted in a number of electronic formats, including tiff and jpeg formats.

DISPOSAL OF CHEMICAL WASTE

General Reference: *Prudent Practices in the Laboratory* National Academy Press, Washington, D.C. 2011.

Effluents from synthetic organic chemistry fall into the following categories:

1. **Gases**
   1a. Gaseous materials either used or generated in an organic reaction.
   1b. Solvent vapors generated in reactions swept with an inert gas and during solvent stripping operations.
   1c. Vapors from volatile reagents, intermediates and products.

2. **Liquids**
   2a. Waste solvents and solvent solutions of organic solids (see item 3b).
   2b. Aqueous layers from reaction work-up containing volatile organic solvents.
   2c. Aqueous waste containing non-volatile organic materials.
   2d. Aqueous waste containing inorganic materials.

3. **Solids**
   3a. Metal salts and other inorganic materials.
   3b. Organic residues (tars) and other unwanted organic materials.
   3c. Used silica gel, charcoal, filter aids, spent catalysts and the like.

The operation of industrial scale synthetic organic chemistry in an environmentally acceptable manner* requires that all these effluent categories be dealt with properly. In small scale operations in a research or academic setting, provision should be made for dealing with the more environmentally offensive categories.

1a. Gaseous materials that are toxic or noxious, e.g., halogens, hydrogen halides, hydrogen sulfide, ammonia, hydrogen cyanide, phosphine, nitrogen oxides, metal carbonyls, and the like.
1c. Vapors from noxious volatile organic compounds, e.g., mercaptans, sulfides, volatile amines, acrolein, acrylates, and the like.

*An environmentally acceptable manner may be defined as being both in compliance with all relevant state and federal environmental regulations *and* in accord with the common sense and good judgment of an environmentally aware professional

2a. All waste solvents and solvent solutions of organic waste.

2c. Aqueous waste containing dissolved organic material known to be toxic.

2d. Aqueous waste containing dissolved inorganic material known to be toxic, particularly compounds of metals such as arsenic, beryllium, chromium, lead, manganese, mercury, nickel, and selenium.

3. All types of solid chemical waste.

Statutory procedures for waste and effluent management take precedence over any other methods. However, for operations in which compliance with statutory regulations is exempt or inapplicable because of scale or other circumstances, the following suggestions may be helpful.

Gases

Noxious gases and vapors from volatile compounds are best dealt with at the point of generation by "scrubbing" the effluent gas. The gas being swept from a reaction set-up is led through tubing to a large trap to prevent suck-back and into a sintered glass gas dispersion tube immersed in the scrubbing fluid. A bleach container can be conveniently used as a vessel for the scrubbing fluid. The nature of the effluent determines which of four common fluids should be used: dilute sulfuric acid, dilute alkali or sodium carbonate solution, laundry bleach when an oxidizing scrubber is needed, and sodium thiosulfate solution or diluted alkaline sodium borohydride when a reducing scrubber is needed. Ice should be added if an exotherm is anticipated.

Larger scale operations may require the use of a pH meter or starch/iodide test paper to ensure that the scrubbing capacity is not being exceeded.

When the operation is complete, the contents of the scrubber can be poured down the laboratory sink with a large excess (10–100 volumes) of water. If the solution is a large volume of dilute acid or base, it should be neutralized before being poured down the sink.

Liquids

Every laboratory should be equipped with a waste solvent container in which *all* waste organic solvents and solutions are collected. The contents of these containers should be periodically transferred to properly labeled waste solvent drums and arrangements made for contracted disposal in a regulated and licensed incineration facility.**

**If arrangements for incineration of waste solvent and disposal of solid chemical waste by licensed contract disposal services are not in place, a list of providers of such services should be available from a state or local office of environmental protection.

Aqueous waste containing dissolved toxic organic material should be decomposed *in situ*, when feasible, by adding acid, base, oxidant, or reductant. Otherwise, the material should be concentrated to a minimum volume and added to the contents of a waste solvent drum.

Aqueous waste containing dissolved toxic inorganic material should be evaporated to dryness and the residue handled as a solid chemical waste.

Solids

Soluble organic solid waste can usually be transferred into a waste solvent drum, provided near-term incineration of the contents is assured.

Inorganic solid wastes, particularly those containing toxic metals and toxic metal compounds, used Raney nickel, manganese dioxide, etc. should be placed in glass bottles or lined fiber drums, sealed, properly labeled, and arrangements made for disposal in a secure landfill.** Used mercury is particularly pernicious and small amounts should first be amalgamated with zinc or combined with excess sulfur to solidify the material.

Other types of solid laboratory waste including used silica gel and charcoal should also be packed, labeled, and sent for disposal in a secure landfill.

Special Note

Since local ordinances may vary widely from one locale to another, one should always check with appropriate authorities. Also, professional disposal services differ in their requirements for segregating and packaging waste.

HANDLING HAZARDOUS CHEMICALS

A Brief Introduction

General Reference: *Prudent Practices in the Laboratory*; National Academy Press; Washington, DC, 2011.

Physical Hazards

Fire. Avoid open flames by use of electric heaters. Limit the quantity of flammable liquids stored in the laboratory. Motors should be of the nonsparking induction type.

Explosion. Use shielding when working with explosive classes such as acetylides, azides, ozonides, and peroxides. Peroxidizable substances such as ethers and alkenes, when stored for a long time, should be tested for peroxides before use. Only sparkless "flammable storage" refrigerators should be used in laboratories.

Electric Shock. Use 3-prong grounded electrical equipment if possible.

Chemical Hazards

Because all chemicals are toxic under some conditions, and relatively few have been thoroughly tested, it is good strategy to minimize exposure to all chemicals. In practice this means having a good, properly installed hood; checking its performance periodically; using it properly; carrying out all operations in the hood; protecting the eyes; and, since many chemicals can penetrate the skin, avoiding skin contact by use of gloves and other protective clothing at all times.

a. Acute Effects. These effects occur soon after exposure. The effects include burn, inflammation, allergic responses, damage to the eyes, lungs, or nervous system (e.g., dizziness), and unconsciousness or death (as from overexposure to HCN). The effect and its cause are usually obvious and so are the methods to prevent it. They generally arise from inhalation or skin contact, so should not be a problem if one follows the admonition "work in a hood and keep chemicals off your hands". Ingestion is a rare route, being generally the result of eating in the laboratory or not washing hands before eating.

b. Chronic Effects. These effects occur after a long period of exposure or after a long latency period and may show up in any of numerous organs.

Of the chronic effects of chemicals, cancer has received the most attention lately. Several dozen chemicals have been demonstrated to be carcinogenic in man and hundreds to be carcinogenic to animals. Although there is no simple correlation between carcinogenicity in animals and in man, there is little doubt that a significant proportion of the chemicals used in laboratories have some potential for carcinogenicity in man. For this and other reasons, chemists should employ good practices at all times.

The key to safe handling of chemicals is a good, properly installed hood, and the referenced book devotes many pages to hoods and ventilation. It recommends that in a laboratory where people spend much of their time working with chemicals there should be a hood for each two people, and each should have at least 2.5 linear feet (0.75 meter) of working space at it. Hoods are more than just devices to keep undesirable vapors from the laboratory atmosphere. When closed they provide a protective barrier between chemists and chemical operations, and they are a good containment device for spills. Portable shields can be a useful supplement to hoods, or can be an alternative for hazards of limited severity, e.g., for small-scale operations with oxidizing or explosive chemicals.

Specialized equipment can minimize exposure to the hazards of laboratory operations. Impact resistant safety glasses are basic equipment and should be worn at all times. They may be supplemented by face shields or goggles for particular operations, such as pouring corrosive liquids. Because skin contact with chemicals can lead to skin irritation or sensitization or, through absorption, to effects on internal organs, protective gloves should be worn at all times.

Laboratories should have fire extinguishers and safety showers. Respirators should be available for emergencies. Emergency equipment should be kept in a central location and must be inspected periodically.

MSDS (Materials Safety Data Sheets) sheets are available from the suppliers of commercially available reagents, solvents, and other chemical materials; anyone performing an experiment should check these data sheets before initiating an experiment to learn of any specific hazards associated with the chemicals being used in that experiment.

PREFACE

This volume of *Organic Syntheses* represents a milestone – the 90th edition. Very few scientific journals or book series can boast of a series that spans 90+ years, and perhaps none, other than *Organic Syntheses*, operates as a non-profit organization. Starting with an 84-page book of 16 checked procedures edited by Roger Adams in 1921, a volume has been published annually every year except three. The longevity of this series speaks to the need for carefully checked and reliable procedures, as well as to the dedication of the editors, contributors, and checkers over this remarkable life span. The 104 former editors, which may be found on the *Organic Syntheses* webpage, are the scientists we recognize as the pioneers of modern organic chemistry. I am honored to be a volume editor from industry, represented by 23 former editors. I was preceded in this role by three process chemistry legends from Merck, Max Tishler (vol 39), Ichiro Shinkai (vol 74), and Ed Grabowski (vol 82). I would also like to acknowledge Dave Mathre, who served the first 3 years of this 8 year term before his unique expertise was called upon to support large molecule development at Merck.

Max McKeown defines innovation as "*new stuff that is made useful*," noting that "*if your invention is useful to someone then the invention has become an innovation.*"[1] This quote aptly summarizes the mission of *Organic Syntheses*, which is to make creative synthetic methodology truly innovative by ensuring it is practical, reliable, and useful for all practitioners. Importantly, the procedures and discussion sections provide critical details and perspective that can often enable readers to apply the chemistry to their particular substrate of interest.

The importance of the publication of reproducible synthetic methods in *Organic Syntheses* is underscored by concerns in recent years regarding the lack of reproducibility of published scientific work. "*In modern research, false findings may be the majority or even vast majority of published research claims.*" Most, if not all, scientists would find the above statement absurd, yet this quote from Professor John Ioannidis comes from the opening paragraph of an article in which he presents compelling evidence that most published biomedical research is false.[2] The concern with the irreproducibility of results has prompted the NIH to consider having independent labs verify key experimental results.[3] While this is largely focused on *in-vivo* studies and molecular biology, most organic chemists have been frustrated by not being able to reproduce synthetic methods, even though most methodology papers include dozens of examples. The lack of reproducibility is not due to negligence or fraud by the authors, but to the vast number of inter-dependent variables that go into each step of a transformation - perhaps dozens for even a simple procedure. In addition, many published procedures are carried out on no more than a few milligrams, where variables are even harder to define, much less control. The procedures in *Organic Syntheses* are run on a scale (generally > 5g) where experimental conditions can be well-controlled and reproduced.

Of the 30 procedures in Vol 90 of *Organic Syntheses*, 24 were contributed from academic labs and 6 from the pharmaceutical industry. While metal-based reactions have been the mainstay of recent volumes, this trend is reversed in the current volume, with only 11 procedures involving base or precious metals. Of the metal-based catalysis methods, 2 involve C—H activation, an area of current keen interest in the synthetic community. Both procedures involve use of an amide group for ortho-directing C—H functionalization of aromatic rings, including:

- Paureau, Besset, and Glorius, *Preparation of (E)-N,N-Diethyl-2-styryl-benzamide by Rh-catalyzed C—H Activation*, p. 41
- Magano, Kiser, Shine, and Chen, *Oxindole Synthesis via Palladium-catalyzed C—H Functionalization*, p. 74.

Also a contrast from recent volumes, only 5 procedures pertain to asymmetric synthetic methodology. Two contributions describing state-of-the-art asymmetric phase transfer chemistry are from the Maruoka labs, including the preparation of both tertiary and quaternary asymmetric centers:

- Shirakawa, Yamamoto, Liu, and Maruoka, *Enantioselective Alkylation of N-(Diphenylmethylene)glycinate tert-Butyl Ester: Synthesis of (R)-2-(Benzhydrylidenamino)-3-Phenylpropanoic Acid tert-Butyl Ester*, p. 112
- Shirakawa, Yamamoto, Liu, and Maruoka, *Enantioselective Alkylation of 2-[(4-Chlorobenzyliden)Amino]Propanoic Acid tert-Butyl Ester: Synthesis of (R)-2-Amino-2-Methyl-3-Phenylpropanoic Acid tert-Butyl Ester*, p. 121.

Preparation of new reagents that facilitate alternate synthetic approaches has always been an important area for *Organic Syntheses* and this volume continues that tradition with 7 such procedures. Three of these are highlighted below.

Prakash and co-workers designed and developed α-fluorobis(phenyl-sulfonyl)methane as a versatile, selective, and mild nucleophilic mono-fluoromethylating reagent.

- Prakash, Shao, Wang, and Ni, *Preparation of α-Fluorobis(phenyl-sulfonyl)methane (FBSM)*, p. 130.

DABSO, a 1:2 complex between DABCO and sulfur dioxide, is a convenient and safe source of sulfur dioxide. However, its preparation still requires handling of gaseous sulfur dioxide. Martial and Bischoff have now developed a safe and practical alternative for the preparation of DABSO using commercially available Karl-Fischer reagent as the source of sulfur dioxide.

- Martial and Bischoff, *Preparation of DABSO from Karl-Fischer reagent*, p. 301

Dimethyldioxirane is an effective reagent for the preparation of sensitive epoxides but has found limited use, in part due to a cumbersome workup and isolation procedure. The Taber labs have developed a practical procedure for preparation of millimolar quantities of DMDO in acetone. DMDO is isolated by simple rotary evaporation of the reaction slurry with collection of the distillate containing DMDO in the bump bulb of a rotary evaporator – likely a first for *Organic Syntheses*.

- Taber, DeMatteo, and Hassan, *Simplified Preparation of Dimethyldioxirane (DMDO)*, p. 350.

Nearly half of the contributions in this volume can be broadly classified under the heading of achiral synthetic methodology, including 6 novel preparations of heterocycles. The following examples give a flavor of the contributions in this area.

N-Acetyl enamides are key substrates for asymmetric hydrogenations to prepare chiral amines. Several literature methods are available but suffer from limited substrate scope, low yields, or cryogenic conditions. Tang and co-workers describe a straightforward 2-step synthesis of *N*-acyl enamides from ketones via conversion to an oxime followed by iron-mediated reductive acylation.

- Tang, Patel, Capacci, Wei, Yee, Senanayake, *Synthesis of N-Acetyl Enamides by Reductive Acetylation of Oximes Mediated with Iron(II) Acetate:N-(1-(4-Bromophenyl)vinyl)Acetamide*, p. 62.

"Click" chemistry has had a profound impact across many areas of organic chemistry due to its extremely wide scope that allows applications to a variety of small and large molecule motifs. The current contribution from the Fokin labs describes the Ru-catalyzed variant, which provides 1,5-disubstituted 1,2,3-triazoles from terminal alkynes (instead of the 1,4-regiosiomers generated from Cu-catalyzed reactions) and fully substituted 1,2,3-triazoles from internal alkynes, generally with high regioselectivity.

- Oakdale and Fokin, *Preparation of 1,5-Disubstituted 1,2,3-Triazoles via Ruthenium-catalyzed Azide Alkyne Cycloaddition*, p. 96.

The Weix group has contributed a Ni-catalyzed sp2-sp3 cross coupling method to directly couple haloalkanes with haloarenes that does not require any additional organometallic reagents. The carbon nucleophiles are synthesized from the corresponding organic halides, thus eliminating one step. Further, excellent functional group compatibility is achieved by avoiding stoichiometric strong nucleophiles or bases.

- Everson, George, and Weix, *Nickel-Catalyzed Cross-Coupling of Aryl Halides with Alkyl Halides: Ethyl 4-(4-(4-methylphenylsulfonamido)-phenyl)butanoate*, p. 200.

The Stahl group has contributed a safe and greener aerobic oxidation protocol of primary alcohols to aldehydes using a Cu/TEMPO catalyst system, a compelling alternative to traditional oxidation methods.

- Hoover and Stahl, *Air Oxidation of Primary Alcohols Catalyzed by Copper(I)/TEMPO. Preparation of 2-Amino-5-bromobenzaldehyde*

Literature methods to prepare cyclic amines through cyclodehydration of amino alcohols typically involve a tedious sequence of protection/activation/cyclization/deprotection. Xu and Simmons describe a straightforward one-pot cyclodehydration transformation, which is achieved by 'inverse' addition of a solution of the free amino alcohol to a solution of SOCl2 to generate the alkyl chloride intermediate, followed by NaOH-catalyzed ring closure.

- Xu and Simmons, *One-Pot Preparation of Cyclic Amines from Amino Alcohol*, p. 251.

In addition to checked procedures, each volume of *Organic Syntheses* includes Discussion Addenda, four of which are included in this volume. These addenda are intended to provide any new information that has become available in the years since the original *Organic Syntheses* procedure was published, such as updated scope and limitations, new experimental conditions, or a discussion of alternate methods.

In closing, I would like to thank the Board of Editors and, in particular, their co-workers who have diligently checked the procedures. As a reader and user of the procedures provided in *Organic Syntheses* prior to becoming an editor, I was unaware of the time and effort required by the checkers and editors to perfect them. Their dedicated efforts have guaranteed the high quality of each contribution, ensuring the continued success of *Organic Syntheses*. Largely working behind the scenes, the

Board of Directors provides wise oversight, expert fiscal management, and a steadfast commitment to the mission. Finally, a special thanks to Rick Danheiser, Editor-in-Chief, and Chuck Zercher, Associate Editor. Without their creative leadership, their dedication and passion for this enterprise over the past decade, *Organic Syntheses* would not be in the excellent shape it is today.

DAVID HUGHES
Rahway, New Jersey

1. McKeown, Max *The Truth About Innovation*. London, UK: Prentice Hall (2008).
2. Ioannidis, J. P. A. *PLoS Medicine* **2005**, *2*, 696–701.
3. Wadman, M. *Nature* **2013**, *500*, 14–16.

CONTENTS

Enantioselective Nitroaldol (Henry) Reaction of p-Nitrobenzaldehyde and Nitromethane Using a Copper (II) Complex Derived from (R,R)-1,2-Diaminocyclohexane: (1S)-1-(4-Nitrophenyl)- 2-nitroethane-1-ol

52

Antoinette Chougnet and Wolf-D. Woggon*

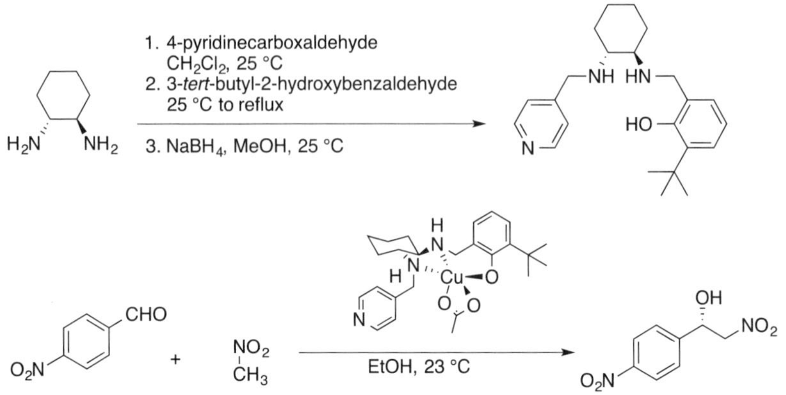

Synthesis of N-Acetyl Enamides by Reductive Acetylation of Oximes Mediated with Iron(II) Acetate: N-(1-(4-Bromophenyl)vinyl)acetamide

62

Wenjun Tang,* Nitinchandra D. Patel, Andew G. Capacci, Xudong Wei, Nathan K. Yee, and Chris H. Senanayake

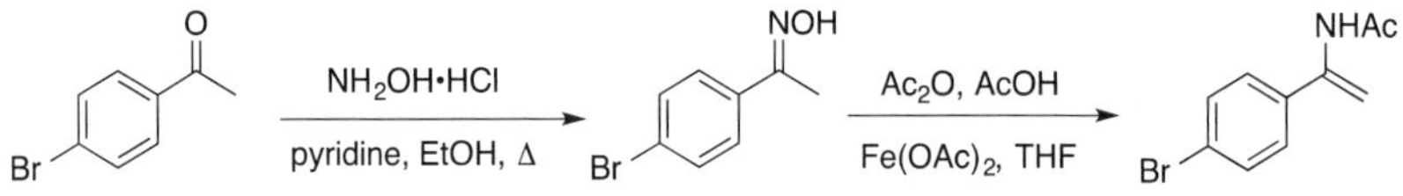

Oxindole Synthesis via Palladium-catalyzed C–H Functionalization

74

Javier Magano,* E. Jason Kiser, Russell J. Shine, and Michael H. Chen

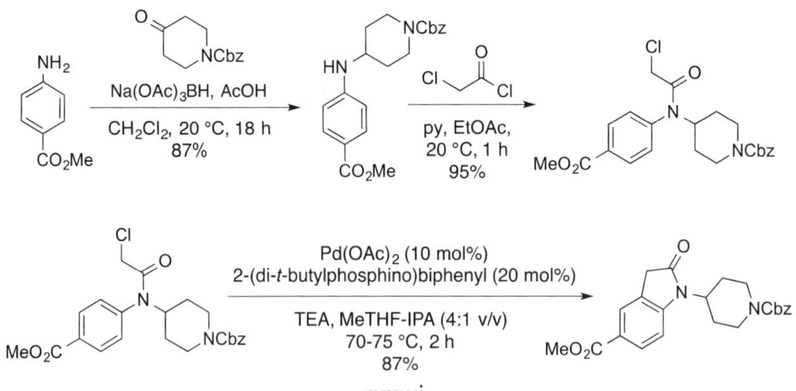

Enantioselective Alkylation of *N*-(Diphenylmethylene) Glycinate *tert*-Butyl Ester: Synthesis of (*R*)-2-(Benzhydryliden-amino)-3-Phenylpropanoic Acid *tert*-Butyl Ester

112

Seiji Shirakawa, Kenichiro Yamamoto, Kun Liu, and
Keiji Maruoka*

Ph–C(N–CH₂–COO*t*-Bu)(Ph) + PhCH₂Br →[(*S*)-**1** (0.1 mol%)] [48% KOH *aq* toluene 0 °C, 20 h] Ph–C(N–CH(COO*t*-Bu)CH₂Ph)(Ph)

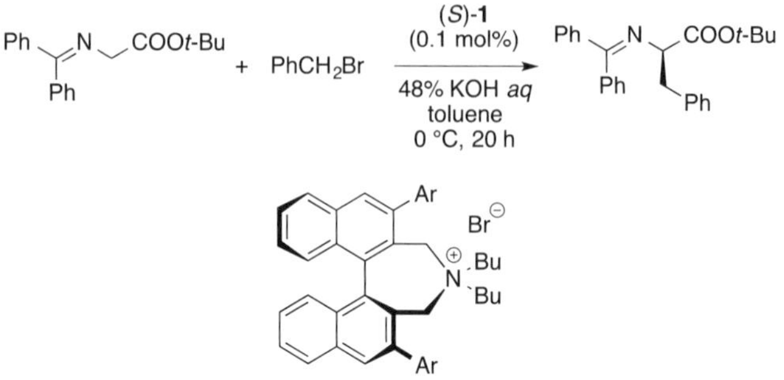

(*S*)-**1** (Ar = 3,4,5-F₃-C₆H₂)

Enantioselective Alkylation of 2-[(4-Chlorobenzyliden) Amino]Propanoic Acid *tert*-Butyl Ester; Synthesis of (*R*)-2-Amino-2-Methyl-3-Phenylpropanoic Acid *tert*-Butyl Ester

121

Seiji Shirakawa, Kenichiro Yamamoto, Kun Liu, and
Keiji Maruoka*

4-Cl-C₆H₄-CHO + Cl⁻ H₃N⁺–C(COO*t*-Bu)(Me) →[NEt₃ toluene 70 °C, 4 h] 4-Cl-C₆H₄-CH=N–C(COO*t*-Bu)(Me)

4-Cl-C₆H₄-CH=N–C(COO*t*-Bu)(Me) + PhCH₂Br →[(*S*)-**1** (0.1 mol%) 80% CsOH *aq* toluene 0 °C, 18 h] →[citric acid THF rt, 3 h] H₂N–C(COO*t*-Bu)(Me)(CH₂Ph)

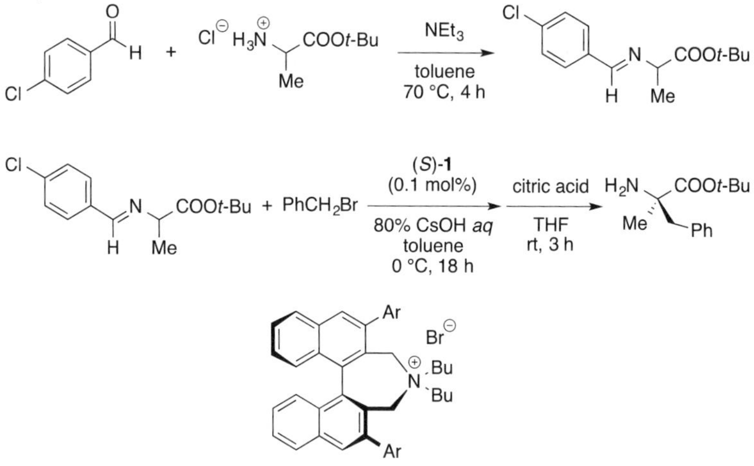

(*S*)-**1** (Ar = 3,4,5-F₃-C₆H₂)

Ar = 4-BrPh

Nickel-Catalyzed Cross-Coupling of Aryl Halides with Alkyl Halides: Ethyl 4-(4-(4-methylphenylsulfonamido)-phenyl)butanoate

Daniel A. Everson, David T. George, and Daniel J. Weix*

Discussion Addendum for: Preparation of 4-Acetylamino-2,2,6,6-tetramethylpiperidine-1-oxoammonium Tetrafluoroborate and the Oxidation of Geraniol to Geranial (2,6-Octadienal, 3,7-dimethyl-, (2e)-)

James M. Bobbitt,* Nicholas A. Eddy, Jay J. Richardson, Stephanie A. Murray, and Leon J. Tilley

Preparation of 3-Oxocyclohex-1-ene-1-carbonitrile

Jesus Armando Lujan-Montelongo and Fraser F. Fleming*

Air Oxidation of Primary Alcohols Catalyzed by Copper(I)/TEMPO. Preparation of 2-Amino-5-bromo-benzaldehyde

Jessica M. Hoover and Shannon S. Stahl*

One-Pot Preparation of Cyclic Amines from Amino Alcohols

Feng Xu* and Bryon Simmons

Preparation of Tetrabutylammonium (4-fluorophenyl)trifluoroborate

Fabrizio Pertusati, Parag V. Jog, and G. K. Surya Prakash*

Allyl Cyanate-To-Isocyanate Rearrangement: Preparation of *tert*-Butyl 3,7-Dimethylocta-1,6-dien-3-ylcarbamate

Yoshiyasu Ichikawa,* Noriko Kariya, and Tomoyuki Hasegawa

Palladium-Catalyzed Triazolopyridine Synthesis: Synthesis of 7-Chloro-3-(2-Chlorophenyl)-1,2,4-Triazolo[4,3-a]Pyridine

Oliver R. Thiel* and Michal M. Achmatowicz

Preparation of DABSO from Karl-Fischer reagent

Ludovic Martial, and Laurent Bischoff*

Low-epimerization Peptide Bond Formation with Oxyma Pure: Preparation of Z-*L*-Phg-Val-OMe

Ramon Subirós-Funosas, Ayman El-Faham,* and Fernando Albericio

W

Synthesis of Koser's Reagent and Derivatives

Submitted by Nazli Jalalian and Berit Olofsson.[1]
Checked by David Yeung, Stephen M. Shaw and Margaret Faul.

1. Procedure

Caution! Reactions and subsequent operations involving peracids and peroxy compounds should be run behind a safety shield. For relatively fast reactions, the rate of addition of the peroxy compound should be slow enough so that it reacts rapidly and no significant unreacted excess is allowed to build up. The reaction mixture should be stirred efficiently while the peroxy compound is being added, and cooling should generally be provided since many reactions of peroxy compounds are exothermic. New or unfamiliar reactions, particularly those run at elevated temperatures, should be run first on a small scale. Reaction products should never be recovered from the final reaction mixture by distillation until all residual active oxygen compounds (including unreacted peroxy compounds) have been destroyed. Decomposition of active oxygen compounds may be accomplished by the procedure described in Korach, M.; Nielsen, D. R.; Rideout, W. H. Org. Synth. 1962, 42, 50 (Org. Synth. 1973, Coll. Vol. 5, 414).

1-[Hydroxy(tosyloxy)iodo]-3-trifluoromethylbenzene (1). A 250-mL single-necked, round-bottomed flask equipped with a magnetic stirring bar (3.5 cm, egg-shaped) is charged with *m*-chloroperbenzoic acid (*m*-CPBA, 4.18 g, 17 mmol, 1.0 equiv) (Note 1) and 2,2,2-trifluoroethanol (TFE, 65 mL) (Note 2). The mixture is dissolved by sonication (Note 3) (approx. 20 min) and the concentration of *m*CPBA is determined by iodometric titration of the solution (Notes 4 and 5). 3-Iodobenzotrifluoride (2.45 mL, 1.71 g, 17.0 mmol, 1.0 equiv) (Note 6) and *p*-toluenesulfonic acid monohydrate (3.23 g, 17.0 mmol, 1.0 equiv) (Note 7) are added to the solution. The flask is loosely capped with a septum, and the reaction mixture is lowered into a 40 °C preheated oil bath and stirred vigorously during 1 h (Notes 8 and 9). The solvent is then removed by distillation to recover the TFE (Note 10), or evaporated under reduced pressure if TFE recovery is not of interest (Notes 11 and 12). After concentration, diethyl ether (65 mL) (Note 13) is added to the residue and the mixture is allowed to stir for 30 min, resulting in trituration of the product as a white solid. The product is collected by suction filtration on a Büchner funnel, followed by washing with diethyl ether (65 mL). The product is vacuum-dried at room temperature (<1 mmHg, 22 °C) to afford compound 1 as a white solid (7.66 g, 97%) (Notes 14 and 15).

2. Notes

1. m-CPBA ($\leq$ 77%) was purchased from Sigma-Aldrich. Typical purity of samples received was 66–69% by weight.
2. TFE ($\geq$99%) was purchased from Sigma-Aldrich and used as received. The reaction is run without precautions to avoid air or moisture.
3. The sonicator used to dissolve m-CPBA was an 85 W ultrasonic cleaner from VWR, model B2500A-DTH.
4. The concentration of active oxidizing agent is determined by iodometric titration.[2] NaI (1.5 g) is dissolved in distilled water (50 mL). Chloroform (5 mL), glacial acetic acid (5 mL) and approximately 0.67 g of m-CPBA solution are added and the mixture is stirred vigorously. The solution is titrated with aq. $Na_2S_2O_3$ (0.100 M) – the brown iodine color fades to yellow and then disappears. As a rough guide, 2 mL $Na_2S_2O_3$ solution is required for 0.5 g of solution. The m-CPBA concentration is calculated using (mol m-CPBA = mol I_2 = 0.5 mol $S_2O_3^{2-}$). Typical concentration of m-CPBA as determined by titration is approximately 3.0 wt%.
5. As an alternative to dissolving the m-CPBA in TFE via sonication for titration, the submitters reported the following procedure: To obtain a homogeneous powder, m-CPBA is dried under vacuum in a round-bottomed flask at room temperature until the flask is no longer cold. The time required depends on the amount of m-CPBA and the vacuum; 5 g requires approximately 1.5 h at 10 mmHg. The dried m-CPBA was stored for prolonged time in a refrigerator. Caution should be exercised, however, as dried m-CPBA is shock sensitive. The reaction performs well without drying of the m-CPBA (86–92% yield), but the titration (see Note 4) of the

heterogeneous *m*-CPBA becomes less reliable than when the dried or sonicated/dissolved material is used.

6. 3-Iodobenzotrifluoride (98%) was purchased from Sigma-Aldrich and used as received.

7. *p*-Toluenesulfonic acid monohydrate (≥98.5%) was purchased from Sigma-Aldrich and used as received.

8. The flask does not need to be sealed. A slightly yellow precipitate is observed after 5 min and the mixture may become difficult to stir as the reaction proceeds and thickens; if this occurs, more TFE may be added. Both **1** and *m*-CBA (the reduced form of *m*-CPBA, which is a by-product of the reaction and present in the starting *m*-CPBA) are poorly soluble in TFE. The disappearance of iodobenzotrifluoride can be followed by TLC using neat pentane (R_f = 0.83 using EMD TLC Silica Gel 60 F_{254}) and visualized by UV.

9. Reactions performed at higher temperatures (60 or 80 °C) provide lower yield due to partial decomposition of **1** (80 °C for 1 h gave 76% yield). The reaction may also be performed at room temperature but requires extended reaction time (2 h reaction time gave 67% yield, 18 h gave 90% yield).

10. TFE is recovered using a short path distillation head and a 25 cm condenser under reduced pressure in a 40 °C oil bath (bp = 28 °C at 86 mmHg). The distillate is collected over ~90 min with the receiving flask immersed in an ice bath. Vacuum should be applied in a controlled manner to prevent foaming. Experiments using microscale distillation resulted in lower recovery of TFE. Higher distillation temperatures should be avoided due to partial decomposition of **1**. The TFE can be recovered almost quantitatively (63 mL, 97%) and can be re-used without loss in yield.

11. The TFE can be removed by evaporation under reduced pressure using a rotary evaporator (25 °C, ~20 mmHg), if recovery of the solvent is not of interest.

12. Unreacted *m*-CPBA, e.g. from charging excess *m*-CPBA, may pose a risk of explosion during evaporation of TFE since dry *m*-CPBA is shock sensitive. It is important that the concentration be carried out using an oil bath or water bath, not a heating mantle, to avoid heating the concentrated solids above 40 °C. Due to the insolubility of **1** in TFE, the isolation is straightforward even if residual TFE remains. Direct filtration of the solid material from the crude reaction, without concentration, gave **1** in 77% yield after washing with diethyl ether (65 mL). Another 18% yield of **1**

was obtained by concentration of the filtrate followed by precipitation with diethyl ether.

13. Diethyl ether (≥99.0%) was purchased from Sigma-Aldrich and used as supplied.

14. The reaction was also performed at 5.0 and 13.0 mmol scale to afford a comparable yield of **1** (2.16 g, 94% and 5.69 g, 95%, respectively).

15. The product is stable to air but somewhat light sensitive and is best stored in a refrigerator in the dark. Analytical data of **1**: mp: 155–157 °C (lit 156–157 °C)[3]; ^{1}H NMR (400 MHz, DMSO-d_6) δ: 2.29 (s, 3 H), 7.12 (d, J = 8.0 Hz, 2 H), 7.48 (d, J = 8.0 Hz, 2 H), 7.85 (t, J = 8.0 Hz, 1 H), 8.08 (d, J = 8.0 Hz, 1 H), 8.50 (d, J = 8.0 Hz, 1 H), 8.63 (s, 1 H), 9.98 (bs, 1 H); ^{13}C NMR (100 MHz, DMSO-d_6) δ: 20.7; 123.1 (q, $^1J_{F\text{-}C}$, J = 273.4 Hz), 123.8, 125.5, 128.6 (q, $^3J_{F\text{-}C}$, J = 2.8 Hz), 130.5 (q, $^3J_{F\text{-}C}$, J = 3.5 Hz), 130.8 (q, $^2J_{F\text{-}C}$, J = 32.5 Hz), 131.9, 138.0, 138.6, 144.0; ^{19}F NMR (376 MHz, DMSO-d_6)[4] δ: –61.5; IR (ATR ν 686, 794, 1199, 1319, 3076 cm^{-1}; HRMS (ESI): calcd for C$_7$H$_5$F$_3$IO ([M-OTs]$^+$): 288.9332; found 288.9330; Anal. Calcd. for C$_{14}$H$_{12}$F$_3$IO$_4$S (460.21): C 36.54; H 2.63. Found: C 36.4; H 2.6.

Handling and Disposal of Hazardous Chemicals

The procedures in this article are intended for use only by persons with prior training in experimental organic chemistry. All hazardous materials should be handled using the standard procedures for work with chemicals described in references such as "Prudent Practices in the Laboratory" (The National Academies Press, Washington, D.C., 2011 www.nap.edu). All chemical waste should be disposed of in accordance with local regulations. For general guidelines for the management of chemical waste, see Chapter 8 of Prudent Practices.

In the development and checking of these procedures, every effort has been made to identify and minimize potentially hazardous steps. The Editors believe that the procedures described in this article can be carried out with minimal risk if performed with the materials and equipment specified, and in careful accordance with the instructions provided. However, these procedures must be conducted at one's own risk. *Organic Syntheses, Inc.,* its Editors, and its Board of Directors do not warrant or guarantee the safety of individuals using these procedures and hereby disclaim any liability for any injuries or damages claimed to have resulted from or related in any way to the procedures herein.

3. Discussion

Hypervalent iodine(III) reagents are mainly used as mild oxidants in organic synthesis. [Hydroxy(tosyloxy)iodo]benzene (HTIB, Koser's reagent) is employed in α-oxidation of carbonyl compounds, oxidation of olefins, ring contractions and expansions, dearomatization of phenols and synthesis of iodonium salts (Scheme 1).[5] The reactivity of the reagent can be varied by adding substituents on the aryl moiety, and compound **1** is more efficient than HTIB in α-oxidation of ketones.[6]

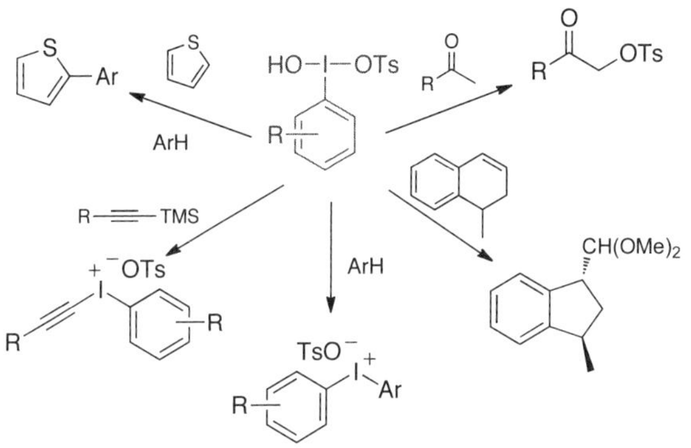

Scheme 1. Application areas of HTIBs.

[Hydroxy(tosyloxy)iodo]arenes (HTIBs) are usually synthesized from other iodine(III) reagents, but Yamamoto and Togo recently reported an efficient one-pot reaction where iodoarenes were treated with *m*-CPBA in chloroform at room temperature to give HTIBs in high yields.[3]

The one-pot synthesis of HTIBs from iodoarenes reported herein utilizes TFE as solvent, which reduces reaction times considerably. In our original report, a large scope of electron-rich and electron-poor HTIBs has been demonstrated (Table 1), and the synthesis of electron-rich HTIBs can be performed directly from iodine and arenes (Table 2).[7] The reaction then employed a 1:1 mixture of CH$_2$Cl$_2$ and TFE to cut the solvent costs. Large scale preparations are benefitted from the use of neat TFE with recovery of the solvent, which is also preferable from an environmental point of view.

Table 1. Synthesis of HTIBs from iodoarenes.

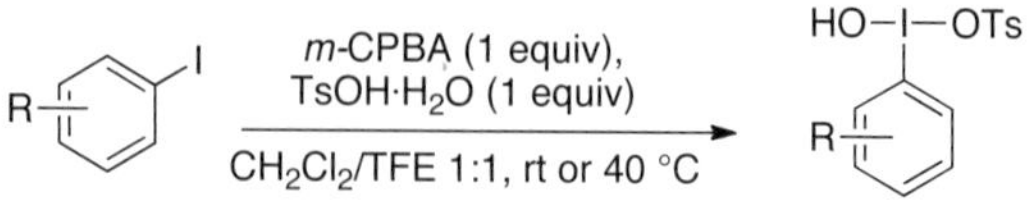

Entry	Product	Yield (%)	Entry	Product	Yield (%)
1	HO—I—OTs (phenyl)	94	6	HO—I—OTs (Cl-phenyl)	91
2	HO—I—OTs (o-methyl phenyl)	87	7	HO—I—OTs (methyl, CF₃ phenyl)	87
3	HO—I—OTs (p-methyl phenyl)	95	8	HO—I—OTs (CF₃, methyl phenyl)	80
4	HO—I—OTs (p-OMe phenyl)	76	9	HO—I—OTs (p-NO₂ phenyl)	57
5	HO—I—OTs (p-Br phenyl)	91			

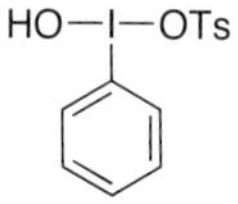

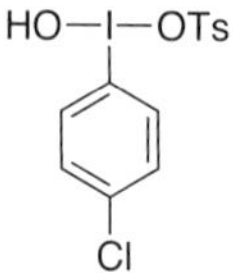

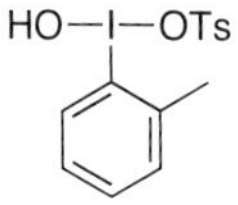

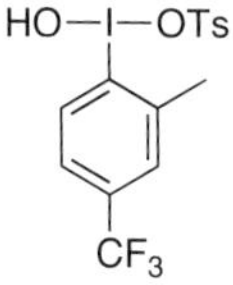

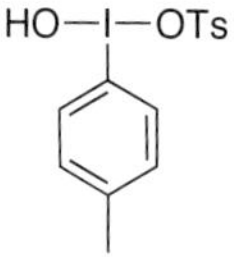

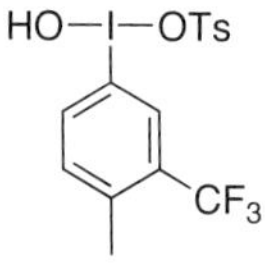

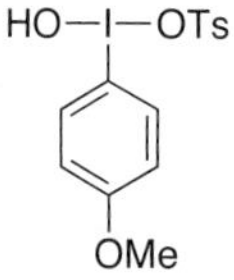

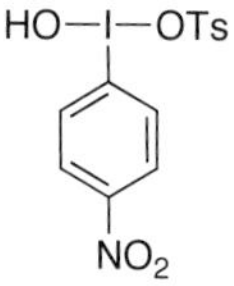

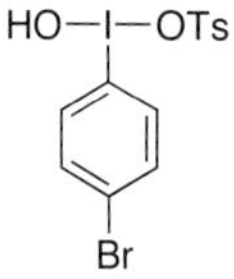

Table 2. Synthesis of HTIBs from iodine and arenes.

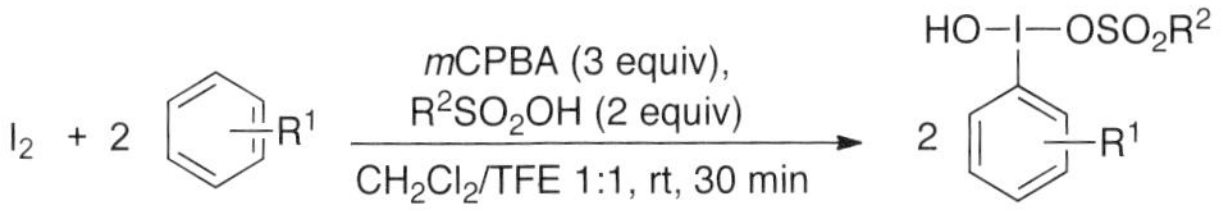

Entry	Product	Yield (%)	Entry	Product	Yield (%)
1	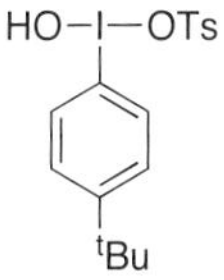 HO—I—OTs, ^{t}Bu	85	5	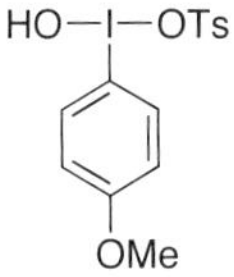 HO—I—OTs, OMe	78
2	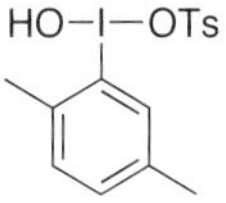 HO—I—OTs	67	6^a	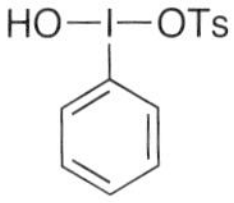 HO—I—OTs	75
3	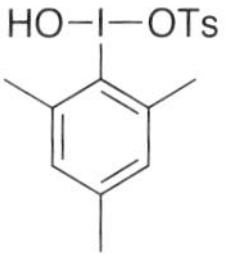 HO—I—OTs	67	7	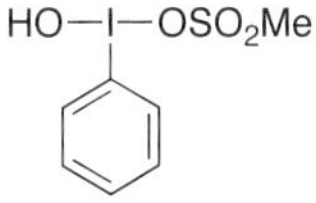 HO—I—OSO$_2$Me	76
4	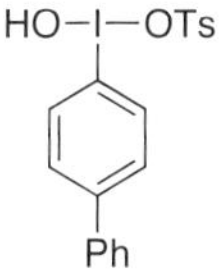 HO—I—OTs, Ph	23	8	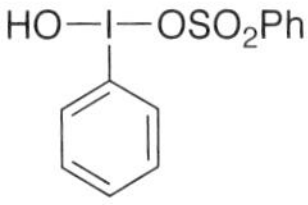 HO—I—OSO$_2$Ph	88

^{a} Sequential reaction with *m*-CPBA and TFOH followed by addition of TsOH.

1. Department of Organic Chemistry, Arrhenius Laboratory, Stockholm University, SE-106 91 Stockholm Sweden. E-mail: berit@organ.su.se. The authors thank Prof. Hans Adolfsson for fruitful discussions. This work was financially supported by the Swedish Research Council and Wenner-Gren Foundations.

2. (a) Vogel, A. I.; Furniss, B. S.; Hannaford, A. J.; Rogers, V.; Smith, P. W. G.; Tatchell, A. R., *Vogel's Textbook of Practical Organic Chemistry* (4th Ed) **1978**, p. 308. (b) *Org. Synth, Coll. Vol. 6,* **1988**, *276; Org. Synth.* **1970**, *50*, 15.

3. Yamamoto, Y.; Togo, H. *Synlett* **2005**, 2486–2488.
4. Fluorobenzene used as internal reference, see Fifolt, M. J.; Sojka, S. A.; Wolfe, R. A.; Hojnicki, D. S.; Bieron, J. F.; Dinan, F. J. *J. Org. Chem.* **1989**, *54*, 3019–3023.
5. (a) Zhdankin, V. V.; Stang, P. J. *Chem. Rev.* **2008**, *108*, 5299–5358; (b) Merritt, E. A.; Olofsson, B. *Synthesis* **2011**, 517-538; (c) Koser, G. F. *Aldrichimica Acta* **2001**, *34*, 89–102.
6. Nabana, T.; Togo, H. *J. Org. Chem.* **2002**, *67*, 4362–4365.
7. Merritt, E. A.; Carneiro, V. M. T.; Silva, L. F. J.; Olofsson, B. *J. Org. Chem.* **2010**, *75*, 7416–7419.

Appendix
Chemical Abstracts Nomenclature; (Registry Number)

1-[Hydroxy(tosyloxy)iodo]-3-trifluoromethylbenzene; (440365-98-0)
m-Chloroperbenzoic acid; Peroxybenzoic acid, *m*-chloro-;
    Benzenecarboperoxoic acid, 3-chloro-; (937-14-4)
2,2,2-Trifluoroethanol; Ethanol, 2,2,2-trifluoro-; (75-89-8)
3-Iodobenzotrifluoride; Benzene, 1-iodo-3-(trifluoromethyl)-; (401-81-0)
p-Toluenesulfonic acid monohydrate; Benzenesulfonic acid, 4-methyl-;
    (104-15-4)

Berit Olofsson was born in Sundsvall, Sweden, in 1972. She got her M.Sc. at Lund University in 1998, and finished her Ph.D. in asymmetric synthesis at KTH, Stockholm in 2002, supervised by Prof. P. Somfai. She then moved to Bristol University, UK for a post doc on methodology and natural product synthesis with Prof. V. K. Aggarwal. Returning to Sweden, she became assistant supervisor in the group of Prof. J.-E. Bäckvall at Stockholm University. In 2006 she was appointed to Assistant Professor, and got a permanent position as Associate Professor in 2010. Her research is focused on the synthesis and application of hypervalent iodine compounds as environmentally benign reagents in asymmetric synthesis.

Nazli Jalalian was born in 1982 in Tehran, Iran, and moved to Sweden at the age of four. She studied Chemistry at Stockholm University, where she also started her Ph.D. studies in 2007, under the supervision of Berit Olofsson. She has developed synthetic routes to diaryliodonium salts and applied the salts in arylation of phenols to diaryl ethers. She is currently working with further applications of the salts in organic synthesis.

David Yeung earned his B.Sc. in Biochemistry in 2000 and M.Sc. in Chemistry in 2004 from Concordia University, Montreal. He joined Amgen's Analytical Research and Development department later that year and, in 2012, moved to the Chemical Process Research and Development group.

Reagent for Divalent Sulfur Protection: Preparation of 4-Methylbenzenesulfonothioic Acid, S-[[[(1,1-Dimethylethyl)-Dimethylsilyl]oxy]methyl] Ester

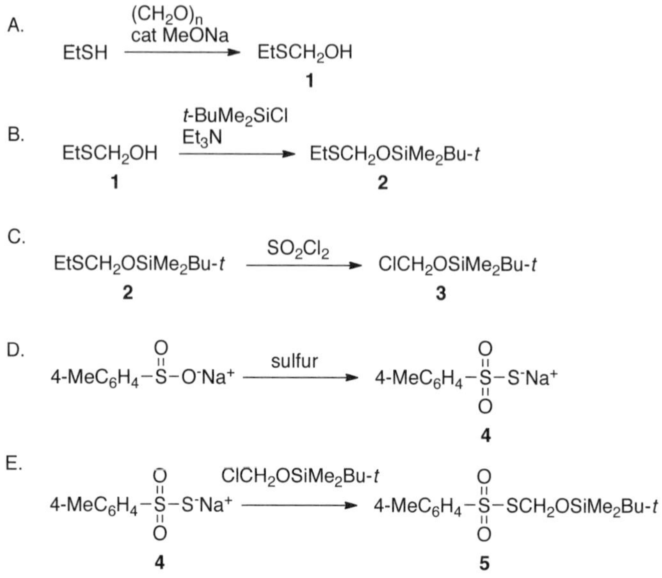

A.
$$EtSH \xrightarrow[\text{cat MeONa}]{(CH_2O)_n} EtSCH_2OH$$
1

B.
$$EtSCH_2OH \xrightarrow[\text{Et}_3\text{N}]{t\text{-BuMe}_2\text{SiCl}} EtSCH_2OSiMe_2Bu\text{-}t$$
1 **2**

C.
$$EtSCH_2OSiMe_2Bu\text{-}t \xrightarrow{SO_2Cl_2} ClCH_2OSiMe_2Bu\text{-}t$$
2 **3**

D.
$$4\text{-MeC}_6H_4-\overset{O}{\underset{}{S}}-O^-Na^+ \xrightarrow{\text{sulfur}} 4\text{-MeC}_6H_4-\overset{O}{\underset{O}{S}}-S^-Na^+$$
4

E.
$$4\text{-MeC}_6H_4-\overset{O}{\underset{O}{S}}-S^-Na^+ \xrightarrow{ClCH_2OSiMe_2Bu\text{-}t} 4\text{-MeC}_6H_4-\overset{O}{\underset{O}{S}}-SCH_2OSiMe_2Bu\text{-}t$$
4 **5**

Submitted by Lihong Wang and Derrick L. J. Clive.[1]
Checked by David Hughes.

1. Procedure

> *Caution: Steps A and B must be conducted in an efficient fumehood.*

A. *(Ethylthio)methanol (1).* An oven-dried 250-mL, 3-necked, round-bottomed flask equipped with a PTFE-coated magnetic stirring bar (3 x 1 cm) is fitted with a gas inlet adapter connected to a nitrogen line and a gas bubbler. The other two necks are capped with rubber septa; a thermocouple probe is inserted through one of the septa (Note 1). Paraformaldehyde (5.58 g, 0.186 mol, 1.07 equiv) (Notes 2 and 3) and ethanethiol (12.8 mL, 10.7 g, 0.173 mol, 1.0 equiv) (Note 4) are added to the flask. The mixture is stirred gently (avoid splashing) and cooled in an ice-water bath to 4 °C. A 25% solution of NaOMe in MeOH (0.07 mL, 0.002 equiv) is added through

10

one of the septa via a 100 µL syringe over 1 min. The temperature rises to 33 °C over 3 min, then cools to 13 °C over 5 min. The solids completely dissolve (Note 5). The ice-water bath is removed and the solution is stirred at 13–16 °C for 15 min. The pale yellow *(ethylthio)methanol* (*1*) (Note 6) is used immediately in the next step (Note 7).

B. *(1,1-Dimethylethyl)[(ethylthio)methoxy]dimethylsilane* *(2)*. Dichloromethane (Note 8) (120 mL) is added to the same flask containing neat (ethylthio)methanol from Step A. The stirred solution is cooled in an ice-water bath to 3 °C, then 4-(dimethylamino)pyridine (0.90 g, 7.4 mmol, 0.04 equiv) and triethylamine (21.9 g, 0.22 mol, 1.27 equiv) are added, followed by chloro(1,1-dimethylethyl)dimethylsilane (30.0 g, 0.20 mol, 1.18 equiv), added in 3 portions over 10 min (Note 9). The ice-water bath is replaced with a water bath and the reaction is stirred for 4 h (Note 10). The mixture is transferred to a 1-L separatory funnel using dichloromethane (60 mL) to rinse the flask. The organic layer is washed with water (2 x 100 mL) and saturated aqueous ammonium chloride (100 mL). The organic phase is filtered through a bed of sodium sulfate (50 g) in a medium porosity sintered glass funnel into a 1-L round-bottomed flask, using dichloromethane (2 x 50 mL) to rinse the filter cake. The solution is concentrated by rotary evaporation (70 mmHg, bath temperature 40 °C). The residue is diluted with hexanes (150 mL), transferred to a 1-L separatory funnel, and washed with water (2 x 200 mL) and brine (75 mL). The organic layer is filtered through a bed of sodium sulfate (50 g) in a medium porosity sintered glass funnel into a 500-mL round-bottomed flask, using hexanes (2 x 50 mL) to rinse the filter cake. The solution is concentrated by rotary evaporation (70 mmHg, bath temperature 40 °C) to afford crude **2** (38 g) as an oil. Vacuum distillation (Note 11) provides (1,1-dimethylethyl)[(ethylthio)methoxy]dimethylsilane (**2**) as a colorless liquid (28.0–28.6 g, 78–80 % yield over Steps A and B) (Notes 12 and 13).

> **WARNING:** *Chloromethyl ethers are potent carcinogens. The preparation and handling of compound* **3** *should be conducted at all times in a hood or ventilated balance enclosure.*

C. *(Chloromethoxy)(1,1-dimethylethyl)dimethylsilane (3)*. An oven-dried 500-mL round-bottomed flask equipped with a PTFE-coated magnetic stirring bar (3 x 1 cm) is capped with a rubber septum pierced with a nitrogen inlet needle connected to a gas bubbler. A thermocouple thermometer probe is also inserted through the septum (Note 1). (1,1-Dimethylethyl)[(ethylthio)methoxy]dimethylsilane (**2**) (12.1 g, 58.6 mmol, 1.00 equiv) and dichloromethane (120 mL) are added to the flask. The stirred solution is cooled to 2 °C using an ice-water bath. Sulfuryl chloride (8.14 g, 60.3 mmol, 1.03 equiv) (Note 14) is added via a 10 mL syringe over 10 min, keeping the temperature <5 °C. During the addition the reaction mixture turns yellow. After the addition, the ice-bath is removed and the solution is stirred for 20 min (Note 15). The stir bar is removed and the solution is concentrated by rotary evaporation (70 mmHg, bath temperature 40 °C) to afford crude **3** (12.4 g). Vacuum distillation provides (chloromethoxy)(1,1-dimethylethyl)dimethylsilane (**3**) as a slightly yellow liquid (8.41–8.66 g, 80–82 % yield) (Notes 16 and 17).

D. *4-Methylbenzenesulfonothioic acid, sodium salt* (**4**). Sodium *p*-toluenesulfinate monohydrate (Note 18) (25.4 g, 0.13 mol. 1.0 equiv), sulfur (4.54 g, 0.14 mol. 1.09 equiv), ethanol (100 mL) and water (100 mL) are added to a 500-mL round-bottomed flask equipped with a PTFE-coated magnetic stir bar (3 x 1 cm). The flask is lowered into a heating mantle and fitted with a Liebig condenser. The mixture is stirred and warmed to reflux for 8 h (Note 19). After the reaction mixture is cooled to room temperature, the residual sulfur is removed by filtration through a 60-mL medium porosity sintered glass funnel into a 1-L round bottom flask, using water (2 x 20 mL) to wash the reaction flask and the collected solid. The filtrate is concentrated by rotary evaporation (50 °C water bath, 70 mmHg) to wet solids (49 g). The flask is equipped with a PTFE stir bar (3 x 1 cm) and water (40 mL) is added. The slightly hazy mixture is stirred at room temperature for 3 h. The solution, which contains some suspended particles, is filtered into a 500-mL round bottomed flask through pad of pre-wetted Celite (3 g) (Note 20) in a 40-mL medium porosity sintered glass funnel, the flask and Celite pad being rinsed with water (4 x 10 mL). The clear filtrate is concentrated by rotary evaporation (50 °C water bath, 70 mmHg) to provide a wet solid (32 g) (Note 21). The flask is equipped with a PTFE stir bar (3 x 1 cm) and a Liebig condenser. Absolute ethanol (80 mL) is added and the heterogeneous mixture is stirred and warmed to reflux over a 30 min period using a heating mantle. Once the mixture reaches reflux, the heating mantle is removed and the mixture is cooled in air to room temperature over the course of 1 h, then is stirred at room temperature for 3 h. The resulting white solid is collected on a 150-mL medium porosity sintered glass funnel, portions of the filtrate being used to rinse all solids out of the flask. The

12

filter cake is washed with absolute ethanol (25 mL), air-dried in the funnel by continued suction (ca 1 h) and then dried for 9 h in a vacuum oven (70 mmHg, 50 °C) to afford 4-methylbenzenesulfonothioic acid, sodium salt (22.4–23.4 g, 82–86 %) (Note 22).

E. *4-Methylbenzenesulfonothioic acid, S-[[[(1,1-dimethylethyl)-dimethylsilyl]oxy]methyl] ester* (**5**). An oven-dried 250-mL round-bottomed flask equipped with a PTFE-coated magnetic stirring bar (3 x 1 cm) is capped with a rubber septum pierced with a nitrogen inlet needle connected to a gas bubbler. A thermocouple thermometer probe is also inserted through the septum (Note 1). The septum is removed momentarily and 4-methylbenzenesulfonothioic acid, sodium salt (**4**) (5.97 g, 28.4 mmol, 1.03 equiv) and acetonitrile (35 mL) (Note 23) are added and the suspension is stirred. (Chloromethoxy)(1,1-dimethylethyl)dimethylsilane (**3**) (4.96 g, 27.4 mmol, 1.00 equiv) is added via a 10-mL syringe over 1 min (Note 24). The mixture is stirred vigorously for 4 h (Note 25), then the septum is replaced with a 200-mL addition funnel and *t*-butyl methyl ether (140 mL) is added dropwise to the stirred mixture over 15 min. The mixture is filtered through a tightly packed pad of Celite (4 cm in diameter x 2.5 cm in height) (Note 26) in a 40-mL sintered glass funnel into a pre-weighed 500-mL round-bottomed flask, using MTBE (3 x 10 mL) as a rinse of the flask and filter cake. The filtrate is concentrated by rotary evaporation (40 °C bath, 70 mmHg) and dried for 3 h at ambient temperature (70 mmHg) to give *4-methylbenzenesulfonothioic acid, S-[[[(1,1-dimethyl-ethyl)dimethylsilyl]-oxy]methyl] ester* (**5**) as a colorless, viscous oil (9.06 g, 99 %) (Notes 27, 28, and 29).

2. Notes

1. The internal temperature was monitored using a J-Kem Gemini digital thermometer with a Teflon-coated T-Type thermocouple probe (12-inch length, 1/8 inch outer diameter, temperature range –200 to +250 °C). The submitters did not monitor the internal temperature in any of the procedure's steps A-E.

2. The following reagents and solvents were obtained from Sigma-Aldrich and used as received for Step A: paraformaldehyde (powder, 95%), ethanethiol (97%), sodium methoxide (25% wt % solution in MeOH).

3. Paraformaldehyde is added in slight excess with the intention to consume all ethanethiol during the reaction to minimize odors.

4. Ethanethiol was transferred to the flask via a 10 mL graduated glass pipette (2 transfers). After transfer, the pipette was immediately rinsed with bleach.

5. If a portion of the mixture has splashed onto the walls of the flask, the material should be rinsed down by gentle swirling.

6. (Ethylthio)methanol (**1**) has the following physical and spectroscopic properties: FTIR (microscope) υ (cm^{-1}): 3395 br, 2968, 2928, 2873, 1453 cm^{-1}; ^{1}H NMR (CDCl$_3$, 500 MHz) δ: 1.32 (t, J = 7.4 Hz, 3 H), 1.82 (br s, 1 H), 2.73 (q, J = 7.4 Hz, 2 H), 4.74 (s, 2 H); ^{13}C NMR (CDCl$_3$, 125 MHz) δ: 15.2, 24.8, 65.8. Low resolution EI m/z calcd for C$_3$H$_8$OS 92, found 92. The submitters determined a purity of 91% by GC/MS (Agilent Technologies 7890 GC with 5975C mass spectrometer; column ZEBRON ZB-5, length 30 m, ID 0.25 mm; film thickness 0.25 μm; initial temperature 35 °C for 2 min, final temperature 290 °C for 2 min; rate of temperature increase 10 °C/min; helium gas; flow 0.6636 mL/min; inlet temperature 200 °C; 50:1 split injection); and 83% by NMR. The checker determined a purity of approximately 85% by ^{1}H NMR.

7. To minimize odors, it is recommended that the material produced in step A be used directly in Step B using the same flask.

8. The following reagents and solvents were used as received for Step B: dichloromethane (Fisher ACS certified, stabilized), chloro(1,1-dimethylethyl)dimethylsilane (Oakwood Products, Inc., West Columbia, SC), triethylamine (Sigma-Aldrich, 99.5% distilled), hexanes (Fisher, ACS reagent, >98.5%).

9. The temperature rises to 10 °C after the addition.

10. The reaction was monitored by ^{1}H NMR. A 0.1 mL aliquot of the reaction mixture was quenched into 1 mL CDCl$_3$/1 mL sat. aq. NH$_4$Cl. The layers were separated, then the CDCl$_3$ layer was concentrated to dryness. The concentrated sample was diluted with CDCl$_3$ for ^{1}H NMR analysis. Diagnostic peaks: product **2**, δ 4.82 (s, 2H, CH_2); starting material **1**, δ 4.74 (s, 2H, CH_2). Approx 1.5 % unreacted **1** remained at the 3 h sampling point.

11. The checker carried out the distillation in a 100-mL pear-shaped flask equipped with a 1-cm oval PTFE stir bar using a 3-cm Vigreaux column at a pressure of 70 mmHg. Three fractions were collected: fr 1, 80 - 120 °C (2.2 g); fr 2, 125-128 °C (28.6 g); and fr 3, 128-134 °C (3.2g). The pot residue was 1.7 g. By GC analysis (conditions in Note 12), fraction 2 was >98 % pure and fraction 3 was 85 - 90% pure. The yield was based on fraction 2. The submitters carried out the distillation at 4.8 mmHg, 58.5 – 61 °C.

14

12. *(1,1-Dimethylethyl)[(ethylthio)methoxy]dimethylsilane (2)* has the following physical and spectroscopic properties: FTIR (microscope) υ (cm^{-1}): 2957, 2930, 2897, 2858, 1472, 1463 cm^{-1}; ^{1}H NMR (CDCl$_3$, 500 MHz) δ: 0.13 (s, 6 H), 0.91 (s, 9 H), 1.30 (t, J = 7.4 Hz, 3 H), 2.68 (q, J = 7.4 Hz, 2 H), 4.82 (s, 2 H); ^{13}C NMR (CDCl$_3$, 125 MHz) δ: –4.8, 15.1, 18.4, 24.8, 26.0, 66.2. GC-MS *m/z* (relative intensity), 149 (92%, M$^+$ - *t*-Bu), 119 (100%, M$^+$ – *t*-BuMe$_2$), 91(40%), 89 (61%), 75 (43%), 73 (63%). Purity = 98 % by GC (t$_R$ = 7.5 min; conditions: Agilent DB35MS column; 30 m x 0.25 mm; initial temp 60 °C, ramp at 20 °C/min to 280 °C, hold 15 min).

13. The submitters report the compound is stable at room temperature for at least 1 month when kept in a stoppered flask.

14. The following reagents and solvents were used as received for Step C: dichloromethane (Fisher ACS certified, stabilized,), sulfuryl chloride (Acros, 97%).

15. The temperature rose to 12 °C.

16. The distillation was carried out in a 50-mL pear-shaped flask containing a PTFE-coated oval magnetic stir bar (1 cm) using a 3-cm Vigreaux column at a pressure of 70 mmHg. Three fractions were collected: fr 1, 30–83 °C (1.18 g); fr 2, 83–87 °C (6.89 g); fr 3, 87–88 °C (1.52 g). GC purity (Note 12) was 96.5% for fraction 2 and 94.5% for fraction 3. Fractions 2 and 3 were combined. The submitters reported distillation at 70.0–72.5 °C (26 mmHg).

17. *(Chloromethoxy)(1,1-dimethylethyl)dimethylsilane (3)* has the following physical and spectroscopic properties: FTIR (neat film, microscope) υ (cm^{-1}): 2958, 2932, 2901, 2860, 1473, 1464 cm^{-1}; ^{1}H NMR (CDCl$_3$, 500 MHz) δ: 0.21 (s, 6 H), 0.92 (s, 9 H), 5.61 (s, 2 H); ^{13}C NMR (CDCl$_3$, 125 MHz) δ: –5.0, 18.0, 25.7, 76.6. GC-MS *m/z* (relative intensity), 125 (9 %, M$^+$ - *t*-Bu) 123 (24%, M$^+$ - *t*-Bu), 95 (35 %, M$^+$ – *t*-BuMe$_2$), 93 (100 %, M$^+$ – *t*-BuMe$_2$), 73 (25%), 57 (45%). Purity = 96 % by GC (t$_R$ = 5.5 min, conditions in Note 12). The checkers noted the compound decomposes at a rate of 15% per week at room temperature when stored in a glass flask with a glass stopper based on ^{1}H NMR analysis. In the refrigerator, a small amount of decomposition (2-3%) occurred over a two-week period. When stored in the freezer, no decomposition occurred over a two-week period.

18. The following reagents and solvents were used as received for Step D: *p*-toluenesulfinate hydrate (Acros, 98%), sulfur (Fisher, sublimed powder, ethanol (Sigma-Aldrich, 99.5%), Celite (Sigma-Aldrich, acid-washed).

19. Reaction progress was monitored by ^{1}H NMR as follows. A 0.1 mL reaction aliquot was evaporated to dryness then dissolved in DMSO-d$_6$. Diagnostic resonances were δ 7.14–7.15 (m, 2 H), 7.61–7.63 (m, 2 H) for product **4** and 7.35–7.37 (m, 2H) for p-TsSO$_2$Na. Approximately 1% p-TsSO$_2$Na remained at the 7 h timepoint.

20. The Celite was pre-wetted by filtering 20-mL water through the Celite cake. The hazy filtrate was discarded and the receiving flask rinsed with water prior to filtering the reaction mixture.

21. The crude material contains 3% sodium p-toluenesulfinate based on ^{1}H NMR analysis.

22. *4-Methylbenzenesulfonothioic acid, sodium salt* (**4**) has the following physical and spectroscopic properties: IR (KBr) υ (cm^{-1}): 3039, 3023, 2981, 2920, 2861, 1934, 1664, 1596, 1494, 1398 cm^{-1}; ^{1}H NMR (500 MHz, DMSO-d$_6$) δ: 2.30 (s, 3 H), 7.14–7.15 (m, 2 H), 7.61–7.63 (m, 2 H); ^{13}C NMR (125 MHz, DMSO-d$_6$) δ: 20.7, 124.0, 128.0, 138.2, 152.6; exact mass (electrospray) m/z calcd for C$_7$H$_7$O$_2$S$_2$ (M-Na) 186.9893, found 186.9895. The material contained 0.2 wt% water based on Karl Fischer titration. Weight percent purities of 94% and 97% (2 runs) were determined based on quantitative ^{1}H NMR analysis (DMSO-d$_6$) using dichloroethane as internal standard, based on 2 weighings (samples 30-60 mg each), 4 acquisitions using a pulse delay of 10 seconds, and 8 independent peak integrations. The material with 94 wt% purity was purified by slurrying as follows: Compound **4** (22.6 g) and 95% ethanol (80 mL) were added to a 500-mL round bottomed flask equipped with a PTFE oval-stir bar (3 x 1 cm). The stirred mixture was heated to reflux (remains a slurry) with a heating mantle, then cooled in air with stirring to ambient temperature over 1 h and stirred for an additional 3 h. The material was filtered into a 60-mL sintered glass funnel, washed with absolute ethanol (25 mL), and dried under vacuum (70 mmHg, 50 °C) to afford product (17.9 g, 79% recovery). NMR quantitative assay indicated 98 wt% purity.

23. The following reagents and solvents were used as received for Step E: acetonitrile (Fisher Optima, water content 0.001%), t-butyl methyl ether (>98.5%, Sigma-Aldrich), Celite (Sigma-Aldrich, acid-washed).

24. The reaction warms to 27 °C over 10 min, then cools slowly to room temperature over 30 min.

25. The reaction remains heterogeneous throughout. The reaction was monitored by ^{1}H NMR as follows. A 0.1 mL reaction aliquot was added to 1 mL MTBE, filtered, and concentrated to dryness. The sample was dissolved in CDCl$_3$ for analysis. Diagnostic peaks were δ 5.61 (s, 2H) for

16

unreacted **3** and 5.41 (s, 2H) for product **5**. Less than 1% starting material remained at the 3 h timepoint.

26. The Celite was pre-wetted by filtering 20-mL MTBE through the Celite cake. The filtrate was discarded and the receiving flask rinsed with MTBE prior to filtering the reaction mixture. The submitters used diethyl ether in this experiment instead of MTBE.

27.   *4-Methylbenzenesulfonothioic   acid,   S-[[[(1,1-dimethylethyl)-dimethylsilyl]oxy]methyl]   ester* (**5**)   has   the   following   physical   and spectroscopic   properties: mp   33–34   °C   (MTBE/hexanes);   FTIR (dichloromethane cast film) υ (cm^{-1}): 2955, 2930, 2886, 2858, 1595, 1493, 1472, 1464, 1444 cm^{-1}; ^{1}H NMR (CDCl$_3$, 400 MHz) δ: –0.03 (s, 6 H); 0.76 (s, 9 H), 2.42 (s, 3 H), 5.41 (s, 2 H), 7.27–7.32 (m, 2 H), 7.84–7.86 (m, 2 H); ^{13}C NMR (100 MHz, CDCl$_3$) δ: –5.3, 18.1, 21.8, 25.6, 71.5, 127.3, 129.8, 144.0, 144.6; exact mass (electrospray) *m/z* calcd for C$_{14}$H$_{24}$NaO$_3$S$_2$Si 355.0828, found 355.0824; GC-MS *m/z* (relative intensity), 275 (6%, M$^+$ - *t*-Bu), 245 (59%, M$^+$ – *t*-BuMe$_2$), 91 (39%), 75 (100%), 73 (30%). Anal. calcd. for C$_{14}$H$_{24}$O$_3$S$_2$Si: C, 50.56; H, 7.27; S, 19.28; found C, 50.44; H, 6.83; S, 19.61. Weight percent purities of 87% and 90% (2 runs) were determined   based   on   quantitative ^{1}H   NMR   analysis   (CDCl$_3$)   using dichloroethane as internal standard, as outlined in note 22.

28. A purified sample of **5** was obtained by crystallization as follows: Compound **5** (7.1 g) was dissolved in MTBE (10 mL). Residual solids were removed   by   filtration   through   a   0.45   μm   PTFE   syringe   filter   (25   mm, Millipore catalog # SLCR025NS) into a 100-mL round-bottomed flask. Hexanes (20 mL) were added and the solution was held in a –20 °C freezer for 20 h. (Obtaining crystals for the first time at small scale required 6 days in   the   freezer;   thereafter,   crystallization   typically   initiated   within   a   few hours). The product was isolated by filtration on a 40-mL sintered glass funnel, using MTBE/hexanes (1:1) (5 mL, –20 °C) as a wash, to provide colorless cubic crystals (4.1 g 59% recovery). Weight percent purities of 97% and 98 % (2 runs) were determined based on quantitative ^{1}H NMR analysis (CDCl$_3$) using dichloroethane as internal standard, as outlined in note 22.

29. The submitters report the compound decomposes on silica gel and on neutral Grade III alumina. The compound has been kept without change (^{1}H NMR) at room temperature for 1 week and is stable in the freezer for at least 2 months.

Handling and Disposal of Hazardous Chemicals

The procedures in this article are intended for use only by persons with prior training in experimental organic chemistry. All hazardous materials should be handled using the standard procedures for work with chemicals described in references such as "Prudent Practices in the Laboratory" (The National Academies Press, Washington, D.C., 2011 www.nap.edu). All chemical waste should be disposed of in accordance with local regulations. For general guidelines for the management of chemical waste, see Chapter 8 of Prudent Practices.

In the development and checking of these procedures, every effort has been made to identify and minimize potentially hazardous steps. The Editors believe that the procedures described in this article can be carried out with minimal risk if performed with the materials and equipment specified, and in careful accordance with the instructions provided. However, these procedures must be conducted at one's own risk. *Organic Syntheses, Inc.*, its Editors, and its Board of Directors do not warrant or guarantee the safety of individuals using these procedures and hereby disclaim any liability for any injuries or damages claimed to have resulted from or related in any way to the procedures herein.

3. Discussion

During the course of synthetic studies directed towards the antitumor agent MPC1001,[2] a need arose for a protected bivalent sulfur unit so constituted that it could be introduced by reaction with a carbanion and later dismantled to release a sulfhydryl group under mild conditions. Although numerous sulfur protecting groups are known,[3] our precise requirements prompted us to investigate the CH_2OSiMe_2Bu-t group for sulfur protection and the reagent 4-$MeC_6H_4SO_2$-SCH_2OSiMe_2Bu-t (**5**) for introducing sulfur protected in this manner.[4] Other sulfonothioic acid esters such as $PhSO_2$-SPh have been used for sulfenylation of carbanions,[5] and the use of a silyl group offered the possibility of controlling the deprotection conditions by changing the substituents on silicon.[6] Although chloromethoxysilanes have been used for protection of alcohols,[7] they do not appear to have been used for sulfur protection.

A direct method for preparing reagent (**5**) appeared to be by way of reaction E above, and so we prepared the two components, salt (**4**) and the chloromethoxysilane (**3**). Both were made by the literature methods[8,9] which

18

are described in detail here. Reaction of the salt with silane (**3**) afforded the desired sulfenylating agent (**5**).

In order to test if the $CH_2OSiMe_2Bu\text{-}t$ group could indeed serve as an effective form of protection for bivalent sulfur, we treated a number of simple thiols with reagent (**3**) so as to generate the sulfides shown in Table 1.[4] DMF is a satisfactory solvent and proton sponge or 2,6-lutidine gave the best results, depending on the particular case.

Table 1. Protection of thiols

	Substrate	Conditions	Product
1	Ph_3CSH	DMF, $ClCH_2OSiMe_2Bu\text{-}t$ proton sponge, rt, 5.5 h, 77%	Ph_3CSPg^a
2	naphthalene-2-SH	DMF, $ClCH_2OSiMe_2Bu\text{-}t$ proton sponge, rt, 2 h, 78%	naphthalene-2-SPg
3	$Ph_3CS(CH_2)_3SH$	DMF, $ClCH_2OSiMe_2Bu\text{-}t$ 2,6-lutidine, 55 °C, 3.5 h, 61%	$Ph_3CS(CH_2)_3SPg$
4	$MeO_2C\!-\!CH(NHCbz)\!-\!SH$	DMF, $ClCH_2OSiMe_2Bu\text{-}t$ 2,6-lutidine, 55 °C, 3 h, 86%	$MeO_2C\!-\!CH(NHCbz)\!-\!SPg$
5	$Me(CH_2)_{11}SH$	DMF, $ClCH_2OSiMe_2Bu\text{-}t$, 2,6-lutidine, rt, 21 h, 80%	$Me(CH_2)_{11}SPg$
6	$HOCH_2CH_2SH$	DMF, $t\text{-BuOK}$, rt, 30 min; rt $ClCH_2OSiMe_2Bu\text{-}t$, 15 min, 63%	$HOCH_2CH_2SPg$
7	$HOCH_2CH_2SPg$	3,5-dimethoxybenzoic acid EDCI.HCl, DMAP, CH_2Cl_2 0 °C, then rt, 24 h, 60%	$ArCO_2CH_2CH_2SPg$

$^aPg = CH_2OSiMe_2Bu\text{-}t$ $\qquad$ $Ar = 3,5\text{-}(MeO)_2C_6H_3$

Several methods for deprotection of the sulfides were examined (Table 2).[4] Treatment with Bu_4NF in THF; Bu_4NF and AcOH in THF, or HF.pyridine in THF were all suitable, and it was usually convenient to oxidize the thiols in situ by addition of iodine. In a few cases the protecting group was removed by reaction with a sulfenyl halide (2-nitrobenzenesulfenyl chloride, α,α-diphenylbenzenemethanesulfenyl chloride, benzenemethanesulfenyl chloride).

Table 2. Deprotection

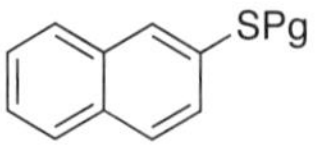

1	Ph_3CSPg^a	Bu$_4$NF, THF, -78 °C, 15 min; to -10 °C (ca. 3 min), 68% ⟶	Ph_3CSH

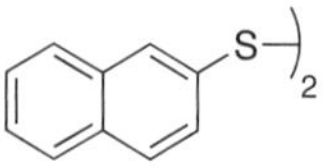

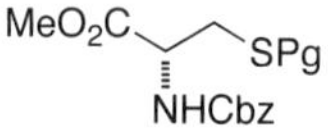

The stability of the protecting group was evaluated[4] by exposing n-$C_{12}H_{25}SCH_2OSiMe_2Bu$-t (6) to a variety of conditions. The compound is stable to H_2/Pd/C in methanol-dichloromethane and to H_2/Rh/Al$_2$O$_3$/ethyl acetate. An O-triethylsilyl ether can be selectively deprotected in the presence of (6), using H_2/Pd/C in methanol-dichloromethane. The protecting group appears to survive typical conditions for removal of a Troc group (zinc dust in acetic acid-diethyl ether) and an Fmoc group can be

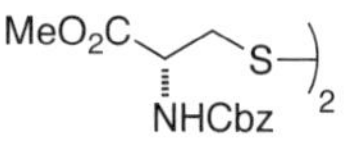

20

removed in its presence by using piperidine. Hydride reducing agents either have no effect (sodium borohydride) or little effect (lithium aluminum hydride, diisobutylaluminum hydride). $(PhS)_2CH_2$ can be deprotonated with butyllithium with very little decomposition (4%) of the test substrate. Acidic reagents (trifluoroacetic acid, p-toluenesulfonic acid hydrate, pyridinium p-toluenesulfonate-MeOH, boron trifluoride etherate) are not compatible with the protecting group, except for pyridinium p-toluenesulfonate in dichloromethane and exposure to silica gel during chromatography. A primary alcohol can be converted into the corresponding bromide (tetrabromomethane, triphenylphosphine) without affecting the protecting group, but oxidizing agents (pyridinium chlorochromate, Dess-Martin periodinane, 2-iodoxybenzoic acid (IBX), Swern conditions) are not compatible with the protecting group. A primary alcohol can be silylated with triethylsilyl triflate in the presence of (6).

The sulfenylating agent (5) has been used to introduce the protected sulfur, as shown in eq 1.[4] Sulfur deprotection is illustrated in eq 2,[10] which represents the result of a single experiment. A more sophisticated use of the sulfenylating agent (5) as well as subsequent deprotection in a synthetically complex setting is summarized in Scheme 1.[10]

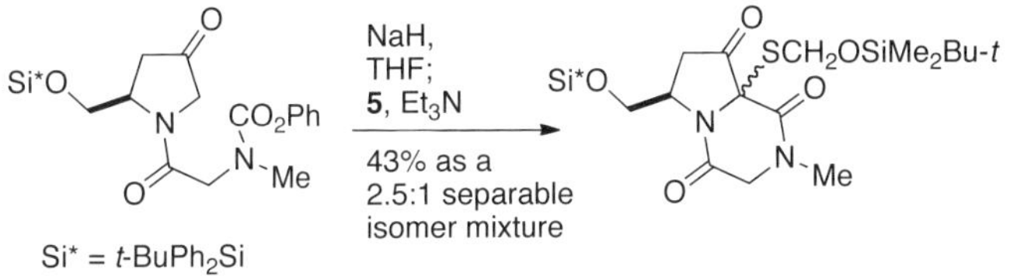

$$eq\ 1$$

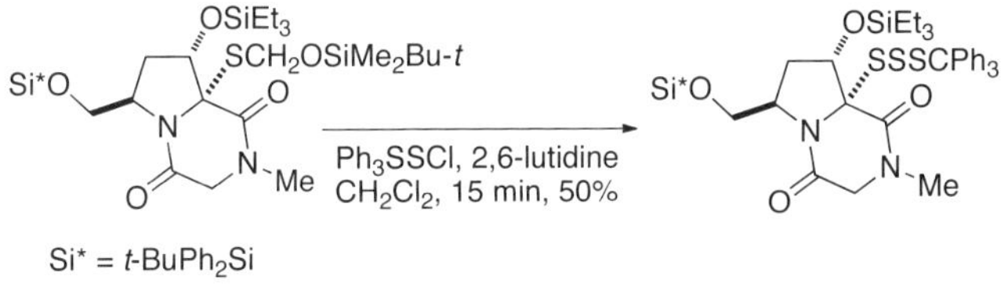

$$eq\ 2$$

Scheme 1. Sulfenylation with reagent **5** and deprotection

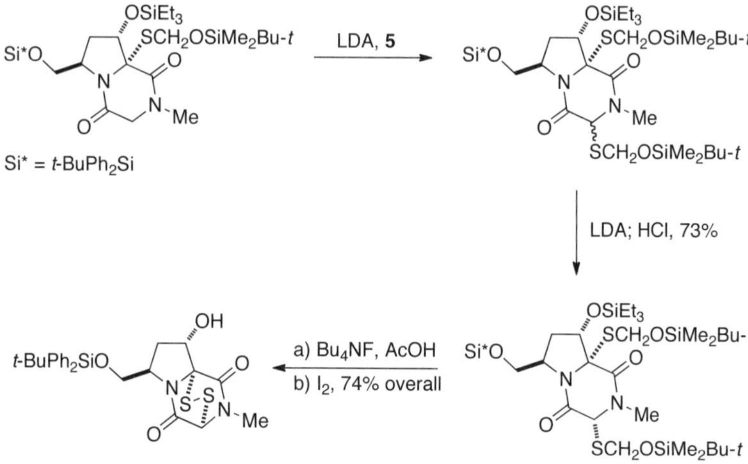

1. Chemistry Department, University of Alberta, Edmonton, Alberta T6G 2G2, Canada; e-mail: derrick.clive@ualberta.ca. We thank the Natural Sciences and Engineering Research Council of Canada for financial support. L.W. held a Marie Arnold Cancer Research Graduate Scholarship and an Alberta Heritage Foundation for Medical Research Graduate Studentship.

2. Peng, J.; Clive, D. L. J. *J. Org. Chem.* **2009**, *74*, 513–519.
3. (a) Wuts, P. G. M.; Greene, T. W. *Greene's Protective Groups in Organic Synthesis*, 4th ed.; Wiley-Interscience: NJ, 2007, p 687. (b) Kocieński, P. J. *Protecting Groups*, 3rd ed.; Thieme: Stuttgart, 2004, p 380.
4. Wang, L.; Clive, D. L. J. *Org. Lett.* **2011**, *13*, 1734–1737.
5. (a) Scholz, D. *Liebigs Ann. Chem.* **1984**, 259–263. Representative recent examples: (b) Trost, B. M.; Salzman, T. N.; Hiroi, K. *J. Am. Chem. Soc.* **1976**, *98*, 4887–4902. (c) Goodridge, R. J.; Hambley, T. W.; Haynes, R. K.; Ridley, D. D. *J. Org. Chem.* **1988**, *53*, 2881–2889. (d) Deng, K.; Chalker, J.; Yang, A.; Cohen, T. *Org. Lett.* **2005**, *7*, 3637–3640. (e) Chen, W.; Pinto, B. M. *Carbohydrate Res.* **2007**, *342*, 2163 2172. (f) Brennan, C.; Pattenden, G.; Rescourio, G. *Tetrahedron Lett.* **2003**, *44*, 8757–8760.
6. i-Pr$_3$SiOCH$_2$Cl is commercially available.

7. (a) Gundersen, L.-L.; Benneche, T.; Undheim, K. *Acta. Chem. Scand.* **1989**, *43*, 706–709. (b) EP 1565479B1 (2006). (c) Pitsch, S.; Weiss, P. A.; Jenny, L.; Stutz, A.; Wu, X. *Helv. Chim. Acta* **2001**, *84*, 3773–3795.
8. Harmon, J. P.; Field, L. *J. Org. Chem.* **1986**, *51*, 5235–5240.
9. Benneche, T.; Gundersen, L.-L.; Undheim, K. *Acta. Chem. Scand.* **1988**, *42B*, 384–389.
10. Wang, L.; Clive, D. L. J. *Tetrahedron Lett.* **2012**, *53*, 1504–1506.

Appendix
Chemical Abstracts Nomenclature (Registry Number)

(Chloromethoxy)(1,1-dimethylethyl)dimethylsilane; (119451-80-8)
Chloro(1,1-dimethylethyl)dimethylsilane; (18162-48-6)
4-(Dimethylamino)pyridine; (1122-58-3)
(1,1-Dimethylethyl)[(ethylthio)methoxy]dimethylsilane; (119451-79-5)
Ethanethiol; (75-08-1)
(Ethylthio)methanol; (15909-30-5)
4-Methylbenzenesulfonothioic acid, *S*-[[[(1,1-dimethylethyl)dimethylsilyl]-oxy]methyl] ester; (1277170-42-9)
4-Methylbenzenesulfonothioic acid, sodium salt (1:1); (3753-27-3)
Paraformaldehyde; (30525-89-4)
Sodium methoxide; (124-41-4)
Sodium *p*-toluenesulfinate hydrate (TolSO$_2$Na·H$_2$O); (207801-20-5)
Sulfur; (7704-34-9)
Sulfuryl chloride; (7791-21-5)
Triethylamine; (121-44-8)

Derrick Clive was born in London and was educated at Imperial College where he obtained a B.Sc. (Special) in Chemistry, and then a Ph.D. in Professor Barton's group. Dr. Jack E. Baldwin (now Sir Jack) assisted in the supervision of these postgraduate studies. Derrick then held a postdoctoral position at Harvard in R. B. Woodward's group. In 1975 he joined the Chemistry Department of the University of Alberta, where he is now Professor of Chemistry. He has published over 200 papers on the development of general synthetic methods — involving mainly selenium chemistry and radical cyclization — and on the total synthesis of complex natural products with significant biological properties.

Lihong Wang was born in Zhoushan, Zhejiang Province, and obtained his B.Sc. at Fudan University. He stayed at Fudan University to begin his graduate studies, but moved to the University of Alberta in 2006 for his Ph.D. under the supervision of Professor Clive. Lihong's research has been supported by a number of Scholarships and is in the area of synthetic methodology and natural product synthesis. After obtaining his Ph.D. in 2011 he joined Professor Nicolaou's group at the Scripps Institute as a postdoctoral fellow.

24

Discussion Addendum for: Applications of (2S)-(–)-3-*exo*-Morpholinoisoborneol [(–)MIB] in Organic Synthesis

Submitted by Mahmud M. Hussain[1,2] and Patrick J. Walsh.[1,*]
Original article: Chen, Y. K.; Jeon, S. -J; Walsh, P. J.; Nugent, W. A. *Org. Synth.* **2005**, *82*, 87–89.

Nugent's morpholinoisoborneol (MIB) is an excellent ligand for the catalytic asymmetric addition of alkyl, vinyl, and aryl groups to aldehydes in the presence of organozinc reagents to furnish a variety of secondary alcohols with high ee.[3,4,5] The major developments have been the significant broadening of the substrate scope and the development of several one-pot methods that streamline the synthesis of synthetically valuable and versatile molecules such as epoxy alcohols, allylic epoxy alcohols, cyclopropyl alcohols, halocyclopropyl alcohols, pyranones and 1,2,4-trioxanes with high enantio-, diastereo-, and chemoselectivity.[6,7,8] In this update, we will summarize many of the advances.

A detailed synthetic procedure for the synthesis of (–)-MIB has been published in *Organic Syntheses*.[9] Starting from either (R)- or (S)-camphor, gram quantities of either enantiomer of MIB can be synthesized in three steps and with only a single purification step. At the time of this report, only (–)-MIB is commercially available.

1. Catalytic Asymmetric Arylation of Aldehydes

Two routes for the catalytic asymmetric arylation of aldehydes to generate highly enantioenriched diarylmethanols and benzylic alcohols were reported (Figure 1).[10] In the first route, diarylzinc intermediates were generated in situ by metalation of unfunctionalized aryl bromides with *n*-BuLi followed by transmetalation with ZnCl$_2$ (Figure 1A). The LiCl formed in the process catalyzes a rapid racemic background reaction. The Lewis acidic LiCl was selectively suppressed by addition of tetraethylethylenediamine (TEEDA). Subsequent addition of catalytic amounts of (–)-MIB and aldehyde to the preformed diarylzinc reagent gave addition products with high enantioselectivities (80–92%) and yields (78–99%). In the second route, mixed alkyl aryl zinc reagents were generated by metalation of aryl bromides with n-BuLi followed by addition to $ZnCl_2$ and subsequent addition of a second equivalent of n-BuLi. The mixed organozinc species was used in situ in the asymmetric addition reaction after addition of 0.8 equivalents of TEEDA (Figure 1B). Higher enantioselectivities (up to 97% ee) were achieved via this route. This methodology was extended to the synthesis of aryl/heteroaryl- and diheteroarylmethanols with high levels of enantioselectivity.[11]

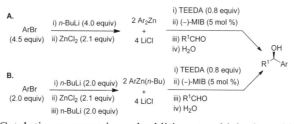

Figure 1. Catalytic asymmetric aryl additions to aldehydes with (A) Ar_2Zn and (B) ArZnBu from aryl bromides.

Charette and coworkers introduced a complementary catalytic asymmetric arylation method by developing a salt-free preparation of diorganozinc reagents using $Zn(OMe)_2$ and alkyl/aryl Grignard reagents (Figure 2).[12] The insoluble salts $Mg(OMe)_2$ and/or NaBr salts were removed by centrifugation or filtration to afford the salt-free diorganozinc reagents. Two examples highlighting the efficiency of this chemistry are illustrated in Figure 2. Ishihara and coworkers adopted Charette's method to synthesize salt-free i-Pr_2Zn and subsequently added it to aldehydes in the presence of 10 mol % (–)-MIB with up to 94% ee.[13,14]

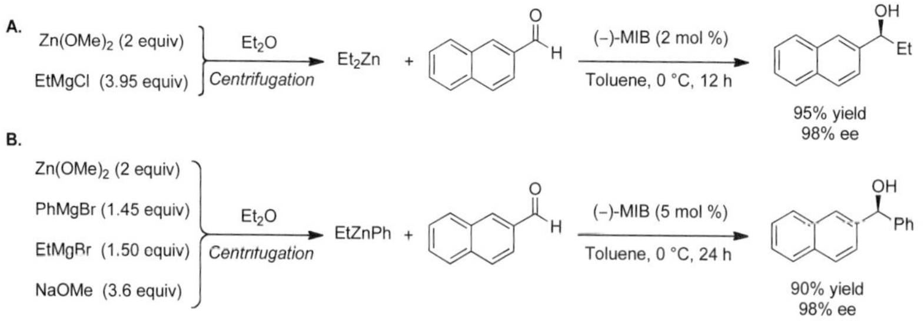

A.

Zn(OMe)₂ (2 equiv)

EtMgCl (3.95 equiv) | Centrifugation → Et₂O → Et₂Zn + [2-naphthaldehyde] → (–)-MIB (2 mol %), Toluene, 0 °C, 12 h → [1-(2-naphthyl)propan-1-ol, OH, Et]

95% yield
98% ee

B.

Zn(OMe)₂ (2 equiv)

PhMgBr (1.45 equiv)

EtMgBr (1.50 equiv) | Centrifugation → Et₂O → EtZnPh + [2-naphthaldehyde] → (–)-MIB (5 mol %), Toluene, 0 °C, 24 h → [phenyl(2-naphthyl)methanol, OH, Ph]

NaOMe (3.6 equiv)

90% yield
98% ee

Figure 2. Charette's catalytic asymmetric (A) alkyl and (B) aryl additions.

2. Synthesis of Chiral Acyclic Epoxy Alcohols

In an effort to streamline the enantio- and diastereoselective synthesis of valuable small molecules,[6] the (–)-MIB-based organozinc catalyzed carbonyl addition step was merged with several other transformations such as epoxidation, cyclopropanation, halocyclopropanation, and various rearrangement reactions. The first tandem reaction developed was for the synthesis of chiral epoxy alcohols.

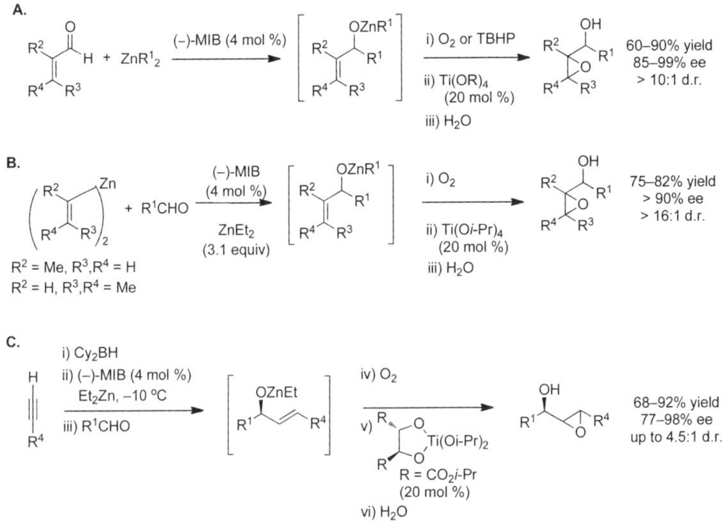

Figure 3. Three one-pot synthesis of epoxy alcohols.

Two complementary methods were developed for one-pot synthesis of highly enantio- and diastereoenriched epoxy alcohols with up to three contiguous stereocenters.[15,16,17] The first route involved highly enantioselective alkyl additions to α,β-unsaturated aldehydes followed by titanium-catalyzed diastereoselective epoxidation with either dioxygen or TBHP (Figure 3A). The second route involves highly enantioselective divinylzinc additions to aliphatic or aromatic aldehydes (Figure 3B). The vinylzinc reagents can be either isolated and purified prior to addition to aldehydes,[18,19] or formed in situ by hydroboration of a terminal alkyne followed by transmetalation to zinc (Figure 3C).[20] The latter method provided access to a synthetically challenging class of secondary *trans*-disubstituted epoxy alcohols with up to 4.5 : 1 dr. Excellent enantioselectivities were obtained with a wide range of aldehyde substitution patterns, except unbranched aldehydes, which undergo additions with up to 85% ee. This alkoxide-based titanium epoxidation catalyst is unique in that the *same* catalyst demonstrated high diastereoselectivity with allylic alkoxides exhibiting either $A^{1,2}$ or $A^{1,3}$ allylic strain in one of the diastereomeric epoxidation transition states.

3. Synthesis of Chiral Allylic Epoxy Alcohols

Asymmetric vinylation of α,β-unsaturated aldehydes in the presence of catalytic amounts of MIB generated unsymmetrical bis(allylic) alkoxide intermediates. In situ alkoxide directed epoxidation afforded densely functionalized allylic epoxy alcohols in good yields and excellent chemo-, diastereo- and enantioselectivity (Figure 4).[16] The vinylzinc reagents may be either prepared and purified (Figure 4A) or generated in situ using Oppolzer's procedure.[20]

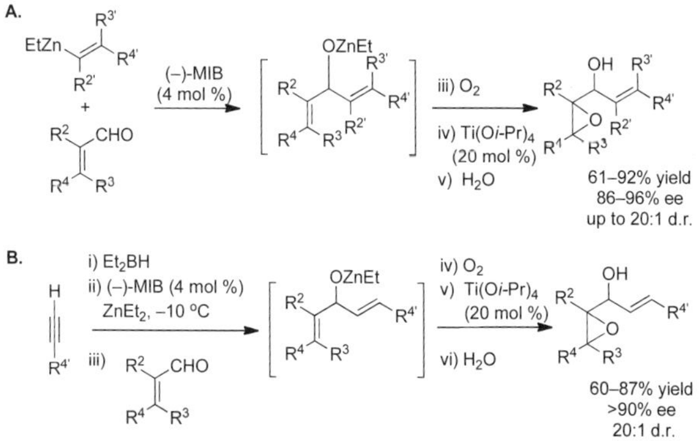

Figure 4. Synthesis of allylic epoxy alcohols using (A) purified vinylzinc reagents, or (B) in situ generated vinylzinc reagents.

The latter procedure afforded only (*E*)-disubstituted vinylzinc reagents whereas the former allowed for more substituted vinylzinc reagents. The enal must bear non-hydrogen substituents in the R^2 or R^3 position so that either $A^{1,2}$ or $A^{1,3}$ strain is present in one of the diastereomeric epoxidation transition states. The unsymmetrical bis(allylic) alkoxide then underwent a highly chemoselective directed epoxidation of the more electron-rich double bond while minimizing $A^{1,2}$ or $A^{1,3}$ strain.

4. Catalytic Asymmetric (*Z*)-Vinylation of Aldehydes

The vinylation procedures above provide (*E*)-allylic alcohols. Walsh and coworkers developed catalytic asymmetric (*Z*)-vinylation of aldehydes

28

with (Z)-vinylzinc reagents via a novel 1,2-metalate rearrangement/transmetalation sequence.[21,22] Thus, hydroboration of 1-halo-1-alkynes followed by reaction with *tert*-butyllithium,[23] transmetalation of the resulting (Z)-vinylboranes to zinc and addition to prochiral aldehydes in the presence of (–)-MIB and the diamine inhibitor TEEDA furnished (Z)-disubstituted allylic alcohols with high enantioselectivity and without contamination by (E)-allylic alcohols (Scheme 1). This transformation is analogous to a net *trans* hydroboration of the corresponding alkynes. Aliphatic aldehydes with α-branching gave products of high enantiopurity whereas β-branched aldehydes gave lower enantioselectivities.

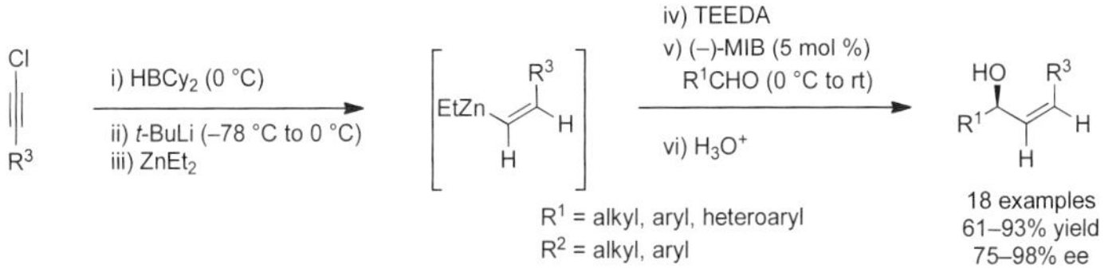

Scheme 1. Catalytic asymmetric synthesis of (Z)-allylic alcohols.

4.1 Tandem Catalytic Asymmetric (Z)-Vinylation of Aldehydes/Diastereoselective Epoxidation and Cyclopropanation

The generation and addition of (Z)-vinylzinc reagents to aldehydes were then applied to two one-pot tandem reactions. In the first tandem reaction, the resulting allylic alkoxides were treated with Et_2Zn, TBHP, and $Ti(Oi\text{-}Pr)_4$ to perform the diastereoselective and/or chemoselective epoxidation to synthesize epoxy alcohols and (Z)-allylic epoxy alcohols (Figure 5A).[21] In the second tandem reaction, the allylic alkoxides were subjected to 5 equiv each of CF_3CH_2OH, Et_2Zn, and CH_2I_2 to provide highly

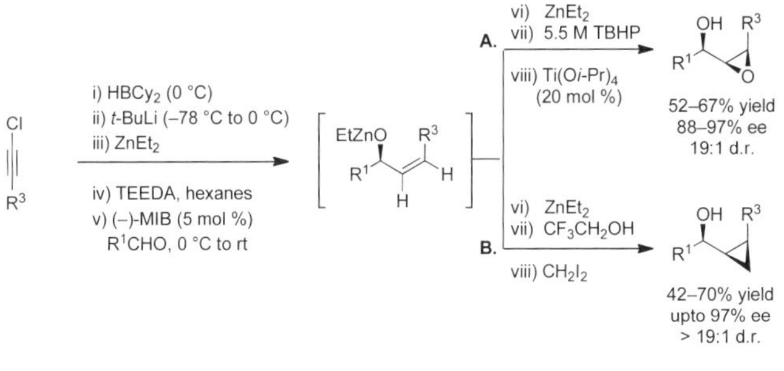

Figure 5. Tandem syntheses of (A) epoxy and allylic epoxy alcohols and (B) *syn-cis*-disubstituted cyclopropyl alcohols.

enantio- and diastereoenriched *syn-cis*-disubstituted cyclopropyl alcohols (Figure 5B).[24] A variety of 1-chloro-1-alkynes and aldehydes (saturated, aromatic, and heteroaromatic) were employed successfully in these tandem reactions.

4.2 Catalytic Asymmetric Synthesis (Z)-Trisubstituted Allylic Alcohols

The (Z)-vinyl zinc reagents generated were *all* disubstituted because the 1,2-metalate rearrangement was executed with a hydride source. A variant of the 1,2-metalate rearrangement/transmetalation sequence for the stereospecific generation of (Z)-*trisubstituted* vinyl zinc reagents was also developed.[25] Hydroboration of 1-bromo-1-hexyne with either diethyl- or dicyclohexylborane followed by 3 equiv of diethylzinc provided the (Z)-trisubstituted vinyl zinc reagents. The dialkylzinc served a two-fold function: it induced a 1,2-metalate shift to form the new C–C bond and promoted the boron to zinc transmetallation. In the presence of TMEDA (to inhibit the zinc halide by-product) and catalytic amounts of (–)-MIB, these reagents were then employed in the catalytic asymmetric addition of (Z)-trisubstituted vinyl zinc reagents to aldehydes to furnish enantioenriched (Z)-trisubstituted allylic alcohols in good yields and excellent enantioselectivity (Scheme 2). Unfortunately, this procedure could not be adapted to the highly enantioselective synthesis of α-methyl-substituted allylic alcohols despite screening several additives (ee <30%, 15–50% yield).

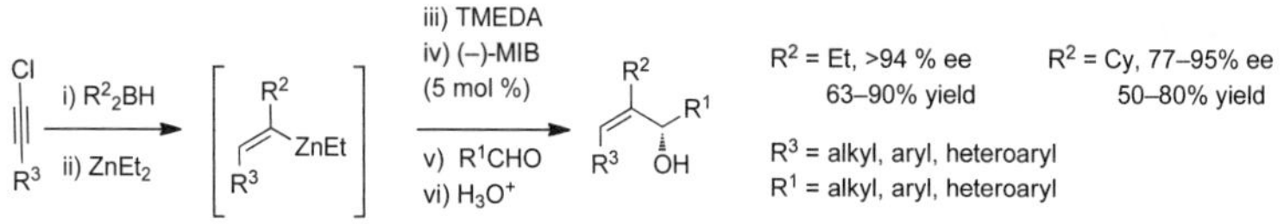

Scheme 2. Catalytic asymmetric synthesis of α-ethyl and α-cyclohexyl (Z)-trisubstituted allylic alcohols.

5. Tandem Catalytic Asymmetric Addition/Diastereoselective Cyclopropanation

5.1. Synthesis of *syn*-Cyclopropyl Alcohols. Two tandem routes have been developed to synthesize highly enantio- and diastereoenriched *syn*-cyclopropyl alcohols.[26] The first route involved enantioselective alkyl addition to α,β-unsaturated aldehydes in the presence of (–)-MIB (Figure 6A) whereas the second route involved addition of vinylzinc reagents to aldehydes to furnish the allylic alkoxide intermediates (Figure 6B). After

30

removal of the volatile materials, the intermediate allylic alkoxide was exposed to either EtZnCH$_2$I or the more reactive CF$_3$CH$_2$OZnCH$_2$I[27] to furnish *syn*-cyclopropyl alcohols with high enantio- and diastereoselectivity. The first route has a broader substrate scope, but is more challenging because unsaturated aldehydes isomerize readily. The second route provides only (*E*)-disubstituted *syn*-cyclopropyl alcohols.

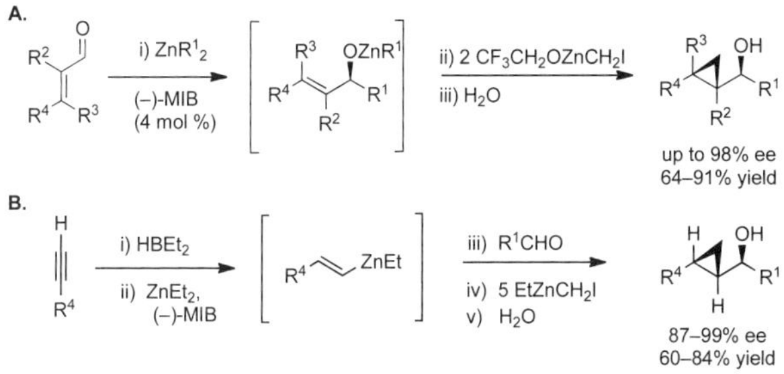

Figure 6. Tandem asymmetric A) alkyl addition to enals followed by diastereoselective cyclopropanation and B) vinylation of aldehydes followed by diastereoselective cyclopropanation.

5.2. Synthesis of Enantioenriched Dienols and *syn*-Vinylcyclopropanes

The catalytic enantioselective vinylation was extended to addition of dienyl groups to aldehydes in the presence of 10 mol % (–)-MIB.[24] The requisite dienyl zinc intermediates were synthesized via chemo- and regioselective hydroboration of enynes followed by transmetallation with diethylzinc. Dienols were obtained in 79–93% yield and 76–94% ee (Figure 7A).

This methodology was further extended to the synthesis of vinylcyclopropanes (VCPs). The dienylzinc alkoxide intermediates were subjected to EtZnCH$_2$I to provide vinyl cyclopropanes with high chemo-, enantio- and diastereoselectivity (Figure 7B). The alkoxide directed cyclopropanation of allylic C=C bonds is faster than remote C=C bonds.[28] A limitation of this method is that aromatic aldehydes were unsuccessful coupling partners.

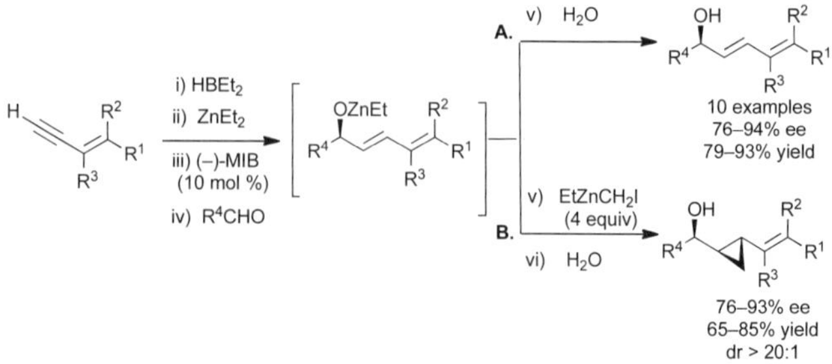

Figure 7. Asymmetric dienylation and diastereoselective cyclopropanation.

5.3. Synthesis of *anti*-Cyclopropyl Alcohols

The inherent bias for *syn*-selectivity in the alkoxide directed cyclopropanation could be switched to synthesize *anti*-cyclopropyl alcohols using a strategy developed by Charette and coworkers[29] wherein allylic alcohols were protected with bulky silyl groups to prevent coordination to zinc carbenoids. In our method for the synthesis of *anti*-cyclopropanes,[19] we silylated the intermediate zinc alkoxide in situ with TMSCl/Et$_3$N and then subjected the silyl ether to cyclopropanation conditions. The cyclopropyl zinc alkoxides were desilylated in situ to furnish *anti*-cyclopropyl alcohols in 60–82% yield with high enantio- and diastereoselectivity (Scheme 3). The Et$_3$N is likely necessary to break up the zinc aggregates, rendering the zinc alkoxides more nucleophilic towards TMSCl.

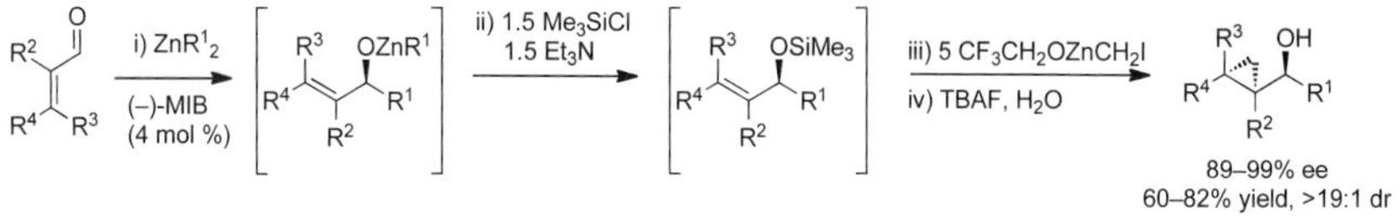

Scheme 3. One-pot tandem asymmetric synthesis of *anti*-cyclopropyl alcohols.

5.4. Synthesis of *syn*-Halocyclopropyl Alcohols

The catalytic enantio- and diastereoselective tandem generation of cyclopropyl alcohols was extended to the synthesis of halo-substituted cyclopropyl alcohols.[26,30] The enantioenriched zinc alkoxide intermediate was subjected to Et$_2$Zn, CF$_3$CH$_2$OH, and either iodoform, bromoform or dichlorobromomethane to furnish iodo-, bromo- or chlorocyclopropyl alcohols respectively in good yields and excellent enantioselectivity (Figure 8A). In these one-pot tandem halocyclopropanation reactions, four consecutive stereogenic centers are established with excellent

32

diastereoselectivity starting from simple achiral α,β-unsaturated aldehyde precursors. Interrogation of the cyclopropyl stereochemistry via ^{1}H NMR and X-ray analyses led to an interesting find; when R^4 = alkyl or H, the halo group was *cis* to the carbinol, whereas when R^4 = Ph, the halo group was *trans*. This switch in stereochemical bias was rationalized by invoking a zinc-phenyl-π interaction.[31]

A complementary approach to iodocyclopropyl alcohols was developed using MIB-catalyzed asymmetric vinyl addition as the first step followed by subjection of the zinc alkoxide intermediate to 3 equiv of $Zn(CHI_2)_2$ (Figure 8B).[30] The iodocyclopropyl alcohols can be further allylated with allyl/methallyl bromide in the presence of $LiCu(n$-Bu$)_2$ to furnish 1,2,3-disubstituted cyclopropanes in good yields as single diastereomers with full retention of the cyclopropane stereochemistry (not shown).

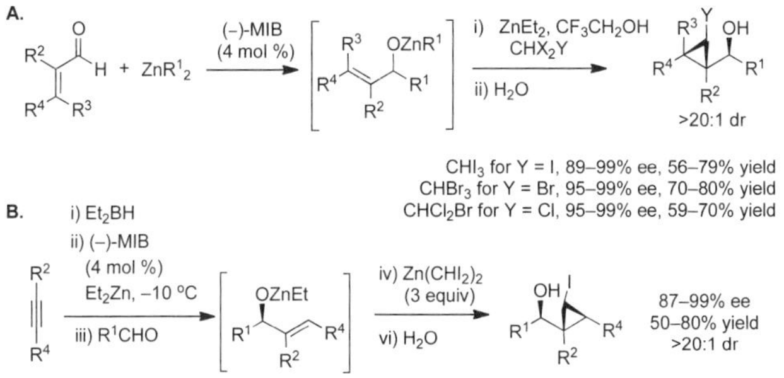

Figure 8. Tandem asymmetric (A) alkyl addition to enals followed by diastereoselective halocyclopropanations and (B) vinyl addition to aldehydes followed by diastereoselective iodocyclopropanation.

6. Catalytic Asymmetric Aminovinylation of Aldehydes: Synthesis of β-Hydroxyenamines, β-Aminoalcohols, and *syn*-Aminocyclopropyl Alcohols

Regioselective hydroboration of ynamides followed by boron to zinc transmetallation and subsequent addition to aldehydes in the presence of 5 mol % (−)-MIB furnished β-hydroxyenamines in moderate yields and high enantioselectivities (up to 98% ee, Figure 9A).[32]

A tandem catalytic asymmetric aminovinylation/diastereoselective cyclopropanation reaction was developed to synthesize *syn*-aminocyclopropyl alcohols with excellent diastereoselectivities (>20:1) in moderate yields (Figure 9B).[32]

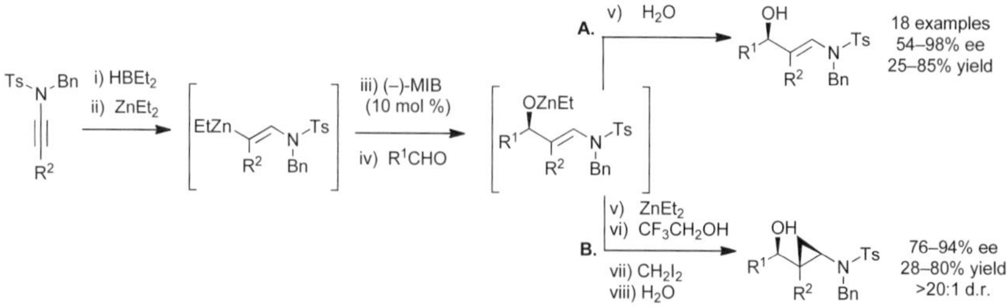

Figure 9. A) Synthesis of β-hydroxyenamines and B) tandem synthesis of *syn*-aminocyclopropyl alcohols.

7. Catalytic Asymmetric Ethoxyvinylation of Aldehydes

Highly enantioselective addition of ethoxyvinyl zinc reagents, generated via hydroboration of ethoxyacetylene followed by in situ transmetalation to zinc and addition to aldehydes in presence of MIB afforded hydroxyenol ethers with high ee (89–95%) and yields (>93%, Scheme 4).[33] Subsequent hydrolysis generated two carbon homologated enantioenriched β-hydroxy aldehydes. In the case of addition to chiral β-hydroxy aldehydes, mismatched and matched catalyst-substrate combinations can be used to achieve moderate to good diastereoselectivities of either the *syn*- (up to 3.8:1 with (–)-MIB) or *anti*-diols (>9:1 with (+)-MIB).

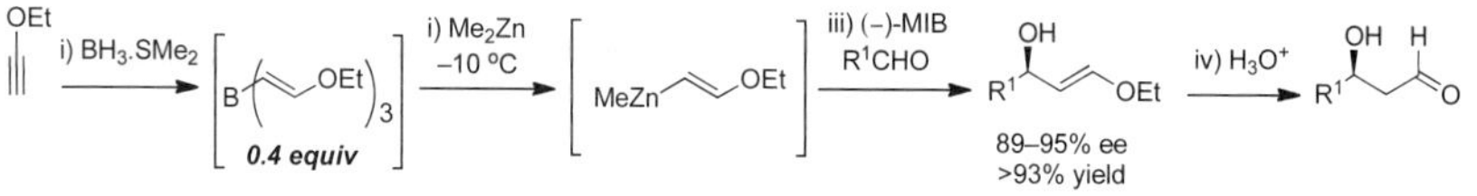

Scheme 4. One-pot tandem catalytic asymmetric ethoxyvinylation of aldehydes.

8. Other uses of MIB in Organic Synthesis

8.1. Synthesis of γ-Unsaturated β-Amino Acid Derivatives

A catalytic enantioselective synthesis of γ-unsaturated β-amino acid derivatives was achieved in three steps from trityl protected 1-butyne-3-ol.[34] The enantioenriched allylic alcohols were transformed into the corresponding allylic amines via a [3,3]-sigmatropic trichloroacetimidate rearrangement, and led to γ-unsaturated β-amino acid derivatives with high ee after a one-pot deprotection-oxidation sequence (Scheme 5). Similar

34

[3,3]-sigmatropic allyl cyanate-to-isocyanate rearrangement reactions were executed to access enantioenriched allylic amines en route to the syntheses of glycocinnasperimicin D and pachastrissamine (Figure 10).[35,36]

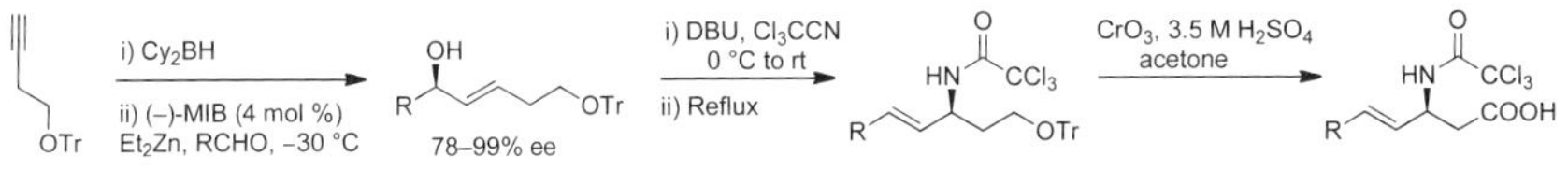

Scheme 5. Asymmetric synthesis of $\tilde{\gamma}$unsaturated β-amino acid derivatives via a [3,3]-sigmatropic rearrangement.

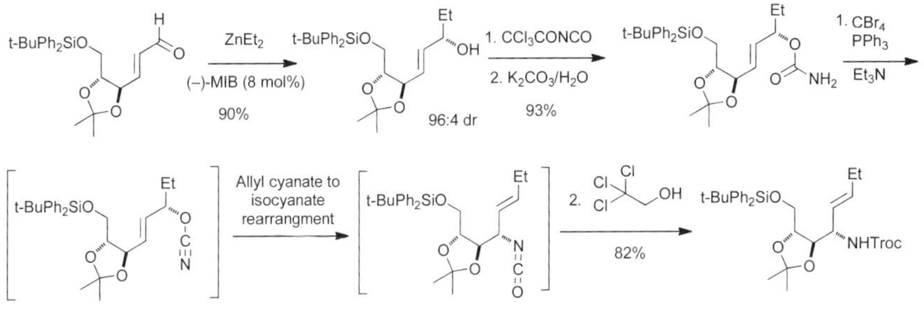

Figure 10. [3,3]-Sigmatropic allyl cyanate-to-isocyanate rearrangement reactions en route to the synthesis of glycocinnasperimicin D.

8.2. Synthesis of Di(allyl) Ether Derivatives

Nelson and coworkers added Et₂Zn to conjugated enals in the presence of MIB followed by O-allylation of the enantioenriched zinc alkoxides to provide di(allyl) ethers in good yields (73–87%) and high ee (88–98%).[37,38,39] Subsequent olefin isomerization and Claisen rearrangement provided access to a variety of enantio- and diastereoenriched Claisen adducts (Scheme 6).

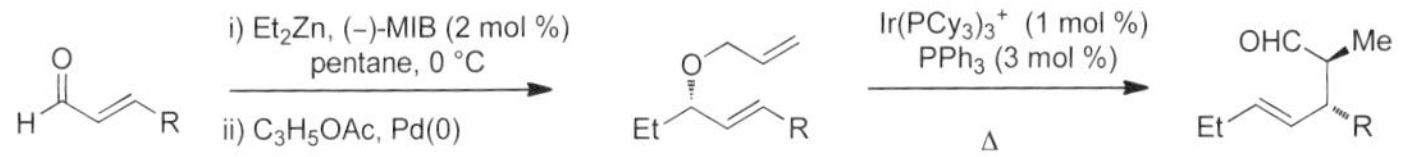

Scheme 6. Asymmetric synthesis of di(allyl) derivatives.

8.3. One-pot Catalytic Asymmetric Synthesis of Pyranones

Enantioenriched pyranones with >90% ee were prepared via a one-pot tandem asymmetric alkylation of 2-furfurals in the presence of catalytic (–)-MIB followed by oxidation with NBS (Scheme 7).[40]

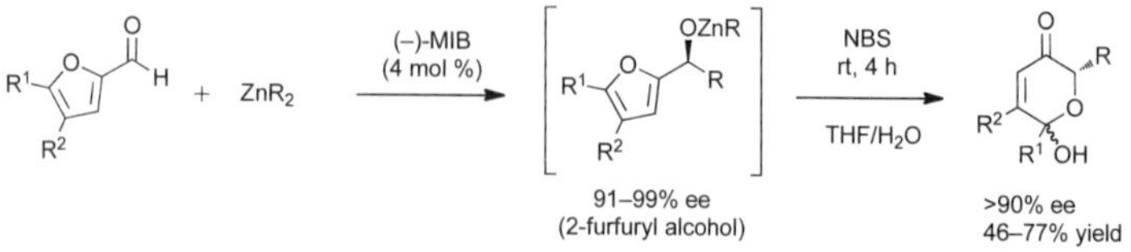

91–99% ee
(2-furfuryl alcohol)

>90% ee
46–77% yield

Scheme 7. One-pot tandem asymmetric synthesis of enantioenriched pyranones.

8.4. Synthesis of Enantiomerically Enriched 1,2,4- Trioxanes

Enantioenriched allylic alcohols can be subjected to a hydroxyl directed regio- and diastereoselective photooxygenation reaction with O_2 in the presence of tetraphenyl porphyrin (TPP) to obtain allylic hydroperoxides in >10:1 dr, with the *threo*-isomer as the major product (Scheme 8).[41] The allylic hydroperoxides were further reacted with cyclic ketones in the presence of catalytic *p*-TsOH to yield enantioenriched 1,2,4-trioxanes (16–78% yield) that exhibited antimalarial activity.

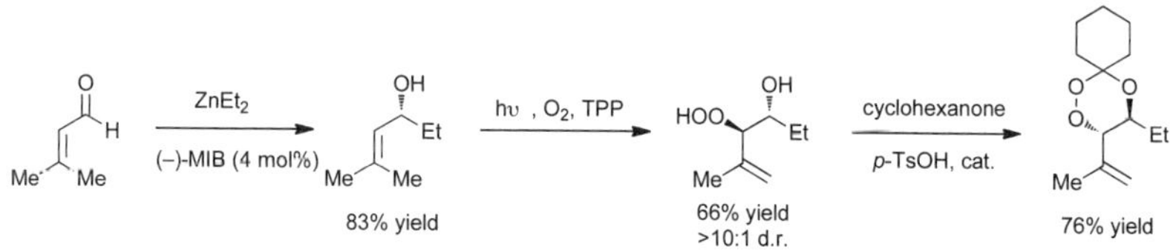

83% yield

66% yield
>10:1 d.r.

76% yield

Scheme 8. Synthesis of enantioenriched β-hydroperoxy alcohols and 1,2,4-trioxanes.

8.5. Polystyrene-supported MIB-derived Ligands

Pericas and coworkers have successfully synthesized and immobilized 3-exopiperazinoisoborneol (PIB), a close analog of MIB, to Merrifield resins (Scheme 9).[42] Polystyrene-supported PIB possessed high catalytic activity and improved chemical stability, and was employed as a ligand (10 mol %) in the asymmetric alkylation of aldehydes with Et_2Zn in batch methods to produce highly enantioenriched alcohols in good yields (50–92% y, 92–99% ee). This method is amenable to continuous flow methods for over 30 h with high conversion and no erosion in enantioselectivity. In this fashion, industrial scale amounts (13.0 g) of enantiopure alcohol were isolated in a single continuous flow operation leading to >30 fold better performance compared to batch conditions (TON = 251 with respect to the product).

36

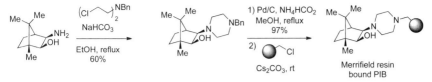

Scheme 9. Synthesis of polystyrene-supported PIB

8.6. MIB in Syntheses of Natural Products

Over the years, MIB has been employed in the synthesis of a number of natural products. In all of these syntheses, MIB has been primarily used in asymmetric alkylation or vinylation reactions to provide diastereo- and enantioenriched alcohols.[35,36,41,43,44] In their formal synthesis of leucascandrolide A, Hong and coworkers employed (–)-MIB in the diastereoselective vinylation of a chiral aldehyde to furnish the requisite allylic alcohol with >32:1 diastereoselectivity (Figure 11).[44] The enantio- and diastereoselective addition step can be further coupled with efficient transformations en route to synthesis of natural products. As outlined earlier, a [3,3]-sigmatropic allyl cyanate-to-isocyanate rearrangement reaction was employed in the synthesis of glycocinnasperimicin D and pachastrissamine (jaspine B) (Figure 10)[35,36] while a diastereoselective Schenk ene reaction with singlet oxygen was utilized in the synthesis of artemisin-type 1,2,4-trioxanes (Scheme 8).[41] It is hopeful that synthetic chemists will adopt the tandem chemo-, regio- and diastereoselective transformations reactions presented herein in efficient synthesis of their natural products.

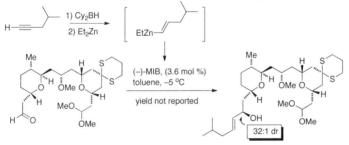

Figure 11. Hong and coworkers diastereoselective vinyl carbonyl addition in the formal synthesis of leucascandrolide A.

Both Nugent's (+)- and (–)-MIB have found significant synthetic utility as a ligand of choice for addition of organozinc groups to carbonyl compounds. The efficient installation of chirality, coupled with tandem chemo-, regio- and diastereoselective transformations, provides high-value added building blocks that augment the synthetic organic chemist's repertoire of enantioenriched small molecules.

1. P. Roy and Diana T. Vagelos Laboratories, University of Pennsylvania, Department of Chemistry, 231 South 34th Street, Philadelphia, PA 19104.
2. Chemical Biology Program, The Broad Institute of Harvard and Massachusetts Institute of Technology, Cambridge, MA 02142.
3. (a) Nugent, W. A. *Chem. Commun.* **1999**, 1369. (b) Nugent, W. A., U. S. Pat. 6,187,918, **2001** (to DuPont Pharmaceutical Company; rights currently held by Bristol-Myers Squibb).
4. White, J. D.; Wardrop, D. J.; Sundermann, K. F. *Org. Synth.* **2002**, *79*, 130.
5. Hussain, M. M.; Walsh, P. J. in The Encyclopedia of Reagents for Organic Synthesis Ed. Pacquette, L. A.; Crich, D.; Fuchs, P. L.; Molander, G. A., John Wiley & Sons, Ltd., Chichester, UK, **2009**, *9*, pp. 7205.
6. Hussain, M. M.; Walsh, P. J. *Acc. Chem. Res.* **2008**, *41*, 883.
7. Kim, H. Y.; Walsh, P. J. *Acc. Chem. Res.* **2012**, DOI: 10.1021/ar300052s.
8. Kim, H. Y.; Walsh, P. J. *J. Phys. Org. Chem.* **2012** DOI: 10.1002/poc.2965.
9. Chen, Y. K.; Jeon, S.; Walsh, P. J.; Nugent, W. A. *Org. Synth.* **2005**, *82*, 87.
10. Kim, J. G.; Walsh, P. J. *Angew. Chem., Int. Ed.* **2006**, *25*, 4175.
11. Salvi, L.; Kim, J. G.; Walsh, P. J. *J. Am. Chem. Soc.* **2009**, *131*, 12483.
12. Côté, A.; Charette, A. B. *J. Am. Chem. Soc.* **2008**, *9*, 2771.
13. Hatano, M.; Mizuno, T.; Ishihara, K. *Chem. Commun.* **2010**, *46*, 5443.
14. Hatano, M.; Gouzu, R.; Mizuno, T.; Abe, H.; Yamada, T.; Ishihara, K. *Catal. Sci. Technol.* **2011**, *1*, 1149.
15. Lurain, A. E.; Maestri, A.; Kelly, A. R.; Carroll, P. J.; Walsh, P. J. *J. Am. Chem. Soc.* **2004**, *42*, 13608.
16. Kelly, A. R.; Lurain, A. E.; Walsh, P. J. *J. Am. Chem. Soc.* **2005**, *42*, 14668.
17. Lurain, A. E.; Carroll, P. J.; Walsh, P. J. *J. Org. Chem.* **2005**, *4*, 1262.
18. Shibata, T.; Nakatsui, K.; Soai, K. *Inorg. Chim. Acta* **1999**, 33.
19. Wooten, A.; Carroll, P. J.; Maestri, A. G.; Walsh, P. J. *J. Am. Chem. Soc.* **2006**, 4624.
20. Oppolzer, W.; Radinov, R. N. *Helv. Chim. Acta* **1992**, 170.
21. Salvi, L.; Jeon, S.; Fisher, E. L.; Carroll, P. J.; Walsh, P. J. *J. Am. Chem. Soc.* **2007**, *51*, 16119.

22. Jeon, S.-J.; Fisher, E. L.; Carroll, P. J.; Walsh, P. J. *J. Am. Chem. Soc.* **2006**, *128*, 9618.
23. Campbell, J. B. J.; Molander, G. A. *J. Organomet. Chem.* **1978**, *156*, 71.
24. Kim, H. Y.; Salvi, L.; Carroll, P. J.; Walsh, P. J. *J. Am. Chem. Soc.* **2010**, *132*, 402.
25. Kerrigan, M. H.; Jeon, S.-J.; Chen, Y. K.; Salvi, L.; Carroll, P. J.; Walsh, P. J. *J. Am. Chem. Soc.* **2009**, *131*, 8434.
26. Kim, H. Y.; Lurain, A. E.; Garcia-Garcia, P.; Carroll, P. J.; Walsh, P. J. *J. Am. Chem. Soc.* **2005**, *38*, 13138.
27. Lorenz, J. C.; Long, J.; Yang, Z.; Xue, S.; Xie, Y.; Shi, Y. *J. Org. Chem.* **2004**, 327.
28. Charette, A. B.; Juteau, H.; Lebel, H.; Molinaro, C. *J. Am. Chem. Soc.* **1998**, *120*, 11943.
29. Charette, A. B.; Lacasse, M.-C. *Org. Lett.* **2002**, *4*, 3351.
30. Kim, H. Y.; Salvi, L.; Carroll, P. J.; Walsh, P. J. *J. Am. Chem. Soc.* **2009**, *131*, 954.
31. Guerrero, A.; Martin, E.; Hughes, D. L.; Kaltsoyannis, N.; Bochmann, M. *Organometallics* **2006**, *25*, 3311.
32. Valenta, P.; Carroll, P. J.; Walsh, P. J. *J. Am. Chem. Soc.* **2010**, *132*, 14179.
33. Jeon, S.; Chen, Y. K.; Walsh, P. J. *Org. Lett.* **2005**, *9*, 1729.
34. Lurain, A. E.; Walsh, P. J. *J. Am. Chem. Soc.* **2003**, *35*, 10677.
35. Nishiyama, T.; Kusumoto, Y.; Okumura, K.; Hara, K.; Kusaba, S.; Hirata, K.; Kamiya, Y.; Isobe, M.; Nakano, K.; Kotsuki, H.; Ichikawa, Y. *Chem.-Eur. J.* **2010**, *16*, 600.
36. Ichikawa, Y.; Matsunaga, K.; Masuda, T.; Kotsuki, H.; Nakano, K. *Tetrahedron* **2008**, *64*, 11313.
37. Nelson, S. G.; Wang, K. *J. Am. Chem. Soc.* **2006**, *13*, 4232.
38. Nelson, S. G.; Bungard, C. J.; Wang, K. *J. Am. Chem. Soc.* **2003**, *43*, 13000.
39. Wang, K.; Bungard, C. J.; Nelson, S. G. *J. Am. Chem. Soc.* **2007**, *12*, 2325.
40. Cheng, K.; Rowley, K. A.; Kohn, R. A.; Dweck, J. F.; Walsh, P. J. *Org. Lett.* **2009**, *11*, 2703.
41. Sabbani, S.; La, P. L.; Bacsa, J.; Hedenstroem, E.; O'Neill, P. M. *Tetrahedron* **2009**, *65*, 8531.
42. Osorio-Planes, L.; Rodriquez-Escrich, C.; Pericas, M. A. *Org. Lett.* **2012**, *14*, 1816.
43. Mori, K.; Tashiro, T.; Zhao, B.; Suckling, D. M.; El-Sayed, A. M. *Tetrahedron* **2010**, *66*, 2642.
44. Lee, K.; Kim, H.; Hong, J. *Org. Lett.* **2011**, *13*, 2722.

Mahmud M. Hussain received his B.A. in Chemistry from Bard College in 2005 and his Ph. D. in synthetic organic chemistry in 2010 with Prof. Patrick J. Walsh from the University of Pennsylvania where he received an *Ahmed Zewail Graduate Fellowship*. Dr. Hussain specializes in new reaction development with particular emphasis on chemoselectivity, stereocontrol and asymmetric catalysis. Dr. Hussain is presently a Howard Hughes Medical Institute postdoctoral fellow in the laboratories of Prof. Stuart L. Schreiber at Harvard University and the Broad Institute. Currently, he aims to identify and optimize novel small molecule modulators of mutant isocitrate dehydrogenase enzymes and autophagy-related diseases.

Patrick J. Walsh received his B.A. from UC San Diego (1986) and Ph.D. with Prof. Robert G. Bergman at UC Berkeley (1991). He was an NSF postdoctoral fellow with Prof. K. B. Sharpless at the Scripps Research Institute. He holds the Alan G. MacDiarmid Chair at the University of Pennsylvania. Walsh's interests are in asymmetric catalysis, development of new methods, reaction mechanisms, and inorganic synthesis. With Prof. Marisa Kozlowski Walsh wrote *Fundamentals of Asymmetric Catalysis*.

Preparation of (*E*)-*N*,*N*-Diethyl-2-styrylbenzamide by Rh-Catalyzed C-H Activation

Submitted by Frederic W. Patureau, Tatiana Besset and Frank Glorius.[1]
Checked by Takaaki Harada and Tohru Fukuyama.

1. Procedure

(E)-N,N-Diethyl-2-styrylbenzamide: A flame-dried, 200-mL, three-necked round-bottomed flask is equipped with 2.5 cm rod-shaped, Teflon-coated, magnetic stirbar, glass stopper, rubber septum and reflux condenser fitted with argon inlet (Note 1). The flask is flushed with argon and charged with [Cp*RhCl₂]₂ (45.8 mg, 0.074 mmol, 0.25 mol%) (Note 2), AgSbF₆ (102 mg, 0.30 mmol, 1.00 mol%) (Note 3), Cu(OAc)₂ (11.3 g, 62.2 mmol, 2.10 equiv) (Note 4) and 1,4-dioxane (30 mL) (Note 5). *N*,*N*-Diethylbenzamide (5.25 g, 29.6 mmol, 1 equiv) (Note 6) in 1,4-dioxane (10 mL) is added via cannula and styrene (5.11 mL, 44.4 mmol, 1.50 equiv) (Note 7) is added via syringe into the resulting blue-green suspension. The rubber septum is changed to a glass stopper and the stirred suspension is heated in an oil bath (125 °C) for 23 h (Note 8). The reaction mixture is allowed to cool to room temperature, gravity filtered through a cotton plug into a 500-mL round bottomed flask and rinsed with toluene (150 mL) (Note 9). The resulting green solution is concentrated by rotary evaporation (40 °C water bath, 40 mmHg,). Toluene (2 x 200 mL) is added and then evaporated twice (Note 10) (40 °C water bath, 40 mmHg). The crude is filtered through a short plug of silica gel (Notes 11 and 12), washed with EtOAc (500 mL), and concentrated by rotary evaporation (40 °C water bath, 70 mmHg) to provide a yellow viscous oil (Note 13). This material is purified by silica gel column chromatography (Notes 14 and 15). The combined eluent is concentrated by rotary evaporation (40 °C water bath, 70 mmHg) and then submitted to high vacuum (0.3 mmHg) at 80 °C for 24 h (Note 16) to furnish

Org. Synth. **2013**, *90*, 41-51
Published on the Web 8/6/2012
© 2013 Organic Syntheses, Inc.

7.62 g (27.3 mmol, 92%) of (*E*)-*N*,*N*-diethyl-2-styrylbenzamide as a light yellow viscous oil (Note 17).

2. Notes

1. The submitters used a dry Schlenk-tube (capacity: 166 mL, diameter: 45 mm) with a screw cap equipped with a 2.5 cm magnetic stir bar under an argon atmosphere. The submitters have shown that air could serve as terminal oxidant; nevertheless, the reaction is less effective in moist conditions, presumably due to the competitive binding of water to the active metal coordination sites. Therefore, use of an argon atmosphere is recommended.

2. The submitters purchased [Cp*RhCl$_2$]$_2$ from Heraeus and stored the catalyst in a glovebox under argon. The checkers purchased [Cp*RhCl$_2$]$_2$ from Aldrich and used the material as received.

3. The submitters purchased AgSbF$_6$ from ABCR and stored the material in a glovebox under argon. The checkers purchased AgSbF$_6$ from Aldrich and used the material as received.

4. The submitters prepared Cu(OAc)$_2$ from Cu(OAc)$_2$(H$_2$O), which was purchased from Aldrich, by heating at 100 °C under high vacuum (approx. 0.1 mmHg) for 48 h. The checkers prepared Cu(OAc)$_2$ from Cu(OAc)$_2$(H$_2$O), which was purchased from Aldrich, by heating at 100 °C under vacuum (2 mmHg) for 72 h.

5. The checkers purchased 1,4-dioxane from Aldrich and the solvent was stored over activated 4Å molecular sieves.

6. The submitters prepared *N*,*N*-diethylbenzamide, R$_f$ = 0.3 (pentane/ethyl acetate 8:2), a light yellow liquid, from a simple condensation of Et$_2$NH with benzoyl chloride. The checkers purchased *N*,*N*-diethylbenzamide (98.0+%) from Wako and used the material as received.

7. The submitters purchased styrene from Acros and used as received. The checkers purchased styrene (>99.0%) from Tokyo Chemical Industry Co., Ltd. and used it as received.

8. The consumption of the starting material was monitored by TLC analysis on Merck silica gel 60 F254 plates (0.25 mm, glass-backed, visualized with 254 nm UV lamp) using 20% acetone in toluene as an eluant. *N*,*N*-Diethylbenzamide (starting material) had R$_f$ = 0.49 (UV active) and (*E*)-*N*,*N*-Diethyl-2-styrylbenzamide had R$_f$ = 0.58 (UV active)

9. Large brown-red precipitate that is typical for Cu^I/Cu^0 species is removed.

10. Azeotropic removal of styrene and 1,4-dioxane was performed in order to simplify the purification process.

11. The crude material is dissolved in EtOAc (20 mL) and then charged onto a silica gel column (diameter = 3 cm, height = 10 cm) of 35 g (80 mL) of silica gel.

12. The checkers purchased silica gel (acidic) from Kanto Chemical Co., Inc. (40–100 μm). The submitters purchased silica gel (0.040–0.063 mm) from Merck.

13. The submitters performed GC-MS analysis of the crude material and detected only the peak corresponding to the product, corresponding to the only major spot (R_f = 0.4) on the TLC plate (pentane/ethyl acetate = 8:2, UV detection)

14. The crude material was dissolved in toluene (10 mL) and then charged onto a column (diameter = 10 cm, height = 11 cm) of 425 g (1000 mL) of silica gel. The column was eluted with n-hexane/EtOAc = 8:1 (7.0 L) to 2:1 (3.6 L) and 100-mL fractions were collected. Fractions 73-108 were combined.

15. The major by-product was a dimer of N,N-diethylbenzamide (R_f = 0.62 (hexane/ethyl acetate = 8:2, UV detection with 254 nm UV lamp), Merck silica gel 60 F_{254} plates (0.25 mm, glass-backed) were used for TLC analysis). Fractions 46-70 were combined to obtain 216 mg of such compound.

16. A very minor condensation may appear eventually in the vacuum line system (< 10 mg). NMR shows it to be exclusively the starting material N,N-diethylbenzamide. This represents a convenient way to remove trace impurities that may still be present in spite of the silica gel purification.

17. (E)-N,N-Diethyl-2-styrylbenzamide has the following physicochemical and spectroscopic properties: R_f = 0.4 (pentane/ethyl acetate = 8:2), R_f = 0.34 (hexane/ethyl acetate = 8:2); ^{1}H NMR (400 MHz,

CDCl$_3$) δ: 1.00 (t, ^3J = 7.1 Hz, 3 H), 1.31 (t, ^3J = 7.1 Hz, 3 H), 3.11 (q, ^3J = 7.1 Hz, 2 H), 3.39 (broad s, 1 H), 3.85 (broad s, 1 H), 7.09 (d, ^3J = 16.5 Hz, 1 H), 7.13 (d, ^3J = 16.5 Hz, 1 H), 7.23–7.47 (m, 8 H), 7.70 (d, ^3J = 7.8 Hz, 1 H); ^{13}C NMR (100 MHz, CDCl$_3$) δ: 12.9 (s, CH$_3$), 13.8 (s, CH$_3$), 38.8 (s, CH$_2$), 42.8 (s, CH$_2$), 124.9 (s, CH), 125.2 (s, CH), 126.1 (s, CH), 126.5 (s, CH), 127.5 (s, CH), 127.8 (s, CH), 128.6 (s, CH), 128.7 (s, CH), 130.8 (s, CH), 133.6 (s, C$_{quat}$), 136.3 (s, C$_{quat}$), 137.0 (s, C$_{quat}$), 170.3 (s, C$_{quat}$); HRMS (ESI): m/z calc. for [(C$_{19}$H$_{21}$NO)Na]$^+$ 302.15208, found 302.15169; IR cm^{-1}: 2974, 2933, 1626, 1428, 1380, 1363, 1314, 1285, 1220, 1113, 1075, 963, 762, 692, 627, 547, 528; Anal. Calc. for C$_{19}$H$_{21}$NO: C, 81.68; H, 7.58; N, 5.01. Found: C, 81.29; H, 7.49; N, 4.94.

Handling and Disposal of Hazardous Chemicals

The procedures in this article are intended for use only by persons with prior training in experimental organic chemistry. All hazardous materials should be handled using the standard procedures for work with chemicals described in references such as "Prudent Practices in the Laboratory" (The National Academies Press, Washington, D.C., 2011 www.nap.edu). All chemical waste should be disposed of in accordance with local regulations. For general guidelines for the management of chemical waste, see Chapter 8 of Prudent Practices.

In the development and checking of these procedures, every effort has been made to identify and minimize potentially hazardous steps. The Editors believe that the procedures described in this article can be carried out with minimal risk if performed with the materials and equipment specified, and in careful accordance with the instructions provided. However, these procedures must be conducted at one's own risk. *Organic Syntheses, Inc.*, its Editors, and its Board of Directors do not warrant or guarantee the safety of individuals using these procedures and hereby disclaim any liability for any injuries or damages claimed to have resulted from or related in any way to the procedures herein.

3. Discussion

The oxidative Heck-type reaction, as pioneered by Fujiwara and Moritani,[2] has emerged as an attractive method for the coupling of arenes and olefins, because in contrast to the traditional Heck reaction,[3] it obviates

44

prior activation of either reaction partner. For this cross-coupling reaction, palladium-based complexes are traditionally privileged catalysts,[4-6] for example in the C-H oxidative olefination of acetanilides with acrylate derivatives reported by de Vries, van Leeuwen et al.,[5] and in the use of remote carboxylic acids as efficient directing groups described by Yu et al.[6] Miura and Satoh et al.,[7] we,[8] and others have looked at other transition metals for these C-H activation processes, specifically rhodium, which often allows lower catalytic loadings, higher selectivities, and broader olefin scope. Whereas acrylates were previously used as the preferred reaction partner, recently we have had some success with the rhodium-catalyzed coupling of unactivated acetanilides (electron-rich arenes) with acrylates, styrenes and even ethylene itself (Scheme 1).[8a]

The resulting olefinated products **1** might serve as valuable building blocks. Still, the C-H activation processes of many difficult-to-activate, *electron-poor* substrates such as common carbonylated arenes have remained underdeveloped,[9,10] especially those with an (often desired) oxidative character. In addition, we have recently found that benzamides and acetophenones constitute excellent classes of substrates for the rhodium-catalyzed oxidative olefination reaction, resulting in the formation of a diverse set of styrene derivatives **2** and **3**.[11] Similarly, vinylic C-H bonds were also utilized in these olefination reactions, providing rapid access to functionalized diene products **4**.[12]

Previously, we found that tertiary benzamides are substantially more efficient for the *ortho* C-H oxidative olefination reaction than primary benzamides such as **5a**. Indeed, **2i** (the title compound) was isolated in 83% yield (1 mmol scale, 1 mol% Rh catalyst), while **2a**, the primary benzamide analogue was only isolated in 60% yield.[11] It seems that the improved electron-richness of the directing group (*tertiary* benzamide **5i**) was beneficial for the reaction, suggesting that the interaction between the directing group and the Rh catalyst is pivotal, and that it can be tuned with the substitution pattern of the amide (Scheme 2). Furthermore, in the case of unbiased benzamides, that is, when positions 2- and 3- of the arene ring are not protected by a bulky substituent, the di-olefination reaction leading to the undesired product **7** can become an issue. While primary benzamides are already very selective in that respect (only traces of **7a** are detected, <10%), tertiary benzamides are found to exclusively form the mono-olefinated

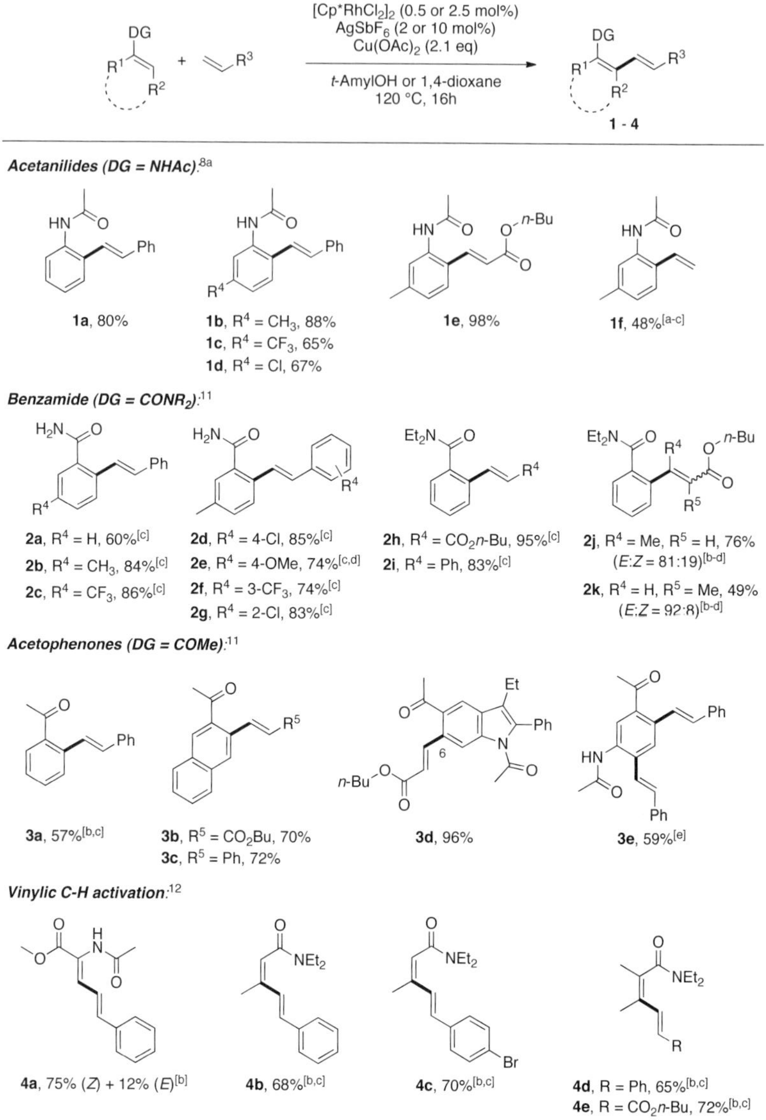

Scheme 1. Rh catalyzed C-H oxidative Heck-type reaction: conditions and isolated yields; selected examples.[8a,11,12] 0.5 mol% of [RhCp*Cl$_2$]$_2$ and 2 mol% AgSbF$_6$; t-AmylOH as solvent. Ratio of coupling partners ranging from 1:1.5 to 1.5:1; run on scales from 0.5 to 1.0 mmol. [a] 2 bar of ethylene. [b] 2.5 mol% of [RhCp*Cl$_2$]$_2$ and 10 mol% AgSbF$_6$. [c] Reaction in 1,4-dioxane. [d] Reaction run at 140 °C. [e] 4.2 equiv of Cu(OAc)$_2$, 3 equiv of styrene and 2.5 mol% of [RhCp*Cl$_2$]$_2$.

46

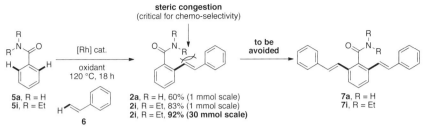

Scheme 2. Chemo-selectivity in the Rh catalyzed C-H oxidative Heck-type reaction.

product (detection limit < 2%). This we think has also to do with the size of the directing group. Indeed, a large directing group will be less prone to provide the *planar* configuration required for an efficient second C-H activation, if it already has a (bulky) styryl substituent on the other *2*-position (see **2a,i**).

While investigating the suitability of our technique on a preparative 30 mmol scale (Scheme 2), we found it experimentally convenient to increase the reaction concentration (from 0.20 M to 0.74 M) and very importantly to decrease the catalyst loading from 0.5 mol% to 0.25 mol% of the $[Cp^*RhCl_2]_2$ precatalyst (see experimental procedure). To our surprise, this afforded a substantial improvement of the reaction, as the desired product, *(E)-N,N*-diethyl-2-styrylbenzamide **2i** (the *Z* stereoisomer was not detected), was obtained in 92% isolated yield.

In conclusion, we have found that tertiary benzamides, such as *N,N*-diethylbenzamide, are excellent substrates for the *mono*-selective dehydrogenative/oxidative olefination reaction catalyzed by Rh, even at very low catalytic loadings (0.25 mol%). Because of its practicality and chemo-selectivity, we expect that this method will find applications in organic synthesis.[13]

1.   Organisch-Chemisches Institut der Westfälischen Wilhelms-Universität Münster, Corrensstrasse 40, 48149 Münster (Germany), E-mail: glorius@uni-muenster.de. The Alexander von Humboldt Foundation (F.P.), the Deutsche Forschungsgemeinschaft (SFB 858), and the European Research Council (ERC starting grant) are gratefully acknowledged for financial support. The research of F.G. was

supported by the Alfried Krupp Prize for Young University Teachers of the Alfried Krupp von Bohlen und Halbach Foundation.

2. (a) Moritani, I.; Fujiwara, Y. *Tetrahedron Lett.* **1967**, *8*, 1119; (b) Jia, C.; Piao, D.; Oyamada, J.; Lu, W.; Kitamura, T.; Fujiwara, Y. *Science* **2000**, *287*, 1992; (c) Jia, C.; Kitamura, T.; Fujiwara, Y. *Acc. Chem. Res.* **2001**, *34*, 633.

3. Recent reviews: (a) The Mizoroki-Heck Reaction (Ed.: M. Oestreich), Wiley, Chichester, **2009**; (b) Bräse, S.; de Meijere, A. in Metal-Catalyzed Cross-Coupling Reactions (Eds. de Meijere A., Diederich, F.) Wiley-VCH, New York, **2004**, p. 217.

4. For intermolecular oxidative Heck couplings, see: (a) Dams, M.; De Vos, D. E.; Celen, S.; Jacobs, P. A. *Angew. Chem. Int. Ed.* **2003**, *42*, 3512; (b) Yokota, T.; Tani, M.; Sakaguchi, S.; Ishii, Y. *J. Am. Chem. Soc.* **2003**, *125*, 1476; (c) Grimster, N. P.; Gauntlett, C.; Godfrey, C. R. A.; Gaunt, M. J. *Angew. Chem. Int. Ed.* **2005**, *44*, 3125; (d) Cai, G.; Fu, Y.; Li, Y.; Wan, X.; Shi, Z. *J. Am. Chem. Soc.* **2007**, *129*, 7666; (e) Cho, S. H.; Hwang, S. J.; Chang, S. *J. Am. Chem. Soc.* **2008**, *130*, 9254; (f) Garcia-Rubia, A.; Arrayas, R. G.; Carretero, J. C. *Angew. Chem. Int. Ed.* **2009**, *48*, 6511; (g) Wu, J.; Cui, X.; Chen, L.; Jiang, G.; Wu, Y. *J. Am. Chem. Soc.* **2009**, *131*, 13888; (h) Nishikata, T.; Lipshutz, B. H. *Org. Lett.* **2010**, *12*, 1972; for intramolecular oxidative Heck couplings, see: (i) Ferreira, E. M.; Stoltz, B. M. *J. Am. Chem. Soc.* **2003**, *125*, 9578; (j) Zhang, H.; Ferreira, E. M.; Stoltz, B. M. *Angew. Chem. Int. Ed.* **2004**, *43*, 6144; (k) Würtz, S.; Rakshit, S.; Neumann, J. J.; Dröge, T.; Glorius, F. *Angew. Chem. Int. Ed.* **2008**, *47*, 7230.

5. (a) Boele, M. D. K.; van Strijdonck, G. P. F.; de Vries, A. H. M.; Kamer, P. C. J.; de Vries, J. G.; van Leeuwen, P. W. N. M. *J. Am. Chem. Soc.* **2002**, *124*, 1586; also: (b) Wang, J.-R. Yang, C.-T. Liu, L. Guo, Q.-X. *Tetrahedron Lett.* **2007**, *48*, 5449.

6. (a) Wang, D.-H.; Engle, K. M.; Shi, B.-F.; Yu, J.-Q. *Science* **2010**, *327*, 315; (b) Wasa, M.; Engle, K. M.; Yu, J.-Q. *J. Am. Chem. Soc.* **2010**, *132*, 3680; (c) Zhang, Y.-H.; Shi, B.-F.; Yu, J.-Q. *J. Am. Chem. Soc.* **2009**, *131*, 5072; (d) Shi, B.-F.; Zhang, Y.-H.; Lam, J. K.; Wang, D.-H.; Yu, J.-Q. *J. Am. Chem. Soc.* **2010**, *132*, 460; (e) Li, J.-J.; Mei, T.-S.; Yu, J.-Q. *Angew. Chem. Int. Ed.* **2008**, *47*, 6452; (f) Engle, K. M.; Wang, D.-H.; Yu, J.-Q. *Angew. Chem. Int. Ed.* **2010**, *49*, 6169.

7. (a) Umeda, N.; Hirano, K.; Satoh, T.; Miura, M. *J. Org. Chem.* **2009**, *74*, 7094; see also: (b) Miura, M.; Tsuda, T.; Satoh, T.; Pivsa-Art, S.;

Nomura, M. *J. Org. Chem.* **1998**, *63*, 5211; (c) Ueura, K.; Satoh, T.; Miura, M. *Org. Lett.* **2007**, *9*, 1407; (d) Ueura, K.; Satoh, T.; Miura, M. *J. Org. Chem.* **2007**, *72*, 5362; (e) Morimoto, K.; Hirano, K.; Satoh, T.; Miura, M. *Org. Lett.* **2010**, *12*, 2068.

8. (a) Patureau, F. W.; Glorius, F. *J. Am. Chem. Soc.* **2010**, *132*, 9982; see also: (b) Rakshit, S.; Grohmann, C.; Besset, T.; Glorius, F. *J. Am. Chem. Soc.* **2011**, *133*, 2350; (c) Rakshit, S.; Patureau, F. W.; Glorius, F. *J. Am. Chem. Soc.* **2010**, *132*, 9585; (d) Patureau, F. W.; Besset, T.; Kuhl, N.; Glorius, F. *J. Am. Chem. Soc.* **2011**, *133*, 2154.

9. For examples of acetophenone directing groups in rhodium catalyzed C-H activation processes, see: (a) Lenges, C. P.; Brookhart, M.; *J. Am. Chem. Soc.* **1999**, *121*, 6616; (b) Tanaka, K.; Otake, Y.; Wada, A.; Noguchi, K.; Hirano, M.; *Org. Lett.* **2007**, *9*, 2203; (c) Tsuchikama, K.; Kuwata, Y.; Tahara, Y.-k.; Yoshinami, Y.; Shibata, T. *Org. Lett.* **2007**, *9*, 3097; Ru catalysts, see: (d) Murai, S.; Kakiuchi, F.; Sekine, S.; Tanaka, Y.; Kamatani, A.; Sonoda, M.; Chatani, N.; *Nature* **1993**, *366*, 529; (e) Kakiuchi, F.; Kan, S.; Igi, K.; Chatani, N.; Murai, S. *J. Am. Chem. Soc.* **2003**, *125*, 1698; (f) Martinez, R.; Chevalier, R.; Darses, S.; Genet, J.-P. *Angew. Chem. Int. Ed.* **2006**, *45*, 8232; (g) Simon, M.-O.; Genet, J.-P.; Darses, S. *Org. Lett.* **2010**, *12*, 3038; (h) Watson, A. J. A.; Maxwell, A. C.; Williams, J. M. *J. Org. Lett.* **2010**, *12*, 3856; iron-catalyzed ortho C-H arylation of protected acetophenones: (i) Yoshikai, N.; Matsumoto, A.; Norinder, J.; Nakamura, E. *Angew. Chem. Int. Ed.* **2009**, *48*, 2925; palladium-catalyzed ortho C-H arylation of acetophenones: (j) Gandeepan, P.; Parthasarathy, K.; Cheng, C.-H. *J. Am. Chem. Soc.* **2010**, *132*, 8569.

10. For examples of the use of benzamide directing groups in rhodium-catalyzed C-H activation processes, see: (a) Guimond, N.; Gouliaras, C.; Fagnou, K. *J. Am. Chem. Soc.* **2010**, *132*, 6908; (b) Mochida, S.; Umeda, N.; Hirano, K.; Satoh, T.; Miura, M. *Chem. Lett.* **2010**, *39*, 744; (c) Hyster, T. K.; Rovis, T. *J. Am. Chem. Soc.* **2010**, *132*, 10565.

11. Patureau, F. W.; Besset, T.; Glorius, F. *Angew. Chem. Int. Ed.* **2011**, *50*, 1064.

12. For the formation of dienes by olefination of vinylic C-H bonds, see: Besset, T.; Kuhl, N.; Patureau, F. W.; Glorius, F. *Chem. Eur. J.* **2011**, *17*, 7167.

13. For reviews on Rh(III)-catalyzed C-H activations, see: (a) Patureau, F. W.; Wencel-Delord, J.; Glorius, F. *Aldrichimica Acta* **2012**, *45*, 31; (b)

Song, G.; Wang, F.; Li, X. *Chem. Soc. Rev.* **2012**, *41*, 3651; (c) Satoh, T.; Miura, M. *Chem. Eur. J.* **2010**, *16*, 11212.

Appendix
Chemical Abstracts Nomenclature; (Registry Number)

(E)-N,N-Diethyl-2-styrylbenzamide; (1269637-39-9).
N,N-Diethylbenzamide; (1696-17-9).
Styrene; (100-42-5).
Pentamethylcyclopentadienylrhodium(III) chloride dimer; (12354-85-7).
Silver Hexafluoroantimonate(V); (26042-64-8).
Copper(II) acetate, anhydrous; (142-71-2).

Frank Glorius was educated in chemistry at the Universität Hannover, Stanford University (Prof. Paul A. Wender), Max-Planck-Institut für Kohlenforschung and Universität Basel (Prof. Andreas Pfaltz) and Harvard University (Prof. David A. Evans). In 2001 he began his independent research career at the Max-Planck-Institut für Kohlenforschung in Germany (Mentor: Prof. Alois Fürstner). In 2004 he became Assoc. Prof. at the Philipps-Universität Marburg and since 2007 he is a Full Prof. for Organic Chemistry at the Westfälische Wilhelms-Universität Münster, Germany. His research program focuses on the development of new concepts for catalysis and their implementation in organic synthesis.

Frederic William Patureau was born in 1982 in St. Germain en Laye, France. He studied biology at the University of Versailles (France, 2000–2002), and obtained his bachelor's degree in chemistry in 2003 from the University of Paris-Sud Orsay (France). There, he completed his master's degree (2005) in the group of Prof. H. B. Kagan and Prof. J.-C. Fiaud (mentors: Dr. E. Schulz and Dr. M. Mellah). He then moved to the Van't Hoff Institute for Molecular Sciences at the University of Amsterdam (Netherlands), where he obtained his doctoral degree in 2009 (mentor: Prof. J. N. H. Reek). He then joined the group of Prof. F. Glorius at the W.W. University of Münster (Germany), where he became an Alexander von Humboldt Fellow. In November 2011, he started his independent career as Junior Professor of Organometallic Catalysis at the Technische Universität Kaiserslautern.

50

Tatiana Besset was educated in chemistry at the University of Grenoble (France) in the group of Dr. Andrew Green, where she obtained her doctoral degree in 2009. She then moved to the Westfälische Wilhelms-Universität Münster (Germany) for a postdoctoral period in the group of Prof. Dr. Frank Glorius. In 2011, she moved to the University of Amsterdam (the Netherlands) in the group of Prof. Dr. Joost N. H. Reek, as postdoctoral fellow. She is currently Assistant Professor (CR2, Chargée de Recherche CNRS) at the University of Rouen, France.

Takaaki Harada was born in 1988 in Osaka, Japan. He graduated in 2010 and received his M. S. degree in 2012 from the University of Tokyo. The same year he started his Ph. D. study under the supervision of Professor Tohru Fukuyama. His current interest is enantioselective total synthesis of complex natural products.

Tohru Fukuyama received his Ph.D. in 1977 from Harvard University with Yoshito Kishi. He remained in Kishi's group as a postdoctoral fellow until 1978 when he was appointed as Assistant Professor of Chemistry at Rice University. After seventeen years on the faculty at Rice, he returned to his home country and joined the faculty of the University of Tokyo in 1995, where he is currently Professor of Pharmaceutical Sciences. He has primarily been involved in the total synthesis of complex natural products of biological and medicinal importance. He often chooses target molecules that require development of new concepts in synthetic design and/or new methodology for their total synthesis.

Enantioselective Nitroaldol (Henry) Reaction of *p*-Nitrobenzaldehyde and Nitromethane Using a Copper (II) Complex Derived from (*R,R*)-1,2-Diaminocyclohexane: (1*S*)-1-(4-Nitrophenyl)-2-nitroethane-1-ol

Submitted by Antoinette Chougnet and Wolf-D. Woggon.[1]
Checked by Larissa Pauli and Andreas Pfaltz.

1. Procedure

A. 2-(1,1-Dimethylethyl)-6-[[[(1R,2R)-2-[(4-pyridinylmethyl)amino]-cyclohexyl]amino]methyl]-phenol (2). An oven dried, 250-mL, two-necked flask equipped with a magnetic stir bar (cylindrical, 2 × 1 cm), a reflux condenser (central neck) with a two-tap Schlenk adapter connected to a bubbler and an argon/vacuum manifold (Note 1) is assembled hot and cooled under a stream of argon. The flask is charged with (1*R*,2*R*)-1,2-diaminocyclohexane (**1**, 350 mg, 3.00 mmol, 1.00 equiv) in dry dichloromethane (30 mL) (Note 2) and the remaining neck is equipped with a rubber septum. A solution of 4-pyridinecarboxaldehyde (328 mg, 3.00 mmol, 1.00 equiv) in dichloromethane (30 mL) (Note 2) is added to the

Org. Synth. **2013**, *90*, 52-61
Published on the Web 8/27/2012
© 2013 Organic Syntheses, Inc.

stirred solution *via* syringe pump over 2 h at 25 °C (Note 3). After complete addition the light yellow reaction mixture is stirred for another hour at ambient temperature. Subsequently a solution of 3-*t*-butyl-2-hydroxybenzaldehyde (558 mg, 3.00 mmol, 1.00 equiv) in dichloromethane (15 mL) (Note 2) is added in one portion and the reaction mixture is refluxed for 19 h (Note 4). The dark yellow solution is concentrated in the reaction flask by rotary evaporation (45 °C water bath, 150 mmHg) and the residue, containing the imines (Note 5), is dissolved in methanol (45 mL) (Note 2). A magnetic stir bar (cylindrical, 2 × 1 cm) and an excess of sodium borohydride (768 mg, 19.9 mmol, 6.6 equiv) (Note 2) is added in small portions over 5 min. Because of the slightly exothermic reaction the reflux condenser is attached, and the solution is stirred for 2 h at ambient temperature (Notes 6 and 7). The reaction is quenched with 1N HCl (3 mL) and after stirring for 5 min 4N NaOH (9 mL) is added and the mixture is transferred to a 500-mL separatory funnel containing 90 mL of saturated aqueous NaHCO$_3$ solution. After the first portion of ethyl acetate (75 mL) is added to the dichloromethane solution, the phases are separated and the aqueous phase is extracted with ethyl acetate (2 × 75 mL). The combined organic phase is washed with distilled water (100 mL), dried over anhydrous magnesium sulfate (4 g), filtered, and concentrated by rotary evaporation (45 °C water bath, 38 mmHg) to afford a pale yellow oil (Note 7), which is chromatographed on silica gel (Note 8). Fractions containing the product were collected, concentrated by rotary evaporation (45 °C water bath, 75 mmHg) and dried under high vacuum (25 °C, 0.2 mmHg) to obtain ligand **2** (579–582 mg, 1.58 mmol, 52–53% yield, > 99% ee) as a white solid (Notes 9, 10 and 11).

　　B. (1S)-1-(4-Nitrophenyl)-2-nitroethane-1-ol (4). A 100-mL, round-bottomed flask equipped with a magnetic stir bar (cylindrical, 2 × 1 cm) is charged with 2-(1,1-dimethylethyl)-6-[[[(1*R*,2*R*)-2-[(4-pyridinylmethyl)-amino]cyclohexyl]amino]methyl]-phenol (**2**, 606 mg, 1.65 mmol, 5.7 mol%) (Note 12), copper acetate (273 mg, 1.47 mmol, 5.1 mol%) and 30 mL of ethanol (Note 12). Within minutes the reaction mixture becomes deep green. After stirring for 45 min at 23 °C to complete the formation of the Cu(II) complex **3**,[2] solid *p*-nitrobenzaldehyde (4.48 g, 29.1 mmol, 1.00 equiv) and nitromethane (16.2 mL, 18.3 g, 295 mmol, 10 equiv) (Note 13) are added. Stirring is continued for an additional 3 h (Note 14). The deep green solution is concentrated by rotary evaporation (21 °C water bath, 24 mmHg) to give 8.01 g of a dark green solid (92% *ee*) (Note 15). The residue is dissolved in

4.4 mL of methanol (Note 13) and immediately purified by column chromatography (Note 16). Fractions containing the product are combined in a 50-mL round-bottomed flask, concentrated by rotary evaporation (21 °C water bath, 24 mmHg) and dissolved in a minimal amount of methanol (4.4 mL, dissolution aided by use of ultrasound). To facilitate the crystallization the flask is placed in a refrigerator (3 °C, 5 h) and freezer (−20 °C, 12 h). The mother liquor is removed with a pasteur pipette and the yellowish crystals are washed with cold dichloromethane (taken from the freezer; 3 × 0.5 mL), which is removed each time with a Pasteur pipette, and dried (23 °C, 0.2 mmHg) to provide product **4** (4.24 g, 69% yield, 95% *ee*). The mother liquor is concentrated by rotary evaporation (21 °C water bath, 8 mmHg), layered with dichloromethane (ca. 0.5 mL) and cooled (3 °C, 2 h then −20 °C, 24 h). The mother liquor is removed with a pasteur pipette and the yellowish crystals are washed with cold dichloromethane (taken from the freezer; 3 × 0.5 mL), which is again removed each time with a Pasteur pipette, and dried (23 °C, 0.2 mmHg) to give a second crop of yellow crystals (0.77 g, 12% yield, 94% *ee*). The two crops were combined to provide *(1S)-1-(4-nitrophenyl)-2-nitroethane-1-ol* *(4)* (5.01 g, 81% yield, 95% *ee*) (Notes 15, 17, 18 and 19).

2. Notes

1. A two-tap Schlenk adapter connected to a bubbler and an argon/vacuum manifold is illustrated in Yu, J.; Truc, V.; Riebel, P.; Hierl, E.; Mudryk, B. *Org. Synth.* **2008**, *85*, 64–71.

2. Reagents and solvents were purchased from companies named in parentheses and used without further purification: (1*R*,2*R*)-1,2-diaminocyclohexane (98%, 99% *ee*, Aldrich), 4-pyridinecarboxaldehyde (98%, Acros), 3-*t*-butyl-2-hydroxybenzaldehyde (96%, Aldrich), sodium borohydride (98+%, Acros), anhydrous dichloromethane (puriss., over molecular sieves, ≥99.5%, Sigma-Aldrich), checkers purchased methanol (puriss., over molecular sieves, ≥99.5%) from Sigma-Aldrich and submitters from J. T. Baker (HPLC Gradient Grade). The number of mmol of reagents given in the procedure are calculated based on the purities listed above.

3. In order to achieve the yield given for ligand **2** it is important to add first 4-pyridinecarboxaldehyde followed by 3-*t*-butyl-2-hydroxybenzaldehyde. In case of reversed addition the C_2 symmetric *bis* phenol diaminocyclohexane ligand is the main product.

Org. Synth. **2013**, *90*, 52-61

4.   Reaction progress can be monitored by ^{1}H NMR (disappearance of the aldehyde signals at 11.8 and 10.1 ppm). The dark yellow color of the reaction is characteristic of imine formation. The submitters report that the reaction mixture is heated under reflux for 1 h and subsequently stirred for 16 h at ambient temperature, which completed the formation of the imines.

5.   A mixture of mono- and diimines is formed; it is not advisable to purify the desired diimine by chromatography since the compound is not stable   under   the   chromatography   conditions   used (dichloromethane/methanol).

6.   The dark yellow solution becomes colorless after reduction of the diimines.

7.   The TLC (checkers used Polygram$^®$ SIL/UV$_{254}$-TLC-plates from Macherey-Nagel and submitters from E. Merck) of the reaction mixture (dichloromethane/methanol, 10%) shows the C_1-symmetric ligand **2** as the main spot R_f = 0.35, with two minor impurities at R_f = 0.62 and R_f = 0.77 (UV 254 nm, KMnO$_4$ stain).

8.   Column chromatography: 4 cm diameter $\times$ 40 cm height of silica gel (145 g) (checkers used "Silica Gel 60" (0.040-0.063 mm) from E. Merck, and submitters used "Silica Gel 60" from Aldrich), eluting with 1.5 L of 3 vol% methanol in dichloromethane and collecting 10 mL fractions. Fraction purity can be assayed by TLC (Note 7). Fractions 42–66 were combined.

9.   *2-(1,1-Dimethylethyl)-6-[[[(1R,2R)-2-[(4-pyridinylmethyl)-amino]-cyclohexyl]amino]methyl]-phenol* **(2)** exhibits the following physical and spectroscopic properties: mp 109–110 °C; $[\alpha]_D^{20}$ –109.8 (*c* 1.01, dichloromethane); >99% *ee*; ^{1}H NMR (400 MHz, CDCl$_3$) δ: 0.97–1.09 (m, 1 H), 1.11–1.30 (m, 3 H), 1.42 (s, 9 H), 1.68–1.81 (m, 2 H), 2.17–2.36 (m, 4 H), 3.72 (d, *J* = 14.4 Hz, 1 H), 3.83 (d, *J* = 13.5 Hz, 1 H), 3.95 (d, *J* = 14.3 Hz, 1 H), 4.04 (d, *J* = 13.4 Hz, 1 H), 6.72 (t, *J* = 7.6 Hz, 1 H), 6.87 (br. dd, *J* = ca. 7.5, 1.6 Hz, 1 H), 7.19 (dd, *J* = 7.8, 1.6 Hz, 1 H), 7.29 (overlaps with CHCl$_3$ residual signal, dd , *J* = 4.4, 1.6 Hz, 2 H), 8.52 (dd, *J* = 4.4, 1.6 Hz, 2 H); ^{13}C NMR (101 MHz, CDCl$_3$) δ: 24.7, 25.2, 29.7, 31.1, 31.9, 34.8, 49.7, 50.7, 60.5, 62.2, 118.3, 123.1, 124.1, 125.9, 126.3, 137.1, 149.7, 150.0, 157.4; IR (ATR) 3287, 2920, 2363, 1600, 1562, 1435, 1429, 1415, 1385, 1355, 1250, 1114, 1085, 991, 967, 848, 797, 752, 668 cm^{-1}; MS (EI, 70 eV): *m/z* (%) 367 ([M$^+$], 11), 276 (19), 275 (100), 178 (11), 163 (31), 162 (20), 147 (47), 121 (11), 119 (27), 113 (52), 107 (15), 96 (48), 93 (85), 92 (22),

91 (11), $C_{23}H_{33}N_3O$ (367.53); Anal. calcd: C, 75.16; H, 9.05; N, 11.43. Found: C, 75.03; H, 8.83; N, 11.37.

10. The submitters report that on a reaction scale of 1 mmol (1R,2R)-1,2-diaminocyclohexane (**1**) and conditions stated in Note 4 the ligand **2** can be isolated in 59% yield exhibiting the following physical properties: mp 111–113 °C; $[\alpha]_D^{20}$ –84.3 (c 0.528, ethanol).

11. Because of the strong peak tailing, the enantiomeric excess of the ligand, 2-(1,1-dimethylethyl)-6-[[[(1R,2R)-2-[(4-pyridinylmethyl)amino]cyclohexyl]amino]methyl]-phenol, could not be determined with high accuracy. However, a more precise measurement was possible for the (1S,2S) enantiomer prepared from (1S,2S)-1,2-diaminocyclohexane by the same procedure ($[\alpha]_D^{20}$ +109.5 (c 1.01, dichloromethane)). The analysis was performed using HPLC with a Chiralcel® OD-H column (0.46 cm × 25 cm) obtained from Daicel Chemical Industries, Ltd. and a diode array detector. The assay conditions were 95:5 n-heptane:i-propanol, 20 °C, 0.5 mL/min flow rate, with detection at 220 nm and 263 nm, retention times: (1R,2R) enantiomer = 31.9 min, (1S, 2S) enantiomer = 35.2 min. The signal of the minor (1R,2R) enantiomer was not visible, implying an enantiomeric excess of >99%.

12. In order to accomplish short reaction times and high enantiomeric excess, it is important to use a small excess of ligand over copper acetate.

13. Reagents and solvents were purchased from companies named in parentheses and used without further purification: p-nitrobenzaldehyde (99%, Acros), nitromethane (puriss., over molecular sieves, ≥98.5%, Sigma-Aldrich), copper acetate (purum, anhydrous, >98%, Fluka, ethanol (puriss., over molecular sieves, ≥99.8% (v/v), Fluka), dichloromethane (HPLC Gradient Grade, J. T. Baker) and hexane (HPLC Gradient Grade, J. T. Baker), submitters purchased methanol (puriss., over molecular sieves, ≥99.5%) from Sigma-Aldrich.

14. Reaction progress can be monitored by TLC (silica gel, dichloromethane) (checkers used Polygram®SIL/UV254-TLC-plates from Macherey-Nagel) R_f p-nitrobenzaldehyde = 0.44, R_f (1S)-1-(4-nitrophenyl)-2-nitroethane-1-ol = 0.12 (UV 254 nm, KMnO$_4$ stain).

15. Enantiomeric excess was determined by HPLC with Chiralcel® OD-H column (0.46 cm × 25 cm) obtained from Daicel Chemical Industries, Ltd. and a diode array detector. The assay conditions were 85:15 n-heptane:i-propanol, 20 °C, 0.8 mL/min flow rate, with detection at 220 nm and

56

254 nm, retention times: (*R*) enantiomer = 20.0 min, (*S*) enantiomer = 24.8 min.

16. Column chromatography: 3 cm diameter × 20 cm height of silica gel (66 g) ("Silica Gel 60" from Aldrich), eluting with 350 mL of 4:1 dichloromethane/hexanes, then 300 mL of 95:5 hexanes/ethyl acetate, collecting 20 mL fractions. The fraction size collected was changed to 10 mL beginning with fraction 19. Fraction purity was assayed by TLC (Note 13). Fractions containing product (13–48) were combined.

17. Two runs at the 3 g scale afforded yields of 77–81%.

18. *(1S)-1-(4-Nitrophenyl)-2-nitroethane-1-ol* (**4**) exhibits the following physical and spectroscopic properties: mp 83–85 °C; $[\alpha]_D^{20}$ + 25.9 (*c* 0.69, ethanol); 95% *ee*; ^{1}H NMR (400 MHz, CDCl$_3$) δ: 3.13 (s, 1 H), 4.58 (dd, *J* = 16, 14 Hz, 1 H), 4.60 (dd, *J* = 20, 14 Hz, 1 H), 5.61 (m, 1 H), 7.63/8.27 (AA'BB' system, 4 H); ^{13}C-NMR (101 MHz, CDCl$_3$) δ: 70.1, 80.7, 124.4, 127.1, 145.0, 148.3; IR (ATR) 3479, 1609, 1545, 1506, 1418, 1377, 1346, 1315, 1290, 1215, 1186, 1105, 1076, 1040, 895, 860, 837, 756, 733, 698, 648 cm^{-1}; MS (EI, 70 eV): *m/z* (%) 165 (66), 152 (25), 151 (99), 152 (100), 105 (23), 104 (16), 92 (11), 91 (26), 77 (51), 76 (15), 65 (11), 61 (18), 51 (34), 50 (17), C$_8$H$_8$N$_2$O$_5$ (212.16); Anal. calcd: C, 45.29; H, 3.80; N, 13.20. Found: C, 45.06; H, 3.78; N, 13.21.

19. The submitters report that on a reaction scale of 10 mmol *p*-nitrobenzaldehyde the product *(1S)-1-(4-nitrophenyl)-2-nitroethane-1-ol* (**4**) is obtained in 86% yield exhibiting the following physical properties: mp 91–93 °C; $[\alpha]_D^{20}$ + 26.6 (*c* 0.519, ethanol); ee 98%.

Handling and Disposal of Hazardous Chemicals

The procedures in this article are intended for use only by persons with prior training in experimental organic chemistry. All hazardous materials should be handled using the standard procedures for work with chemicals described in references such as "Prudent Practices in the Laboratory" (The National Academies Press, Washington, D.C., 2011 www.nap.edu). All chemical waste should be disposed of in accordance with local regulations. For general guidelines for the management of chemical waste, see Chapter 8 of Prudent Practices.

These procedures must be conducted at one's own risk. *Organic Syntheses, Inc.*, its Editors, and its Board of Directors do not warrant or guarantee the safety of individuals using these procedures and hereby disclaim any liability for any injuries or damages claimed to have resulted from or related in any way to the procedures herein.

3. Discussion

Commercially available chiral 1,2-diaminocyclohexanes have been extensively used to synthesize salen ligands in order to prepare their metal complexes for various catalytic reactions.[2] In recent years the corresponding and less explored diamine ligands were investigated.[3] Except for a few cases[4] most of the diamine ligands were C2 symmetric due to their facile synthesis. In the course of our investigation it was discovered[5] that if aldehydes are employed displaying considerable difference in reactivity C1 symmetric ligands can be easily prepared if the less reactive aldehyde is reacted first to form the monoimine. The two-step, high yielding procedure reported here is not an exception as we have prepared more than 50 C1 symmetric ligands in the same manner since our first paper in this area was published.[5] Hence, this method appears to be of general interest and application.

The Henry (nitro aldol) reaction is an important carbon-carbon bond forming process yielding α-hydroxy nitro compounds that are versatile intermediates in organic synthesis.[6] Asymmetric, catalytic versions of this reaction are of particular interest and significant progress has been achieved using chiral complexes of Zn(II),[7] Cr(III),[8] La(III)[9] and Co(II).[10] However, most investigations were pursued with chiral Cu(II) complexes containing a C2 symmetric four-dentate ligand sphere.[11]

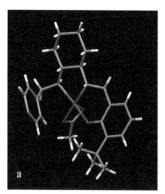 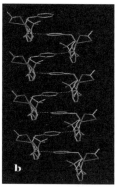

Solid state structure of Cu(II) complex **3**.

The nitro aldol (Henry) reaction described herein is catalyzed by the C1 symmetric Cu(II) complex **3** generated *in situ* from Cu(OAc)$_2$ and ligand **2** derived from commercially available (*R,R*)-1,2-diaminocyclohexane. According to X-ray crystallography of complex **3** a helical, supramolecular structure is formed, see **3a** and **3b**, that most likely is maintained in solution.[5] The complex has a high catalytic activity and the nitroaldol reaction can be performed without additives such as organic bases. The procedure provides a valuable method to prepare enantiomerically enriched β-nitro alcohols from nitromethane and various aromatic and aliphatic aldehydes (Table 1).

Table 1. Asymmetric Henry Reactions of Aromatic and Aliphatic Aldehydes with Nitromethane in the presence of catalyst **3**.[5]

entry[a]	aldehyde (R)	time (h)	yield (%)[d]	ee (%)[e]
1[b]	2-NO$_2$C$_6$H$_4$	3	94	98
2[b]	3-NO$_2$C$_6$H$_4$	3	95	96
3[b]	3-pyridyl	3	94	96
4[b]	4-pyridyl	3	96	95
5[b]	Ph	48	75	90
6[b]	4-ClC$_6$H$_4$	24	65	90
7[b]	4-FC$_6$H$_4$	24	72	87
8[c]	2-thiophenyl	60	58	93
9[c]	2-MeC$_6$H$_4$	60	63	88
10[c]	4-MeC$_6$H$_4$	60	54	93
11[c]	1-naphthyl	60	75	93
12[c]	4-MeOC$_6$H$_4$	60	51	85
13[c]	*n*-butyl	60	88	98
14[c]	*t*-butyl	60	91	99
15[c]	cyclohexyl	60	92	98

[a] All reactions were performed at 0.5M concentration of aldehyde in EtOH using 5.4 mol% of ligand **2**, 5 mol% Cu(OAc)$_2$ and 10 equiv. of nitromethane. [b] Reactions were run at r.t. [c] Reactions were run at 0 °C. [d] Isolated yield. [e] Enantiomeric excess was determined by chiral HPLC using a Chiralcel® OD-H column. The absolute configuration of products was determined as *S* by comparison of their optical rotations with literature values.

1. Department of Chemistry, University of Basel, St. Johann-Ring 19, CH-4056 Basel, Switzerland.
2. (a) Kleij, A. W., *Eur. J. Inorg. Chem.* **2009**, 193–205; (b) Katsuki, T., *Synlett* **2003**, 281–297.
3. (a) Bandini, M.; Cabbidu, S.; Cadoni, E.; Olivelli, P.; Sinisi, R.; Umani-Ronchi, A.; Usai. M., *Chirality* **2009**, *21*, 239–244; (b) Kowalczyk, R.; Sidorowicz, L.; Skarzewski, J., *Tetrahedron Asymmetry* **2008**, *19*, 2310–2315; (c) Xiong, Y.; Wang, F.; Huang, X.; Wen, Y.; Feng, X., *Chem. Eur. J.* **2007**, *13*, 829–833; (d) Matsumoto, K.; Saito, B.; Katsuki, T., *Chem. Commun.* **2007**, 3619–3627.
4. (a) Steurer, M.; Bolm, C., *J. Org. Chem.* **2010**, *75*, 3301–3310; (b) Arai, T.; Watanabe, M.; Fujiwara, A.; Yokoyama, N.; Yanagisawa, A., *Angew. Chem. Int. Ed.* **2006**, *45*, 5978–5981.
5. Zhang, G.; Yashima, E.; Woggon, W.-D., *Adv. Synth. Catal.* **2009**, *351*, 1255–1262.
6. Palomo, C.; Oiarbide M.; Laso A. *Eur. J. Org. Chem.*, **2007**, 2561–2574.
7. Bulut, A.; Aslan A.; Dogan, O. *J. Org. Chem.*, **2008**, *73*, 7373–7375.
8. Zulauf, A.; Mellah, M.; Schulz E., *J.Org. Chem.*, **2009**, *74,* 2242–2245.
9. Sasai, H.; Suzuki, T.; Arai, S.; Shibazaki, M., *J. Am. Chem. Soc.*, **1992**, *114*, 4418–4420.
10. Kogami, Y.; Nakajima,T.; Ikeno,T.; Yamada, T., *Synthesis,* **2004**, 1947–1950.
11. Evans, D. A.; Rueping, M.; Lam, H.W.; Shaw, J. T.; Downey, C. W., *J. Am. Chem. Soc.*, **2003**, *125*, 12692–12693.

Appendix
Chemical Abstracts Nomenclature; (Registry Number)

(1*R*,2*R*)-1,2-Diaminocyclohexane; (20439-47-8)
4-Pyridinecarboxaldehyde; (872-85-5)
3-*t*-Butyl-2-hydroxybenzaldehyde: Benzaldehyde, 3-(1,1-dimethylethyl)-2-hydroxy-; (24623-65-2)
Sodium borohydride: Borate(1-), tetrahydro-, sodium (1:1); (16940-66-2)
Copper acetate: Acetic acid, copper(2+) salt (2:1); (142-71-2)
p-Nitrobenzaldehyde: Benzaldehyde, 4-nitro-; (555-16-8)
Nitromethane: Methane, nitro-; (75-52-5)

2-(1,1-Dimethylethyl)-6-[[[(1*R*,2*R*)-2-[(4-pyridinylmethyl)amino]-cyclohexyl]amino]methyl]-phenol; (1200403-97-9)
(1*S*)-1-(4-Nitrophenyl)-2-nitroethane-1-ol; (454217-09-5

Wolf-D. Woggon was born in 1942 in Berlin, Germany. He completed his undergraduate education in Geology and Chemistry at the Freie Universität Berlin and his PhD in Organic Chemistry at the University of Zurich with the late Hans Schmid. After a postdoctoral stay with Alan Battersby in Cambridge (UK) he started his independent research at the University of Zurich and completed his habilitation in 1985. In 1995 he became professor at the University of Basel where he stayed ever since. His research interests include the synthesis of enzyme models, enzymatic reaction mechanisms, the isolation of new enzymes, the synthesis of vitamins, and the development of new ligand-scaffolds for metallo-organic catalysis.

Antoinette Chougnet (born 1949 in France) has completed her undergraduate studies and her PhD+habilitation in 1980 with Andrée Marquet at the University of Paris. She spent one year at Collège de France as a post-doctoral researcher in the group of J. Glowinski. She became professor at the University of Paris in 1988. From 1990 to 2000, she worked at Hoffmann-La Roche (Basel) as head of the group Drug metabolism. In 2000 she joined the University of Basel where her research interests include the study of several cytochromes P450 isoenzymes and the total synthesis of tocopherol together with the biosynthesis of that natural compound.

Larissa Pauli was born in Karkaralinsk (Kasachstan) in 1981. She studied chemistry at the University of Basel (Switzerland) where she obtained her B.Sc. in Chemistry in 2008 and her M.Sc. in 2010. She started her Ph.D. studies in 2010 under the supervision of Prof. Andreas Pfaltz at the University of Basel and is currently working in the field of Ir-catalyzed enantioselective hydrogenation.

Synthesis of *N*-Acetyl Enamides by Reductive Acetylation of Oximes Mediated with Iron(II) Acetate: *N*-(1-(4-Bromophenyl)vinyl)acetamide

Submitted by Wenjun Tang,[1] Nitinchandra D. Patel, Andew G. Capacci, Xudong Wei, Nathan K. Yee, and Chris H. Senanayake.
Checked by Christopher Haley, Hendrik Klare, and Brian Stoltz.

1. Procedure

A. 1-(4-Bromophenyl)ethanone oxime (**1**). A 1000-mL single-necked, round-bottomed flask equipped with a 7 cm magnetic stir bar, a 50-mL pressure-equalizing dropping funnel fitted with a nitrogen inlet adapter, and an ice-water bath, is charged with 1-(4-bromophenyl)ethanone (50.0 g, 251 mmol, 1 equiv) (Note 1), hydroxylamine hydrochloride (37.0 g, 532 mmol, 2.1 equiv) (Note 2) and denatured ethanol (375 mL) (Note 3). Pyridine (37.5 mL, 465 mmol, 1.85 equiv) (Note 4) is charged through the pressure-equalizing dropping funnel to the suspension over 5 min. Once the addition is complete, the dropping funnel is replaced by a water-cooled condenser fitted with a nitrogen inlet adapter. The ice-water bath is replaced by an oil bath and the reaction mixture is heated at reflux under nitrogen for 4 h (Note 5). After this period, the flask is removed from the oil bath and is placed in an ice-water bath until the internal temperature cools to approximately 20 °C Temperature of the mixture was determined by addition of a thermometer to the flask after approximately 10 min of cooling. The magnetic stir bar is removed and the reaction mixture is concentrated by rotary evaporation (60 °C, 35–40 mmHg) to remove ethanol and other volatiles (~370 mL) to

62

provide a solid residue. The flask is equipped with 7 cm magnetic stir bar and a 500-mL pressure equalizing dropping funnel. Water (375 mL) is charged to the mixture through the dropping funnel over 15 min. The resulting slurry is stirred at room temperature for 1 h, then vacuum-filtered on a 9-cm Büchner funnel (fitted with Whatman-1 filter paper, 90 mm), washed with water (2 x 25 mL) and suction dried (approx. 10 min). The solid is transferred to a 250-mL beaker and dried under vacuum (~130 mmHg) at 65 °C for 12 h to afford the title compound **1** as a white solid (53.0–53.1 g, 99%) (Note 6).

    B.   N-(1-(4-Bromophenyl)vinyl)acetamide (**2**). An oven-dried 1000-mL, three-necked, round-bottomed flask equipped with 7 cm magnetic stir bar, an internal thermometer, a rubber septum and a reflux condenser fitted with a nitrogen inlet, is charged 1-(4-bromophenyl)ethanone oxime (**1**, 20.0 g, 93.4 mmol, 1 equiv) and THF (125 mL) (Note 7) to give a clear solution. To the stirred solution are added acetic anhydride (19.1 g, 17.6 mL, 187 mmol, 2.0 equiv) (Note 8) and acetic acid (16.8 g, 16.0 mL, 280 mmol, 3.0 equiv) (Note 9). The resulting solution is sparged with nitrogen for 20 min (Note 10) and then iron(II) acetate (32.5 g, 187 mmol, 2.0 equiv) (Note 11) is charged in one portion. The resulting brown slurry is heated to approximately 65 °C with a heating mantle and stirred for 12 h under a nitrogen atmosphere (Note 12). The heating mantle is then replaced with an ice-water bath and the resulting mixture is cooled to an internal temperature of approximately 5 °C. To this solution is charged ethyl acetate (100 mL) (Note 13) and then water (200 mL) slowly over 10 min while keeping the internal temperature below 20 °C (Note 14). The pH of the solution is adjusted to ~5 by the addition of 25 mL of 15 wt% aqueous NaOH solution (Note 15). The contents are transferred to a 1-L separatory funnel and the layers are separated. The bottom aqueous layer is extracted with ethyl acetate (125 mL). The combined organic layers are washed sequentially with 1 N aqueous NaOH solution (120 mL) followed by 8 wt% aqueous sodium bicarbonate solution (150 mL) (Notes 16 and 17). The organic layer is dried over anhydrous sodium sulfate (5 g) (Note 18) and filtered. The filtrate is transferred into a 500-mL round-bottomed flask and concentrated by rotary evaporation (40 °C, 25 mmHg) to dryness (Note 19). To the residue is charged ethyl acetate (10 mL) followed by hexanes (95 mL) using a 250 mL addition funnel over a period of 30 min while stirring with a 7 cm magnetic stir bar. The resulting slurry is stirred for 3-4 h at ambient temperature, and then vacuum-filtered on a 6-cm Buchner funnel (fitted with Whatman-1 filter paper, 60 mm). The cake is rinsed sequentially with 30 mL of hexanes/ethyl acetate (5:1 v/v) (Note 20) and 15 mL of hexanes and is suction dried (ca. 10 min). The solid is transferred into a 50-mL round-bottomed flask and further dried under vacuum (~130 mmHg) at 50 °C for 12 h. The title compound **2** is obtained as an off-white solid (17.25–17.90 g, 77–80%) (Note 21).

2. Notes

1.   1-(4-Bromophenyl)ethanone (98%) was purchased from Sigma Aldrich Co. and used as received.
2.   Hydroxylamine hydrochloride (99%) was purchased from Alfa Aesar and used as received.
3.   Denatured ethanol (contains ~5% methanol and ~5% isopropanol as denaturants) was purchased from Sigma-Aldrich Co. and used as received.
4.   Pyridine (anhydrous, 99.8%) was purchased from Sigma-Aldrich Co. and used as received.
5.   The oil bath was heated at 85 °C. All solids dissolve and give a clear solution when the temperature of the oil bath reached approx. 35 °C. The progress of the reaction was monitored by HPLC, (Hewlett Packard 1100, Halo C_8 4.6 mm x 150 mm, 2.7 μm column) by using a liner gradient of B (H_2O with 0.4 % $HClO_4$) and A (acetonitrile) at a flow rate of 1.2 mL/min with UV detection at 220 nm and column temperature of 20 °C; liner gradient, 85-2% B, 7 min).

Compound	Retention time (min)
1-(4-Bromophenyl)ethanone	5.28
1-(4-Bromophenyl)ethanone oxime (**1**)	6.19
N-(1-(4-Bromophenyl)vinyl)acetamide (**2**)	4.70

6.   The product displayed the following physical and spectroscopic properties: mp 120–122 °C; Lit.[2] mp 128–130 °C; ^{1}H NMR (400 MHz, DMSO-d_6) δ: 2.13 (s, 3 H), 7.55–7.63 (m, 4 H), 11.33 (s, 1 H); ^{13}C NMR (100 MHz, DMSO-d_6) δ: 11.36, 121.94, 127.57, 131.30, 136.14, 152.09; [M + H]$^+$ calcd for C_8H_8NOBr: 213.9862. Found: 213.9857. Anal. Calcd. For C_8H_8NOBr C, 44.89, H, 3.77, N, 6.54; found C, 44.97, H, 3.80, N, 6.48.

7.   Anhydrous THF ($\geq$99.9 %) was purchased from Sigma-Aldrich Co. and used as received.

8.   Acetic anhydride (reagent ACS grade, 97+ %) was purchased from ACROS and used as received.

9.   Glacial acetic acid (min 99.7 %) was purchased from EMD Chemicals Inc. and used as received.

10.   It is recommended to sparge the solution with nitrogen thoroughly for approx. 20 min before charging iron(II) acetate into the solution.

11.   Iron(II) acetate (anhydrous, 97 %) was purchased from Strem Chemical Co. and used as received. Iron (II) acetate is air and moisture sensitive; hence, it must be stored under nitrogen or argon and handled with minimum exposure with air.

12.   Keep reaction under nitrogen atmosphere all the times.

13.   Ethyl acetate (ACS reagent grade, $\geq$ 99.5 %) was purchased from EMD Chemicals Inc. and used as received.

14.   An exotherm is observed during water addition ($\sim$10 °C).

15.   Sodium hydroxide (pellets, 99.1 %) was purchased from Fisher Scientific and used as received to prepare 15 wt% and 1 N NaOH aq. solutions.

16.   Sodium bicarbonate (white powder, min 99.7 %) was purchased from EMD Chemicals Inc. and used as received to prepare 8 wt % aqueous solution.

17.   The pH value of the aqueous solution should be $\sim$8 after wash. If the aqueous layer remains acidic, wash again with 8 wt % sodium bicarbonate solution until the pH of the aqueous solution reaches $\sim$8. Caution: Carbon dioxide is released during the wash.

18.   Anhydrous sodium sulfate powder is purchased from EMD Chemicals Inc. and used as received.

19.   The pale orange oil solidified upon cooling to give a beige solid.

20.   Hexanes (ACS grade) was purchased from Sigma-Aldrich Co. and used as received.

21.   The product displayed the following physical and spectroscopic properties: mp 106–113 °C; ^{1}H NMR (400 MHz, CDCl$_3$) δ: 2.06 (s, 3 H), 5.07 (s, 1 H), 5.73 (s, 1 H), 7.11 (s, 1 H), 7.25 (d, J = 8.3 Hz, 2 H), 7.46 (d, J = 8.4 Hz, 2 H), ^{13}C NMR (75 MHz, CDCl$_3$) δ: 24.48, 103.82, 122.74, 127.78, 131.83, 137.18, 139.75, 169.32. [M + H]$^+$ calcd for C$_{10}$H$_{10}$NOBr: 240.0019. Found: 240.0009; Anal. Calcd. For C$_{10}$H$_{10}$NOBr C, 50.02, H,

4.20, N, 5.83; found C, 49.99, H, 4.19, N, 5.70. Note: It was found that compound **2** was prone to tautomerize in sufficiently acidic CDCl$_3$.

Handling and Disposal of Hazardous Chemicals

The procedures in this article are intended for use only by persons with prior training in experimental organic chemistry. All hazardous materials should be handled using the standard procedures for work with chemicals described in references such as "Prudent Practices in the Laboratory" (The National Academies Press, Washington, D.C., 2011 www.nap.edu). All chemical waste should be disposed of in accordance with local regulations. For general guidelines for the management of chemical waste, see Chapter 8 of Prudent Practices.

These procedures must be conducted at one's own risk. *Organic Syntheses, Inc.*, its Editors, and its Board of Directors do not warrant or guarantee the safety of individuals using these procedures and hereby disclaim any liability for any injuries or damages claimed to have resulted from or related in any way to the procedures herein.

3. Discussion

Syntheses of *N*-acetyl enamides have gained increasing attention due to their importance as the starting materials for making chiral amines by asymmetric hydrogenation. Several preparative methods are available, including: 1) direct condensation of acetamides with ketones;[3] 2) addition of an organometallic reagent to a nitrile followed by quench of the resulting imine with an electrophile;[4] 3) transition metal-catalyzed coupling of vinyl derivatives such as vinyl halides,[5] triflates,[6] or tosylates[7] with amides; 4) Heck arylation of vinylacetamide;[8] 5) reductive acylation of ketoximes.[9] While most of these methods suffer from a limited substrate scope, low yields, high cost, or the requirement of cryogenic conditions, the reductive acylation of ketoximes with Fe/Ac$_2$O/AcOH proved to be economical and tolerant of various functionalities, and is often the method of choice at a laboratory scale.[9a-d] However, this method is not amenable to scale-up. The uncontrollable initiation of iron powder often leads to inconsistent yields and impurities. Singh and coworkers[9f] reported a method in which pyrophoric phosphine is used as the alternative reducing reagent. Hydrogen was also reported as the reducing source, primarily for the synthesis of cyclic enamides.[9g] A CuI-catalyzed method with NaHSO$_3$ as the reducing reagent was also reported.[9h]

The procedure presented herein employing iron(II) acetate as the reducing reagent offers several advantages over the Fe/Ac$_2$O/AcOH method.[10] First of all, its mild reaction profile is reliable and amenable to scale-up activities. Additionally, the purification procedure is simpler since both iron(II) acetate as the reagent and iron(III) acetate as the side-product are water-soluble. Higher yields are achieved for most substrates because of the mild reaction conditions (**Table 1** and **Table 2**).

Table 1. Synthesis of acyclic *N*-acetyl α-arylenamides[a]

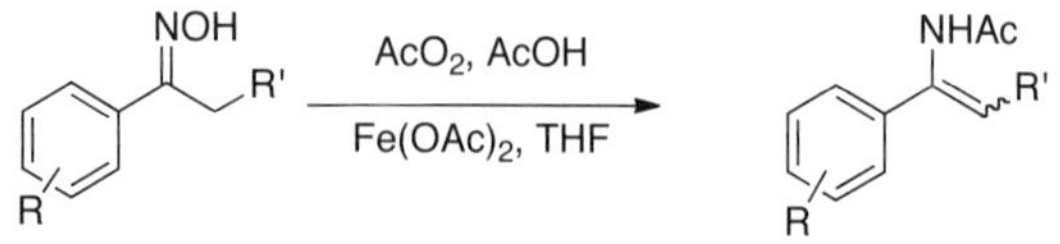

entry	ketoxime	enamide	yield (%)[b]
1	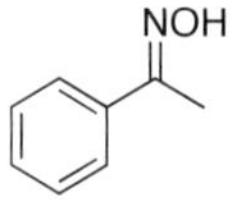	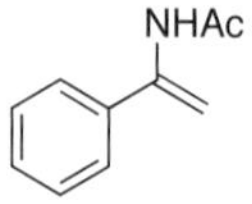	78
2	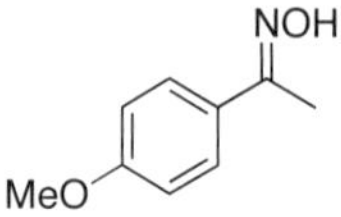	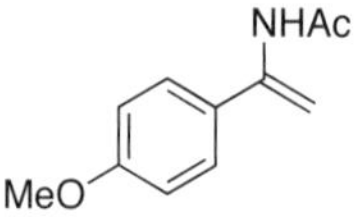	82
3	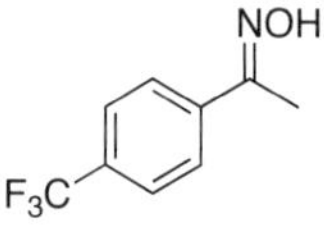	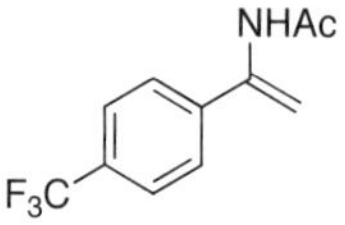	87
4	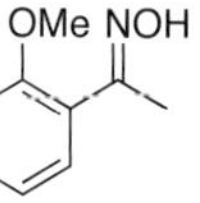	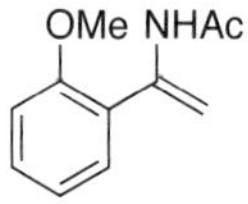	77
5	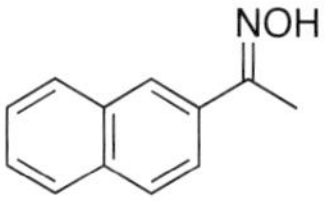	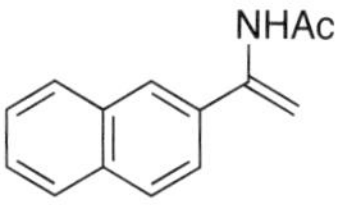	77
6	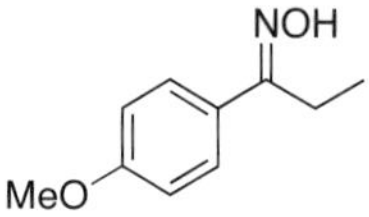	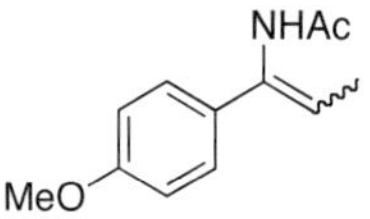	80[c]
7	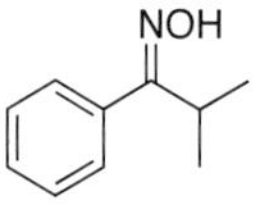	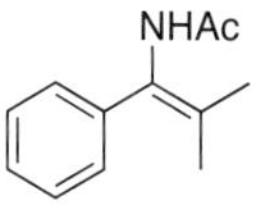	66
8	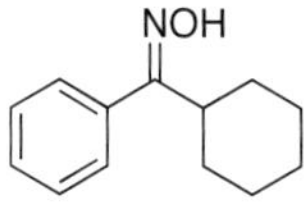	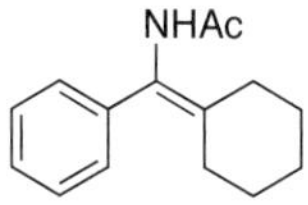	67

[a]Reaction conditions: ketoxime (4 mmol, 1 equiv), Fe(OAc)$_2$ (8 mmol, 2 equiv), AcOH (12 mmol, 3 equiv), Ac$_2$O (8 mmol, 2 equiv), THF (20 mL), at 65-67 °C for 7-15 h under nitrogen; [b]Unoptimized isolated yields. [c]*E/Z* ratio ~1/1.5.

Org. Synth. **2013**, *90*, 62-73

Table 2. Synthesis of acyclic *N*-acetyl α-arylenamides[a]

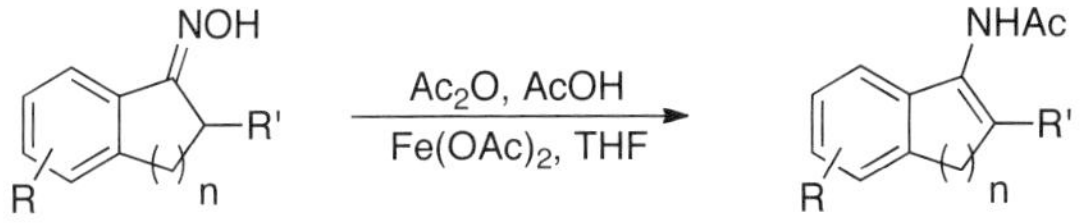

entry	ketoxime	enamide	yield (%)
1	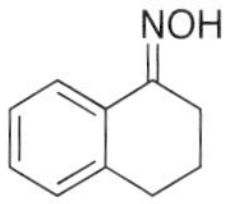	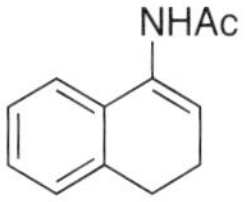	80
2	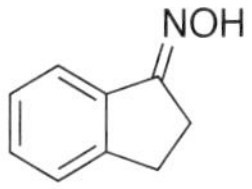	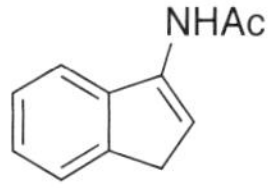	50
3	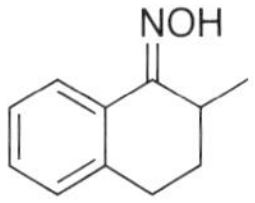	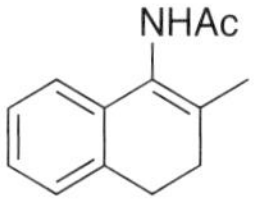	55
4	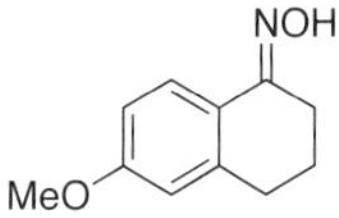	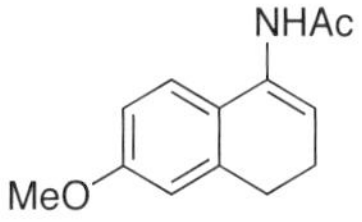	82
5	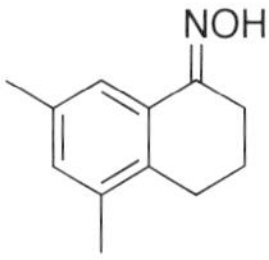	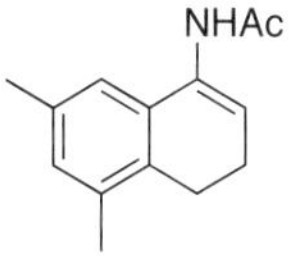	74

[a]The reaction conditions were similar as shown in Table 1. Yields were not optimized.

1. Chemical Development, Boehringer Ingelheim Pharmaceutical Inc., Ridgefield, CT 06877, USA. Email: wenjun.tang@boehringer-ingelheim.com
2. Prateeptongkum, S.; Jovel, I.; Jackstell, R.; Vogl, N.; Weckbecker, C.; Beller, M. *Chem. Comm.* **2009**, 1990.
3. (a) Tschaen, D. M.; Abramson, L.; Cai, D.; Desmond, R.; Dolling, U. H.; Frey, L.; Karady, S.; Shi, Y.-Y.; Verhoeven, T. R. *J. Org. Chem.* **1995**, *60*, 4324. (b) Dupau, P.; Le Gendre, P.; Bruneau, C.; Dixneuf, P. H. *Synlett* **1999**, 1832. (c) Chen, J.; Zhang, W.; Geng, H.; Li, W.; Hou, G.; Lei, A.; Zhang, X. *Angew. Chem., Int. Ed.* **2009**, *48*, 800.

4. (a) Savarin, C. G.; Boice, G. N.; Murry, J. A.; Corley, E.; DiMichele, L.; Hughes, D. *Org. Lett.* **2006**, *8*, 3903, and references therein. (b) Enthaler, S.; Hagemann, B.; Junge, K.; Erre, G.; Beller, M. *Eur. J. Org. Chem.* **2006**, 2912. (c) Van den Berg, M.; Haak, R. M.; Minnaard, A. J.; de Vries, A. H. M.; Vries, J. G.; Feringa, B. L. *Adv. Synth. Catal.* **2002**, *344*, 1003. (d) Burk, M. J.; Lee, J. R.; Wang, Y. M. *J. Am. Chem. Soc.* **1996**, *118*, 5142.

5. (a) Pan, X.; Cai, Q.; Ma, D. *Org. Lett.* **2004**, *11*, 1809. (b) Coleman, R. S.; Liu, P.-H. *Org. Lett.* **2004**, *6*, 577. (c) Jiang, L.; Job, G. E.; Klapars, A.; Buchwald, S. L. *Org. Lett.* **2003**, *5*, 3667. (d) Shen, R.; Porco, J. A., Jr. *Org. Lett.* **2000**, *2*, 1333. (e) Ogawa, T.; Kiji, T.; Hayami, K.; Suzuki, J. *Chem. Lett.* **1991**, 1443.

6. Wallace, D. J.; Klauber, D. J.; Chen, C.-y.; Volante, R. P. *Org. Lett.* **2003**, *5*, 4749.

7. Klapars, A.; Campos, K. R.; Chen, C.-y.; Volante, R. P. *Org. Lett.* **2005**, *7*, 1185.

8. (a) Liu, Z.; Xu, D.; Tang, W.; Xu, L.; Mo, J.; Xiao, J. *Tetrahedron Lett.* **2008**, *49*, 2756. (b) Lindhardt, A.; Skrydstrup, T. *J. Org. Chem.* **2005**, *70*, 5997. (c) Harrison, P.; Meek, G. *Tetrahedron Lett.* **2004**, *45*, 9277.

9. (a) Burk, M. J.; Casey, G.; Johnson, N. B. *J. Org. Chem.* **1998**, *63*, 6084. (b) Zhu, G.; Zhang, X. *J. Org. Chem.* **1998**, *63*, 9590. (c) Laso, N. M.; Quiclet-Sire, B.; Zard, S. Z. *Tetrahedron Lett.* **1996**, *37*, 1605. (d) Le, J. C.-D.; Pagenkopf, B. L. *J. Org. Chem.* **2004**, *69*, 4177. (e) Allwein, S. P.; McWilliams, J. C.; Secord, E. A.; Mowrey, D. R.; Nelson, T. D.; Kress, M. H. *Tetrahedron Lett.* **2006**, *47*, 6409. (f) Zhao, H.; Vandenbossche, C. P.; Koenig, S. G.; Singh, S. P.; Bakale, R. P. *Org. Lett.* **2008**, *10*, 505. (g) Guan, Z.-H.; Huang, K.; Yu, S.; Zhang, X. *Org. Lett.* **2009**, *11*, 481. (h) Guan, Z.-H.; Zhang, Z.-Y.; Ren, Z.-H.; Wang, Y.-Y.; Zhang, X. *J. Org. Chem.* **2011**, *76*, 339.

10. Tang, W.; Capacci, A.; Sarvestani, M.; Wei, X.; Yee, N. K.; Senanayake, C. H. *J. Org. Chem.* **2009**, *74*, 9528.

Appendix
Chemical Abstracts Nomenclature; (Registry Numbers)

1-(4-Bromophenyl)ethanone: Ethanone, 1-(4-bromophenyl)-; (99-90-1)

Hydroxylamine hydrochloride: (5470-11-1)

1-(4-Bromophenyl)ethanone oxime: Ethanone, 1-(4-bromophenyl)-, oxime; (5798-71-0)

Iron(II) acetate: Acetic acid, iron(2+) salt (2:1); (3094-87-9)

N-(1-(4-Bromophenyl)vinyl)acetamide: Acetamide, N-[1-(4-bromophenyl)ethenyl]-; (177750-12-8)

Dr. Wenjun Tang was born in 1974 in Zhejiang Province, China. He received his B.Eng. degree in 1995 from East China University of Sciences and Technology and his M.S degree in 1998 in Chemistry from Shanghai Institute of Organic Chemistry, Chinese Academy of Sciences, where he worked with Professor Dawei Ma. He was awarded his Ph.D. in 2003 from The Pennsylvania State University for his work in asymmetric catalysis with Professor Xumu Zhang. After two-year postdoctoral research with Professor K. C. Nicolaou at the Scripps Research Institute, he joined the Department of Chemical Development at Boehringer Ingelheim Pharmaceuticals Inc, where he is currently Principal Scientist. His research interests include development of efficient chemical processes, asymmetric synthesis and catalysis, and chiral ligand design.

Nitinchandra D. Patel was born in 1980 in Ahmedabad, India. He received his B.Eng. degree in 2001 in Chemical Engineering from Nirma Institute of Technology, Ahmedabad, India. He then moved to USA and joined M.S. program at Texas A & M University-Kingsville, where he obtained his M.S. degree under the guidance of Dr. Apurba Bhattacharya in 2005. Since then, he has been a process chemist at Boehringer Ingelheim Pharmaceuticals Inc. in Ridgefield, Connecticut. His research focuses on the design and development of practical synthetic methods to drug candidates.

Andrew Capacci was born in Maryland in 1987 and moved with his family to Alaska in 1988. He received his bachelor's degree from the University of Toronto in 2009 where he performed undergraduate research with Professor Mark Lautens. He then undertook an internship at Boehringer Ingelheim Pharmaceuticals, Inc. under the supervision of Dr. Wenjun Tang. He is currently a graduate student at Princeton University with Professor David MacMillan.

Dr. Xudong Wei is currently a Senior Principal Scientist and project leader in Boehringer-Ingelheim Pharmaceuticals working on the large-scale synthesis of active pharmaceutical ingredients. He received his BS and PhD degrees from Nanjing University in China and subsequently worked there as an Associate Professor for two years. He did postdoctoral research at Tübingen University, the University of York and Emory University before he joined Boehringer-Ingelheim Pharmaceuticals in Ridgefield, Connecticut in 2001. He is the co-author of >70 research publications and patents.

Dr. Chris H. Senanayake was born in Sri Lanka. He completed his M.S. at Bowling Green State University, and his Ph.D. under the guidance of Professor James H. Rigby at Wayne State University in 1987. He then undertook a postdoctoral fellow with Professor Carl R. Johnson and worked on the total synthesis of polyol systems such as amphotericin B and compactin analogous, and the synthesis of C-nucleoside precursors. In 1989, he joined the Department of Process Development at Dow Chemical Co. In 1990, he joined the Merck Process Research Group. After six years at Merck, he accepted a position at Sepracor, Inc. in 1996 where he was promoted to Executive Director of Chemical Process Research. In 2002, he joined Boehringer Ingelheim Pharmaceuticals. Currently, he is the Vice President of Chemical Development and leading a group of highly talented scientists, engineers, and administrative staff located in Ridgefield, CT.

72

Christopher K. Haley was born in Boston, MA in 1987 and received his B.S. degree in chemistry in 2010 from New York University where he conducted research with Professor David I. Schuster. He subsequently joined the laboratory of Professor Brian M. Stoltz at the California Institute of Technology in 2010 where he is currently pursuing his Ph.D. His research interests focus on the methodologies that exploit aryne reactive intermediates in the preparation of benzannulated heterocycles and their application toward the synthesis of biologically active natural products.

Hendrik Klare, born in Ankum, Germany in 1981, studied chemistry at the Westfälische Wilhelms-Universität Münster where he received his diploma in 2007 and his Ph.D. degree in 2011 under the supervision of Professor Martin Oestreich. His graduate research included the exploitation of diverse modes of silane activation for catalysis. As part of its DFG predoctoral fellowship, he also spent six months as a visiting scholar in the group of Professor Kazuyuki Tatsumi at Nagoya University in Japan. After postdoctoral studies with Professor Gerhard Erker at Münster focusing on frustrated Lewis pair chemistry, Hendrik joined the laboratories of Professor Brian M. Stoltz at the California Institute of Technology as a DAAD postdoctoral fellow and is currently involved in the development of novel methodologies for natural product synthesis.

Oxindole Synthesis via Palladium-catalyzed C–H Functionalization

A.

B.

C.

Submitted by Javier Magano,[1]* E. Jason Kiser, Russell J. Shine, and Michael H. Chen.

Checked by David Hughes.

1. Procedure

A. Benzyl 4-(4-(methoxycarbonyl)phenylamino)piperidine-1-carboxylate (3). A 1-L 3-necked round-bottomed flask equipped with a PTFE-coated magnetic stirring bar (3 x 1 cm) is fitted with a gas inlet adapter connected to a nitrogen line and a gas bubbler. The other two necks are capped with rubber septa; a thermocouple probe is inserted through one of the septa (Note 1). The flask is charged with methyl 4-aminobenzoate (**1**, 25.6 g, 169 mmol, 1 equiv), 1-benzyloxycarbonyl-4-piperidone (**2**, 47.0 g, 201 mmol, 1.2 equiv) and dichloromethane (300 mL) (Note 2). Glacial acetic acid (9.5 mL, 10.0 g, 167 mmol, 1 equiv) is added in one portion to the stirred solution using a graduated pipette. The flask is immersed in a room temperature water bath. Sodium triacetoxyborohydride (53.1 g, 250 mmol, 1.5 equiv) is added in 4 portions (12–14 g each) at 30-min intervals,

74

Published on the Web 9/10/2012
© 2013 Organic Syntheses, Inc.

keeping the internal temperature below 27 °C (Notes 3 and 4). After 17 h at 22–23 °C (Notes 5, 6, and 7), one septum is replaced with a 125-mL dropping funnel which is charged with 2 N aq. NaOH (125 mL, 0.25 mol, 1.5 equiv). The NaOH solution is added to the flask over 5 min (Note 8), keeping the internal temperature below 30 °C. The biphasic mixture is vigorously stirred for 1 h, then the contents are transferred to a 1-L separatory funnel and the layers separated. The organic phase is washed with water (2 × 125 mL). The dichloromethane extracts are filtered through a bed of anhydrous sodium sulfate (50 g) into a tared 1-L round-bottomed flask and concentrated by rotary evaporation (bath temperature: 40 °C; 200–250 mmHg) to 130 g. Methyl *t*-butyl ether (125 mL) is added and the contents are concentrated by rotary evaporation (bath temperature: 40 °C; 100 mmHg) to 130 g. Two additional methyl *t*-butyl ether flushes are carried out (Note 9). The flask is equipped with a PTFE-coated magnetic stirring bar (3 x 1 cm) and a 500-mL addition funnel. Methyl *t*-butyl ether (250 mL) is added to the flask in one portion and the resulting clear solution is stirred at 22 °C. Crystallization initiates within 10 min. The addition funnel is charged with *n*-heptane (250 mL), which is added drop wise over 1 h. The resulting white suspension is stirred at 20–22 °C for 1 h, then vacuum-filtered through a 350-mL medium-porosity sintered-glass funnel. The solid is washed with 1:1 (vol/vol) *n*-heptane/methyl *t*-butyl ether (50 mL) and *n*-heptane (50 mL). The solid is dried in a vacuum oven (70 mmHg) at 50 °C for 24 h to afford benzyl 4-(4-(methoxycarbonyl)phenylamino)piperidine-1-carboxylate (**3**) (52.0 g, 84% yield) (Notes 10 and 11).

B. *Benzyl 4-(2-chloro-N-(4-(methoxycarbonyl)phenyl)acetamido)-piperidine-1-carboxylate (5).* A 1-L 3-necked round-bottomed flask equipped with a PTFE-coated magnetic stirring bar (3 x 1 cm) is fitted with a gas inlet adapter connected to a nitrogen line and a gas bubbler. The other two necks are capped with rubber septa; a thermocouple probe is inserted through one of the septa (Note 1). The flask is charged with benzyl 4-(4-(methoxycarbonyl)phenyl-amino)piperidine-1-carboxylate (**3**, 25.0 g, 68 mmol, 1 equiv), ethyl acetate (250 mL), and pyridine (8.3 mL, 103 mmol, 1.5 equiv) (Note 12). The flask is placed in an ice-water bath and cooled to an internal temperature of 3 °C. Chloroacetyl chloride (10.1 g, 89 mmol, 1.3 equiv) is added via a 12-mL syringe over 5 min, keeping the internal temperature below 10 °C. The ice-bath is removed and the orange slurry is stirred for 1.5 h at ambient temperature (Note 13). One of the septa is replaced with a 125-mL addition funnel and charged with 5% aq. KH_2PO_4

(125 mL), which is added to the reaction contents over 5 min while keeping the internal temperature below 25 °C. After stirring for 10 min (Note 14), the contents of the flask are transferred to a 1-L separatory funnel. The layers are separated and the organic layer is washed with half-saturated brine, then dried by filtration through 50 g anhydrous sodium sulfate into a 1-L round-bottomed flask. The contents are concentrated by rotary evaporation (bath temperature: 40 °C; 100 to 50 mmHg, foaming) (38 g). Methyl t-butyl ether (150 mL) is added to the flask and concentrated (bath temperature: 40 °C; 100 to 50 mmHg, foaming) to an orange oil (35 g). The flask is equipped with a PTFE-coated magnetic stirring bar (3 x 1 cm) and methyl t-butyl ether (150 mL) is added. The contents are warmed in a 50 °C water bath to dissolve the oil, then cooled to ambient temperature with stirring. Within 15 min the mixture turns turbid and a white solid begins to form (Note 15). After stirring for 4 h, the solid is filtered using a 150-mL medium porosity sintered-glass funnel and washed with methyl t-butyl ether (75 mL). The solid is dried in a vacuum oven (70 mmHg) at 40 °C for 2 days to afford benzyl 4-(2-chloro-N-(4-(methoxycarbonyl)phenyl)acetamido)-piperidine-1-carboxylate (**5**) (28.5 g, 94% yield) as an off-white solid (Note 16).

C. *Methyl 1-(1-(benzyloxycarbonyl)piperidin-4-yl)-2-oxoindoline-5-carboxylate (6).* A 500-mL 3-necked round-bottomed flask equipped with a PTFE-coated magnetic stirring bar (3 x 1 cm) is fitted with a reflux condenser with a gas inlet adapter connected to a nitrogen line and a gas bubbler. The other two necks are capped with rubber septa; a thermocouple probe is inserted through one of the septa (Note 1). The flask is charged with benzyl 4-(2-chloro-*N*-(4-(methoxycarbonyl)phenyl)acetamido)-piperidine-1-carboxylate (**5**, 20.0 g, 44.6 mmol, 1 equiv), palladium acetate (1.01 g, 4.5 mmol, 0.1 equiv), 2-(di-t-butylphosphino)biphenyl (2.74 g, 9.2 mmol, 0.2 equiv), 2-methyltetrahydrofuran (130 mL), and 2-propanol (32 mL) (Note 17). The light brown suspension is sparged subsurface with nitrogen gas for 15 min (Note 18). Triethylamine (6.93 g, 68 mmol, 1.5 equiv) is added via a 12-mL syringe and the mixture is sparged for an additional 5 min. The flask is placed in an oil bath at 80 °C and heated to an internal temperature of 74–76 °C for 2 h (Note 19). The hot mixture (Note 20) is vacuum-filtered through a pad of Celite (25 g, pre-wetted with 2-MeTHF) in a 150-mL medium-porosity sintered glass funnel into a 1-L round-bottomed flask. The flask employed to carry out the reaction is rinsed with hot (75 °C) 2-MeTHF (100 mL) and used to wash the Celite pad. The filtrate is

76

concentrated by rotary evaporation (bath temperature: 40 °C; 40 mmHg) to give an orange-brown solid (34 g). The flask is equipped with a PTFE-coated magnetic stirring bar (3 x 1 cm) and 2-propanol (270 mL) is added. The stirred suspension is heated to a gentle reflux with a heating mantle to dissolve all solids, generating a nearly black solution. The heating mantle is turned off and the contents are allowed to cool to ambient temperature over 2 h with stirring to give thick crystallization (Note 21). The suspension is stirred for 4 h, then vacuum-filtered using a 150-mL medium-porosity sintered-glass funnel, washed with 2-propanol (2 x 30 mL, Note 22), then dried in a vacuum oven (70 mmHg) at 40 °C for 2 days to afford oxindole **6** as a gray solid (15.9 g, 87% yield) (Notes 23 and 24).

2. Notes

1.   The internal temperature was monitored using a J-Kem Gemini digital thermometer with a Teflon-coated T-Type thermocouple probe (12-inch length, 1/8 inch outer diameter, temperature range –200 to +250 °C).

2.   The following reagents and solvents were used as received for Step A: methyl 4-aminobenzoate (Sigma-Aldrich, 98%), 1-benzyloxycarbonyl-4-piperidone (Sigma-Aldrich, 99%), sodium triacetoxyborohydride (Sigma-Aldrich, 95%), dichloromethane (Fisher, certified ACS reagent, stabilized), glacial acetic acid (Fisher, certified ACS plus), *t*-butyl methyl ether (Sigma-Aldrich, >98.5%), and heptanes (Sigma-Aldrich, Chromasolv, >99%).

3.   The first two $Na(OAc)_3BH$ portions produced 3–5-degree exotherms.

4.   The addition of $Na(OAc)_3BH$ afforded a pale yellow/green, slightly cloudy mixture.

5.   The submitters monitored the progress of the reaction after each $Na(OAc)_3BH$ addition (5 additions) by quenching an aliquot with water after 30 min, diluting with 9/1 MeCN/water and analyzing by UPLC (Note 6). The results were (limiting reagent/product ratio): 1st portion: 42/58; second portion: 28/72; third portion: 13/87; fourth portion: 8/92. One and two h after the 5th portion had been added, the ratios were 3/97 and 2.4/97.6%, respectively. The checker monitored the reaction by ^1H NMR ($CDCl_3$) by quenching a reaction aliquot into dichloromethane/water, separating the layers, and concentrating the dichloromethane layer to dryness. The diagnostic resonances were δ 6.54–6.57 (m, 2H) for product **3** and 6.63-6.66

(m, 2H) for starting material **1**. Two h after the final addition of Na(OAc)₃BH, 10% unreacted **1** remained; after 16 h, the level was 3%.

6.   UPLC conditions: column, ACQUITY UPLC HSS T3 1.8µm, 2.1 × 50 mm; wavelength: 210 nm; column temperature: 45 °C; eluent A) water (0.05% TFA) B) MeCN; gradient: 0 min: A) 95%, B) 5%; 2.9 min: A) 0% B) 100%; 3.15 min: A) 0% B) 100%; 3.25 min: A) 95% B) 5%; 4.0 min: A) 95% B) 5%.

7.   UPLC analysis by the submitters after 16 h showed 11% of benzyl 4-hydroxypiperidine-1-carboxylate (byproduct from the reduction of 1-benzyloxycarbonyl-4-piperidone; retention time: 1.35 min), 84.7% of desired product **3** (retention time: 2.08 min), and 4.4% of an unidentified byproduct (retention time: 0.88 min).

8.   The flask should be kept under a nitrogen atmosphere during the quench since hydrogen gas is produced upon quenching unreacted Na(OAc)₃BH.

9.   Three co-evaporations with MTBE is an efficient way to remove most of the dichloromethane and maximize the yield in the subsequent crystallization. Concentration of the dichloromethane phase to very small volumes affords a foamy oil, making further removal of dichloromethane by co-evaporation with MTBE less efficient. After the co-evaporations, ¹H NMR (CDCl₃) of the residue showed 5 mol % residual dichloromethane.

10.   Amine **3** has the following physical and spectroscopic properties: Mp: 90–92 °C. ¹H NMR (400 MHz CDCl₃) δ: 1.35–1.43 (m, 2 H), 2.06 (d, *J* = 10.9 Hz, 2 H), 3.03 (t, *J* = 11.8 Hz, 2 H), 3.50–3.55 (m, 1 H), 3.85 (s, 3 H), 4.10–4.17 (m, 3 H), 5.15 (s, 2 H), 6.54–6.57 (m, 2 H), 7.31–7.38 (m, 5 H), 7.85–7.88 (m, 2 H). ¹³C NMR (100 MHz, CDCl₃) δ: 32.2, 42.9, 49.7, 51.7, 67.4, 112.0, 118.8, 128.1, 128.2, 128.7, 131.8, 136.72, 150.7, 155.4, 167.3. IR (ATR cell) cm⁻¹: 3357, 2951, 1707, 1675, 1604, 1531, 1500, 1471, 1436, 1363, 1353, 1309, 1274, 1226, 1197, 1172, 1149, 1095, 1008, 983, 951, 839, 788, 769, 750, 693. LC-MS *m/z* calcd for [M]⁺ (C₂₁H₂₄N₂O₄) 368.4, found, 368.7; Anal. Calcd for C₂₁H₂₄N₂O₄: C, 68.46; H, 6.57; N, 7.60. Found: C, 68.54; H, 6.50; N, 7.57. HPLC area % purity: 97-98% (HPLC method: fused-core C-18 column, 4.6 x 100 mm, 2.7 µm particle size; mobile phase, A = 0.1 % H₃PO₄/H₂O, B = MeCN, gradient 10-95% B in 6 min and hold at 95% B for 2 minutes, detection at 210 nm, flow 1.8 mL/min, temp 40 °C; t$_R$ = 4.93 min).

11.   A yield of 85% was obtained at half scale.

12. The following reagents and solvents were used as received for Step B: ethyl acetate (Fischer Optima, 99.9%, water level <0.002%), pyridine (Sigma-Aldrich, Reagent Plus, >99%), chloroacetyl chloride (Fluka purum ≥ 99%), and *t*-butyl methyl ether (Sigma-Aldrich, >98.5%).

13. The submitters followed the reaction by UPLC as outlined in Note 6. The checker monitored the reaction by ^{1}H NMR as follows: A 0.1 mL reaction aliquot was quenched into 1 mL brine/1 mL EtOAc. The organic layer was separated and dried by filtering through a plug of sodium sulfate. After concentrating to dryness, the sample was dissolved in CDCl$_3$. NMR analysis showed no starting material resonances at 3.85 (s, 3 H) or 6.54-6.57 (m, 2 H).

14. The reaction is quenched with phosphate buffer at pH 10 and stirred for 10 min to fully quench excess chloroacetyl chloride that will otherwise inhibit the subsequent crystallization of the product.

15. Crystallization for the first run required vigorous scratching of the flask with a glass rod. In subsequent runs, the crystallization occurred spontaneously.

16. Chloride **5** has the following physical and spectroscopic properties: Mp: 103–105 °C. ^{1}H NMR (400 MHz, CDCl$_3$) δ: 1.23–1.28 (m, 2 H), 1.84 (d, J = 11.6 Hz, 2 H), 2.88 (br s, 2 H), 3.68 (s, 2 H), 3.96 (s, 3 H), 4.23 (br s, 2 H), 4.74–4.80 (m, 1 H), 5.04 (br s, 2 H), 7.22 (d, J = 8.4, 2H), 7.28–7.35 (m, 5 H), 8.13 (d, J = 8.7 Hz, 2 H). ^{13}C NMR (100 MHz, CDCl$_3$) δ: 30.4, 42.3, 43.5, 52.7, 53.7, 67.4, 128.1, 128.2, 128.6, 130.4, 131.3, 131.4, 136.7, 141.4, 155.1, 165.7, 166.0; IR (ATR cell) cm^{-1}: 2951, 1714, 1705, 1690, 1673, 1604, 1510, 1496, 1449, 1440, 1427, 1391, 1362, 1329, 1293, 1276, 1246, 1231, 1208, 1195, 1176, 1131, 1118, 1105, 1086, 1062, 1020, 985, 968, 956, 934, 921, 870, 834, 814, 797, 788, 779, 769, 753, 715, 707, 678, 665; LC-MS *m/z* calcd for [M]$^+$ (C$_{23}$H$_{25}$ClN$_2$O$_5$) 444.4, found, 444.6; Anal. Calcd for C$_{23}$H$_{25}$ClN$_2$O$_5$: C, 62.09; H, 5.66; Cl, 7.97; N, 6.30. Found: C, 61.69; H, 5.39; Cl, 7.81; N, 6.17; HPLC area % purity: 98% (conditions in Note 10; t_R = 4.73 min

17. The following reagents and solvents were used as received for Step C: palladium acetate (Strem, 98%), 2-(di-*t*-butylphosphino)biphenyl (Acros, 99%), 2-methyltetrahydrofuran (Sigma Aldrich, Reagent Plus ≥99.5%, inhibited with 150-400 ppm BHT), 2-propanol (Sigma-Aldrich, ACS reagent >99.5%).

18. Nitrogen sparging was carried out using a 1-mL plastic syringe with a 10 cm needle with a steady stream of bubbling for 15 min. The heterogeneous mixture darkened during the sparging.

19. The submitters monitored the reaction by UPLC analysis using the conditions in Note 6. The checker monitored the reaction by ^{1}H NMR by diluting a 0.1 mL aliquot from the reaction mixture into 1 mL CDCl$_3$ and filtering through a 0.25 µM filter. At the 1 h time point, 2.5% starting material remained based on resonances integrated at δ 5.04 (br s, 2 H) and 8.13 (d, J = 8.7 Hz, 2 H).

20. The mixture must be filtered while still hot since the product crystallizes upon cooling.

21. Thick solids formed when the internal temperature reached 60–65 °C. The stirring speed had to be increased for efficient mixing.

22. Care was taken to avoid cake cracking prior to the 2-propanol wash, which allowed for the efficient removal of highly colored impurities.

23. Oxindole **6** has the following physical and spectroscopic properties: Mp: 158–159 °C. ^{1}H NMR (400 MHz, CDCl$_3$) δ: 1.73–1.76 (m, 2 H), 2.29–2.39 (m, 2 H), 2.92 (br s, 2 H), 3.56 (s, 2 H), 3.93 (s, 3 H), 4.39–4.47 (m, 3 H), 5.19 (s, 2 H), 6.97 (d, J = 8.4 Hz, 1 H), 7.33–7.42 (m, 5 H), 7.92–7.93 (m, 1 H), 7.96–7.98 (m, 1 H). ^{13}C NMR (100 MHz, CDCl$_3$) δ: 28.3, 35.7, 43.9, 50.4, 52.3, 67.7, 109.3, 124.3, 124.8, 126.1, 128.3, 128.4, 128.8, 130.5, 136.8, 147.8, 155.4, 166.9, 175.1. IR (ATR cell) cm^{-1}: 2948, 1699, 1616, 1588, 1489, 1453, 1428, 1386, 1358, 1332, 1320, 1291, 1272, 1256, 1241, 1227, 1192, 1169, 1139, 1126, 1097, 1077, 1021, 986, 968, 957, 938, 902, 890, 869, 838, 802, 771, 763, 731, 696, 683, 654; LC-MS m/z calcd for [M]$^+$ (C$_{23}$H$_{24}$N$_2$O$_5$) 408.5, found, 408.7; Anal. Calcd for C$_{23}$H$_{24}$N$_2$O$_5$: C, 67.63; H, 5.92; N, 6.86. Found: C, 67.72; H, 5.75; N, 6.77. HPLC area % purity: 98% (conditions in Note 10; t_R = 4.60 min).

24. A reaction at half scale afforded an 84% yield.

Handling and Disposal of Hazardous Chemicals

The procedures in this article are intended for use only by persons with prior training in experimental organic chemistry. All hazardous materials should be handled using the standard procedures for work with chemicals described in references such as "Prudent Practices in the Laboratory" (The National Academies Press, Washington, D.C., 2011 www.nap.edu). All chemical waste should be disposed of in accordance with local regulations. For general guidelines for the management of chemical waste, see Chapter 8 of Prudent Practices.

These procedures must be conducted at one's own risk. *Organic Syntheses, Inc.*, its Editors, and its Board of Directors do not warrant or guarantee the safety of individuals using these procedures and hereby disclaim any liability for any injuries or damages claimed to have resulted from or related in any way to the procedures herein.

3. Discussion

Oxindoles are an important class of compounds with ubiquitous presence in both natural products[2] and pharmaceuticals.[3] In addition, oxindoles can be employed as immediate precursors for the preparation of indoles. Numerous methods have been reported in the literature for the synthesis of oxindoles, such as the derivatization of isatin and indoles, radical cyclizations, the cyclization of *o*-aminophenylacetic acids and α-halo or α-hydroxyacetanilides, cyanoamidation reactions, palladium-catalyzed Heck couplings, and others.[4] Alternatively, Buchwald and co-workers have reported a palladium-catalyzed alkylation reaction via C–H functionalization[5] and Hartwig and co-workers a palladium-catalyzed arylation reaction via amide α-arylation[6] (Figure 1). These two technologies represent direct and unprecedented approaches to the oxindole functionality from readily accessible precursors.

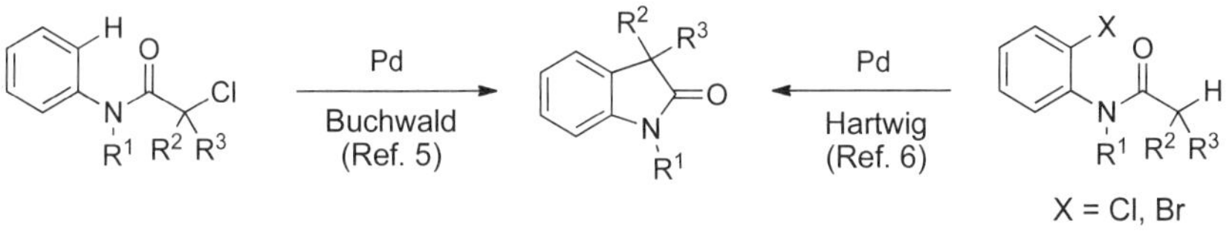

Figure 1. Buchwald's and Hartwig's palladium-catalyzed methodologies for oxindole formation

Compound **7** (Figure 2) is a serine palmitoyl transferase (SPT) enzyme inhibitor candidate for the potential treatment of heart disease. This molecule contains an oxindole functionality and was originally prepared by our Medicinal Chemistry group in nine steps from acid **8**. A key intermediate in this synthesis is oxindole **6**, which can be generated in five steps from **8**; however, this route has several disadvantages, such as the use of unsafe reagents (NaH) and the need for several chromatographic purifications. For the preparation of large quantities of **6** (hundreds of g to several kg), we wanted to avoid those issues and also identify a shorter synthesis. We focused our attention on Buchwald's protocol since no halogenated precursor is needed to install the 5-membered oxindole ring and one precedent was found in the literature on kg-scale.[7]

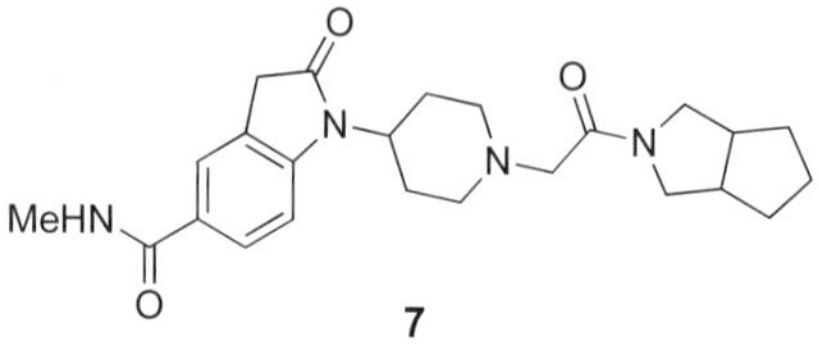

Figure 2. Structure of serine palmitoyl transferase (SPT) enzyme inhibitor **7**

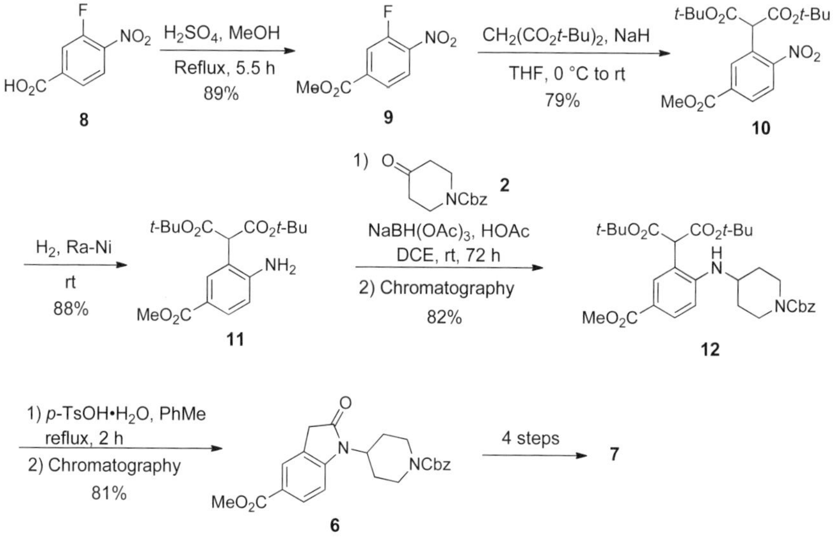

Scheme 1. Medicinal Chemistry synthesis of drug candidate **7**

The new and shorter route to oxindole **6** starts with the reductive amination between methyl 4-aminobenzoate (**1**), a considerably cheaper starting material than 3-fluoro-4-nitrobenzoic acid (**8**), and 1-benzyloxycarbonyl-4-piperidone (**2**) in dichloromethane in the presence of 1 equiv of AcOH. The addition of Na(OAc)$_3$BH in several portions resulted in

Org. Synth. **2013**, *90*, 74-86

complete conversion of **1** to secondary amine **3** and gave acceptable levels of alcohol benzyl 4-hydroxypiperidine-1-carboxylate resulting from the reduction of **2**. The crystallization of **3** from heptane/MTBE gives analytically pure material in 84% yield.

The acylation reaction between **3** and chloroacetyl chloride in anhydrous EtOAc and pyridine to afford amide **5** is fast (1 h) and clean. Pyridine gave a cleaner impurity profile compared to other bases such as triethylamine. Schotten-Baumann conditions (CH_2Cl_2 or EtOAc and aqueous K_2CO_3) were also investigated, but incomplete reaction was observed due to acid chloride hydrolysis. After an aqueous workup, amide **5** is crystallized from MTBE in 94% yield.

The cyclization step to produce oxindole **6** was first attempted under the conditions reported by Buchwald in his original publication (Pd(OAc)$_2$/2-(di-*t*-butylphosphino)biphenyl (1:2 ratio), triethylamine, toluene, 80 °C)[5] but sticky solids were obtained due to the low solubility of **6** in this medium. A solvent (THF, IPA, MeCN, DMF) and base (triethylamine, *n*-Bu$_3$N) screen identified THF/IPA 4:1 as the optimal combination to dissolve both starting material and product in the presence of triethylamine to give complete conversion in 1 h at 74–76 °C. The Pd catalyst and ligand loadings were 10 and 20 mol%, respectively, and optimization experiments with lower loadings resulted in incomplete reactions.[8] The reasons for the high catalyst and ligand loadings were not fully investigated and remain unclear.

In summary, a short and high yielding protocol for the preparation of oxindole **6** has been demonstrated on multi-gram scale from inexpensive and readily available starting materials. All intermediates can be isolated after crystallization in high purity and chromatographic purifications are no longer required. This chemistry has been scaled up in our laboratories to produce multi kg-quantities of **6**.[9]

1. Chemical Research & Development, Pharmaceutical Sciences, Pfizer Worldwide Research & Development, Eastern Point Road, Groton, Connecticut 06340, United States. Javier.Magano@Pfizer.com.
2. (a) Galliford, C. V.; Scheidt, K. A. *Angew. Chem., Int. Ed.* **2007**, *46*, 8748. (b) Marti, C.; Carreira, E. M. *Eur. J. Org. Chem.* **2003**, 2209.
3. (a) Sun, L.; Tran, N.; Tang, F.; App, H.; Hirth, P.; McMahon, G.; Tang, C. *J. Med. Chem.* **1998**, *41*, 2588. (b) Wood, E. R.; Kuyper, L.; Petrov,

K. G.; Hunter, R. N., III.; Harris, P. A.; Lackey, K. *Bioorg. Med. Chem. Lett.* **2004**, *14*, 953. (c) Masamune, H.; Cheng, J. B.; Cooper, K.; Eggier, J. F.; Marfat, A.; Marshall, S. C.; Shirley, J. T.; Tickner, J. E.; Umland, J. P.; Vazquez, E. *Bioorg. Med. Chem. Lett.* **1995**, *5*, 1965. (d) Robinson, R. P.; Reiter, L. A.; Barth, W. E.; Campeta, A. M.; Cooper, K.; Cronin, B. J.; Destito, R.; Donahue, K. M.; Falkner, F. C.; Fiese, E. F.; Johnson, D. L.; Kuperman, A. V.; Liston, T. E.; Malloy, D.; Martin, J. J.; Mitchell, D. Y.; Rusek, F. W.; Shamblin, S. L.; Wright, C. F. *J. Med. Chem.* **1996**, *39*, 10.

4. (a) Ueda, S.; Okada, T.; Nagasawa, H. *Chem. Commun.* **2010**, *46*, 2462 and references therein. (b) Jia, Y.-X.; Kündig, E. P. *Angew. Chem., Int. Ed.* **2009**, *48*, 1636 and references therein.

5. Hennessy, E. J.; Buchwald, S. L. *J. Am. Chem. Soc.* **2003**, *125*, 12084.

6. (a) Lee, S.; Hartwig, J. F. *J. Org. Chem.* **2001**, *66*, 3402. (b) Shaughnessy, K. H.; Hamann, B. C.; Hartwig, J. F. *J. Org. Chem.* **1998**, *63*, 6546.

7. Choy, A. C.; Colbry, N.; Huber, C.; Pamment, M.; Van Duine, J. *Org. Process Res. Dev.* **2008**, *12*, 884.

8. The Supporting Information in reference 5 (page S11) has the ratios of Pd and ligand reversed.

9. Kiser, E. J.; Magano, J.; Shine, R. J.; Chen, M. H. *Org. Process Res. Dev.* **2012**, *16*, 255.

Appendix
Chemical Abstracts Nomenclature (Registry Number)

Methyl 4-aminobenzoate: Benzoic acid, 4-amino-, methyl ester (619-45-4)

1-Benzyloxycarbonyl-4-piperidone: 1-Piperidinecarboxylic acid, 4-oxo-, phenylmethyl ester (19099-93-5)

Benzyl 4-(4-(methoxycarbonyl)phenylamino)piperidine-1-carboxylate: 1-Piperidinecarboxylic acid, 4-[[4-(methoxycarbonyl)phenyl]amino]-, phenylmethyl ester (1037834-44-8)

Benzyl 4-(2-chloro-N-(4-(methoxycarbonyl)phenyl)acetamido)piperidine-1-carboxylate: 1-Piperidinecarboxylic acid, 4-[(2-chloroacetyl)[4-(methoxycarbonyl)phenyl]amino]-, phenylmethyl ester (1037834-45-9)

Methyl 1-(1-(benzyloxycarbonyl)piperidin-4-yl)-2-oxoindoline-5-carboxylate: 1H-Indole-5-carboxylic acid, 2,3-dihydro-2-oxo-1-[1-

[(phenylmethoxy)carbonyl]-4-piperidinyl]-, methyl ester (1037834-34-6)

Sodium triacetoxyborohydride: Borate(1-), tris(acetato-κO)hydro-, sodium (56553-60-7)

Chloroacetyl chloride: Acetyl chloride, 2-chloro- (79-04-9)

Palladium acetate: Acetic acid, palladium(2+) salt (2:1) (3375-31-3)

2-(Di-*tert*-butylphosphino)biphenyl: Phosphine, [1,1'-biphenyl]-2-ylbis(1,1-dimethylethyl)- (224311-51-7)

Triethylamine: Ethanamine, *N,N*-diethyl- (121-44-8)

2-Methyltetrahydrofuran: Furan, tetrahydro-2-methyl- (96-47-9)

Javier Magano was born in Madrid, Spain. He received his B.S. in organic chemistry from Complutense University in Madrid in 1987 and a M.S. degree in chemistry from the University of Michigan in 1990. After working for the oil industry in Spain for three years, he obtained a M.S. degree in rubber and polymer science from the School of Plastics and Rubber at the Center for Advanced Scientific Research in Madrid. In 1995 he moved back to the United States to carry out graduate work at the University of Michigan and in 1998 he joined Pfizer Inc. to work in the early process group in Ann Arbor, MI, and currently in Groton, CT. He has also worked in the area of biologics for 1.5 years.

E. Jason Kiser was born in Chatham, Ontario (Canada) in 1971. He obtained his Bachelors of Science degree (Honors Chemistry) at the University of Windsor (Canada) in 1995. Jason went on to obtain a Master's of Science degree (Chemistry) under the direction of Dr. John M. McIntosh. Jason has over 14 years of synthesis and scale-up experience. He worked for Pfizer for over 10 years at the Ann Arbor, Michigan and Groton, Connecticut research sites as well as the Kalamazoo manufacturing site. He is currently a senior scientist in the process development group at Ash Stevens in Riverview, Michigan.

Russell J. Shine pursued his undergraduate studies at University of Pennsylvania, where he received his B.A. degree in Biology in 1981. After joining Pfizer Inc. in 1983, he joined Professor Phyllis Brown's research group at the University of Rhode Island, where he received his M.S. in Chemistry in 1991. He is currently working in the API manufacturing group within the Pharmaceutical Science Development Division of Pfizer Inc.

Michael Chen was born in Shanghai, China. He received his B.S. of chemistry from Shanghai University, and a M.S. and Ph.D. from the University of Michigan in 1988. After holding post-doctoral positions with Professor Paul Knochel at the University of Michigan and Peter Wuts and Tomi Sawyer at the Upjohn Company, Michael joined Parke-Davis Pharmaceutical Research/Pfizer where he worked until 2007 as a process chemist. He currently is the Chief Scientific Officer at MasTeam Biotech Research Institute in China.

Intermolecular retro-Cope Type Hydroxylamination of Alkynes with NH₂OH: (*E*-1-(1-Hydroxycyclohexyl)ethanone oxime)

Submitted by Francis Loiseau and André M. Beauchemin.[1]
Checked by Nikolas Huwyler and Erick M. Carreira.

1. Procedure

A.   *E-1-(1-Hydroxycyclohexyl)ethanone oxime.* A 100-mL one-necked (TS 14/23) round-bottomed flask is successively charged with 1-ethynylcyclohexanol (5.00 g, 40.3 mmol, 1 equiv) (Notes 1 and 2), 2-methyl-1-propanol (40 mL) (Note 3) and aqueous hydroxylamine (Note 4), (50 wt. %, 6.18 mL, 101 mmol, 2.5 equiv). The flask is equipped with a Teflon-coated magnetic stir bar (6 × 15 mm) and a reflux condenser, which is sealed with a rubber septum. Argon is introduced by means of a syringe needle through the septum and the system is flushed with argon for 5 min before the flask is immersed in an oil-bath at 125 °C (Note 5). The reaction mixture is refluxed for 18 h under an atmosphere of argon (Note 6). The solution stays clear and colorless for the duration of the reaction. After cooling to ambient temperature, the reaction mixture is transferred to a 250 mL round-bottomed flask and concentrated by rotary evaporation (45 °C, 25 mmHg). The resulting oil is azeotroped twice by adding hot toluene (2 x 50 mL) and concentrating by rotary evaporation (45 °C, 25 mmHg) (Note 7). The crude white solid (6.36–6.42 g) is transferred to a 50-mL Erlenmeyer flask and dissolved in 18 mL of near-boiling 1,2-dichloroethane (80 °C). The resulting solution is left to cool to ambient temperature over 2 h, allowing crystallization to occur, and is further cooled at 0 °C for 2 h. The resulting crystals are collected by suction filtration on a Büchner funnel, and suction-dried for 30 min. The crystals are rinsed with 5 mL of ice-cold hexanes, suction-dried again for 30 min, transferred to a 50-mL round-bottomed flask and dried for 1 h at 0.1 mmHg at ambient temperature to provide *E*-1-(1-hydroxycyclohexyl)-ethanone oxime as white needle crystals (5.27–5.42 g, 83–86% yield) (Notes 8 and 9).

2. Notes

1. The submitters used the following materials as received: 1-ethynylcyclohexanol (≥99%, Sigma-Aldrich), 2-methyl-1-propanol (ACS reagent grade, ≥99%, GFS Chemicals), toluene (ACS grade, 99.9%, Fisher Scientific), 1,2-dichloroethane (ACS grade, 99.9%, Fisher Scientific) and hexanes (ACS grade, 99.9%, Fisher Scientific).

2. The checkers obtained 2-methyl-1-propanol (ACS grade, ≥99.0%), 1-ethynylcyclohexanol (≥99%), and hexane (HPLC grade, ≥95%) from Sigma-Aldrich Co., while toluene (ACS grade, ≥99.7%) and 1,2-dichloroethane (ACS grade, ≥99.5%) were purchased from Fluka. All reagents and solvents were used as received.

3. 2-Methyl-1-propanol was chosen as a solvent for the reaction because its boiling point was in the temperature range needed for the desired reactivity and reaction times. 1-Propanol can be alternatively used as a solvent, using microwave heating (4 h, 140 °C) or heating in a sealed tube placed in an oil bath (18 h, 110 °C), yielding similar results.

4. Aqueous hydroxylamine (50 wt. %) was purchased from Aldrich Chemical Co. and used as received. *Caution:* Due to the high energy content of NH_2OH, appropriate care should be taken when conducting these experiments. The hydroxylamine concentration should not be increased beyond 5–10 wt% (i.e. the typical reaction conditions) and appropriate safety controls should be performed before scaling up this chemistry above the gram-scale, especially at very high temperatures. Note: NH_2OH has been used on the 2 mol scale under similar conditions (reflux in EtOH) in a previously reported procedure. (Steiger, R. E. *Org. Synth., Coll. Vol. III*, **1955**, 91).

5. The success of the reaction does not depend on having previously dried the glassware or on adding the reagents under an inert atmosphere.

6. The progress of the reaction can be followed by TLC analysis on silica gel with 30% EtOAc-hexane as eluent and visualized with potassium permanganate, ensuring proper evaporation of the reaction solvent by heating the plate pre-elution. The propargylic alcohol starting material has $R_f = 0.55$ and the α-hydroxyoxime product has $R_f = 0.37$.

88

7.   The high H-bonding affinity of the oxime product causes binding to alcoholic solvents. Toluene is used to azeotropically distill off the remaining 2-methyl-1-propanol. In cases when the oxime is obtained as a solid after the initial or second concentration, the mixture, after addition of hot toluene, can be heated and swirled until complete dissolution before concentrating again. (The crude oxime melts below the boiling point of toluene and is miscible in the solvent when in the liquid state).

8.   The product displayed the following spectroscopic and physiochemical properties: mp 101–103 °C; ^{1}H NMR (400 MHz, CDCl$_3$) δ: 1.14-1.30 (m, 1 H), 1.48-1.77 (m, 9 H), 1.90 (s, 3 H), 2.85 (br s, 1 H), 8.44 (br s, 1 H); ^{13}C NMR (101 MHz, CDCl$_3$) δ: 9.7, 21.4, 25.4, 35.0, 73.7, 162.4; IR (film) cm^{-1}: 3460, 3202, 3147, 2932, 2852, 1466, 1456, 1445, 1431, 1420, 1371, 1312, 1258, 1177, 1156, 1132, 1068, 1054, 998, 974, 936, 884, 846, 748; HRMS (EI) m/z: [M]$^+$ calcd for C$_7$H$_{15}$NO$_2$: 157.1098. Found: 157.1104; m/z: [M-OH]$^+$ for C$_7$H$_{12}$NO: 140.1070. Found: 140.1072; Elemental anal. calcd for C$_7$H$_{15}$NO$_2$: C, 61.12; H, 9.62; N, 8.91. Found: C, 61.08; H, 9.53; N, 8.91.

8.   The product was found to be stable upon storage for at least 2 weeks at room temperature and ambient air.

Handling and Disposal of Hazardous Chemicals

The procedures in this article are intended for use only by persons with prior training in experimental organic chemistry.   All hazardous materials should be handled using the standard procedures for work with chemicals described in references such as "Prudent Practices in the Laboratory" (The National Academies Press, Washington, D.C., 2011 www.nap.edu).   All chemical waste should be disposed of in accordance with local regulations.   For general guidelines for the management of chemical waste, see Chapter 8 of Prudent Practices.

These procedures must be conducted at one's own risk.   *Organic Syntheses, Inc.*, its Editors, and its Board of Directors do not warrant or guarantee the safety of individuals using these procedures and hereby disclaim any liability for any injuries or damages claimed to have resulted from or related in any way to the procedures herein.

3. Discussion

Oximes are useful building blocks or key intermediates for a variety of organic transformations, including the Beckman rearrangement (yielding amides), the Neber reaction (yielding α-amino carbonyl moieties), and in diverse amination strategies with applications ranging from amination of organometallic reagents (electrophilic amination) to heterocyclic synthesis (via electrophilic, nucleophilic, radical-based or amino-Heck amination manifolds).[2] O-Protected oximes are also good electrophiles providing access to amine derivatives upon reaction with various nucleophiles, and stereoselective variants have been investigated.[3] As such, oximes provide access to various nitrogen-containing functional groups that are prevalent in medicinal chemistry and in natural products.

A variety of approaches are available to access oximes efficiently. The simple condensation of carbonyl compounds and NH_2OH (typically formed from its salt $NH_2OH•HCl$ and a mild base such as NaOAc) is often simple, efficient and practical.[4] However, several routes have been developed to complement this method, which are especially useful when the desired carbonyl compound is not readily available or when alternative starting materials are commercially or more readily available. Such routes include (1) the direct oxidation of primary amines to oximes;[5] (2) the reduction of nitroalkanes or nitroalkenes;[6] (3) a 2-step Mitsunobu approach on primary and secondary alcohols;[7] and (4) the nitrosation reactivity of alkenes.[8]

Recently, numerous important developments have extended the synthetic reach of alkene and alkyne hydroamination to a variety of nitrogen-containing molecules.[9] In particular, metal-catalyzed alkyne hydroamination is emerging as a direct approach to a variety of imine derivatives, through intra- and intermolecular reactions.[10] As part of efforts to enable oxime formation using alkyne hydroamination, and to develop the use of hydroxylamines and hydrazines in metal-free Cope-type hydroaminations,[11] we previously reported that oximes are formed directly upon heating alkynes and NH_2OH (eq 1).[12,13]

$$R^1\!\!=\!\!\!=\!\!\!=\!\!-R^2 \xrightarrow[\textit{i-PrOH, 90-140 °C}]{\text{aq. } NH_2OH} \underset{\substack{H \quad H \\ \textit{29 examples} \\ \textit{up to 99\% yield}}}{R^1\overset{\text{NOH}}{\underset{}{\|}}R^2} \quad (1)$$

After several rounds of reaction optimization, we developed several conditions that allow both terminal and internal alkynes to afford the Markovnikov hydroamination products effectively. The original conditions (Table 1, *conditions A*) involve conventional heating done in sealed tubes with 1,4-dioxane used as the solvent, and an excess of aqueous hydroxylamine (2.5 equiv). Optimized and preferred reaction conditions (*conditions B*) feature the use of a microwave reactor, which allows slightly increased yields and significantly shorter reaction times (<10 h), and use alcoholic solvents.

The present procedure describes one representative example for the synthesis of oximes from alkynes and NH_2OH, and produces *E*-1-(1-hydroxycyclohexyl)ethanone oxime (>5 g) in greater than 83 % yield. The procedure uses a simple reflux in 2-methyl-1-propanol (*conditions C*) and is complementary to our previously published conditions. Further modifications can be necessary in some cases when product degradation occurs under standard reflux conditions (notably when using the propargylamine-based substrates). These last reaction conditions (*conditions D*) are milder, involve conventional reflux for a longer reaction time at a lower temperature, and feature the use of a reduced excess of aqueous hydroxylamine (1.5 equiv rather than 2.5 equiv, as reaction optimization revealed that the reaction is higher yielding under those conditions).[14]

As shown in Table 1, the reaction is compatible with a variety of alkynes (aromatic, aliphatic, enynes, sterically hindered), standard protecting groups, free hydroxyl groups, basic nitrogen atoms (sp^2 and sp^3), and amides. Internal alkynes can also be reacted, albeit resulting in reduced yields. In most cases, the unreacted starting alkynes can be recovered, and the products can be conveniently isolated by recrystallization or chromatography.

Table 1. Synthesis of Oximes from the Cope-type Hydroamination of Alkynes with NH$_2$OH

$$R \equiv R' \xrightarrow[110 - 140\ ^\circ C]{\text{aq. NH}_2\text{OH}} \underset{\text{major}}{\overset{\text{NOH}}{R \diagdown \diagup R'}} + \underset{\text{minor}}{\overset{\text{NOH}}{R \diagdown \diagup R'}}$$

Entry	Substrate	Conditions[a]	Major Product	Yield (%)[b]
1	R = Ph, R' = H	A		87 (5)
2	R = 4-OMeC$_6$H$_4$, R' = H	A	R$\diagup$(NOH)	83 (3)
3	R = 4-FC$_6$H$_4$, R' = H	A[c]		71 (8)
4	R = 2-MeC$_6$H$_4$, R' = H	A		45 (11)
5	(2-ethynylpyridine)	A[c]		73 (15)
6	(1-ethynylcyclohexene)	B		72 (3)
7	R = n-C$_6$H$_{13}$, R' = H	B	R$\diagup$(NOH)	86 (2)
8	R = c-C$_6$H$_{12}$, R' = H	B	R$\diagup$(NOH)	72
9	R" = H	B		91
10	R" = TBS	B		86
11	R"O–(–)$_3$ R" = Bn	B	R"O–(–)$_3$ (NOH)	98
12	R" = Piv	B		90
13	R" = PMB	B		98
14	R" = THP	B		0[d]
15	R" = H	B		89 (<1)
16	HO R" = Me	B	HO	89 (<1)
17	R"R" R" = Ph	B	R"R" (NOH)	89
18	R" = -(CH$_2$)$_2$-	C		81
19	R" = -(CH$_2$)$_3$-	C		87
20	R'",R" R"', R'" = Me	D	R'"-N(R") (NOH)	71
21	R"'–N R" = Bz, R'" = H	B		58
23	Ph≡Me	B	Ph (NOH) Me	53 (5)
24	n-Pr≡n-Pr	B	n-Pr (NOH) n-Pr	12

[a] Reaction conditions: Alkyne (1 equiv.), aq. NH$_2$OH (2.5 equiv.) A: dioxane (1M), sealed tube (behind a blast shield), 113 °C, 16-18 h; B: i-PrOH, (1M), 140 °C (Biotage Initiator or CEM microwave), 5-10 h; C: (*this work*) i-BuOH, reflux (107 °C), 18 h; D: Alkyne (1 equiv.), aq. NH$_2$OH (1.5 equiv.) n-PrOH, reflux (97 °C), 40 h. [b] Yield of isolated products. Yield of minor regioisomer shown in parentheses. [c] dioxane (2M). [d] Led to isolation of unprotected oxime in 65 % yield.

The reaction is most efficient when protic solvents (such as 1-propanol, 2-propanol or 2-methylpropan-1-ol) are used. The increased reactivity observed under these conditions is consistent with such alcohols both stabilizing the expected enamine N-oxide intermediate (**I**) and facilitating the proton transfer step required to access a neutral N-hydroxyenamine (**III**), that can subsequently tautomerize to the oxime product (**IV**). Computational results support a bimolecular proton transfer pathway in which the alcohol mediates the proton transfer through a five-membered, cyclic transition state (**II**), over an intramolecular proton transfer or stepwise alternatives.

Scheme 1

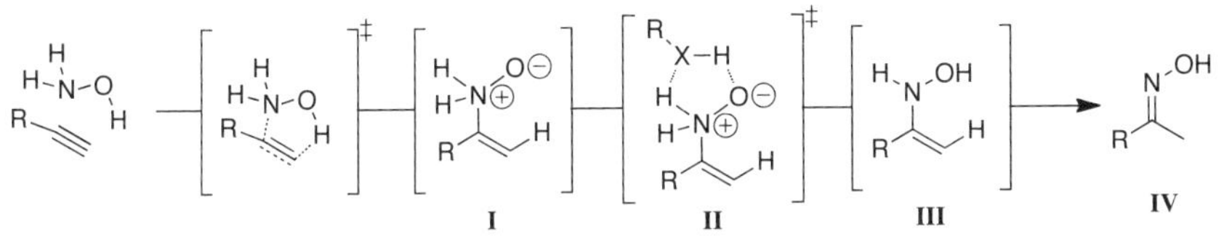

1. Centre for Catalysis Research and Innovation, Department of Chemistry, University of Ottawa, 10 Marie-Curie – D'Iorio Hall, Ottawa, Canada, K1N 6N5. andre.beauchemin@uottawa.ca. The submitters thank the University of Ottawa and NSERC for financial support, FQRNT for a doctoral scholarship (F.L.), and Prof. M. Murugesu (University of Ottawa) and his group for a fruitful collaboration.

2. For selected reviews, see: (a) Yamane, M.; Narasaka, K. In Science of Synthesis; Padwa, A., Ed.; Georg Thieme Verlag: Stuttgart, 2004; Vol. 27, pp. 605-647. (b) Narasaka, K.; Kitamura, M. *Eur. J. Org. Chem.* **2005**, 4505.

3. Moody, C. J. *Chem. Commun.* **2004**, 1341 (and references cited therein).

4. For an example: Bousquet, E.W. *Org. Synth., Coll. Vol. II*, **1943**, 313.

5. Suzuki, K.; Watanabe, T.; Murahashi, S.-I. *Angew. Chem., Int. Ed.* **2008**, *47*, 2079.

6. (a) Mourad, M. S.; Varma, R. S.; Kabalka, G. W. *J. Org. Chem.* **1985**, *50*, 133. (b) Kabalka, G. W.; Laila, G. M. H.; Varma, R. S. *Tetrahedron* **1990**, *46*, 7443. (c) Ghosh, A. K.; Gong, G. *Org. Lett.* **2007**, *9*, 1437.

7. Kitahara, K.; Toma, T.; Shimokawa, J.; Fukuyama, T. *Org. Lett.* **2008**, *10*, 2259.

8. Prateeptongkum, S.; Jovel, I.; Jackstell, R.; Vogl, N.; Weckbecker, C.; Beller, M. *Chem. Commun.* **2009**, 1990.

9. For selected reviews, see: (a) Müller, T. E.; Hultzsch, K. C.; Yus, M.; Foubelo, F.; Tada, M. *Chem. Rev.* **2008**, *108*, 3795 (and reviews cited therein). (b) Aillaud, I.; Collin, J.; Hannedouche, J.; Schulz, E. *Dalton Trans.* **2007**, 5105. (c) Hultzsch, K. C. *Adv. Synth. Catal.* **2005**, *347*, 367. (d) Nobis, M.; Drießen-Hölscher, B. *Angew. Chem., Int. Ed.* **2001**, *40*, 3983. (e) Müller, T. E.; Beller, M. *Chem. Rev.* **1998**, *98*, 675.

10. (a) Severin, R.; Doye, S. *Chem. Soc. Rev.* **2007**, *32*, 1407. (b) Alonso, F.; Beletskaya, I. P.; Yus, M. *Chem. Rev.* **2004**, *104*, 3079. (c) Pohlki, F.; Doye, S. *Chem. Soc. Rev.* **2003**, *32*, 104.

11. Such reactions are also referred to as reverse Cope eliminations in the literature. For a review, see: Cooper, N. J.; Knight, D. W. *Tetrahedron* **2004**, *60*, 243.

12. (a) Beauchemin, A. M.; Moran, J.; Lebrun, M.-E.; Séguin, C.; Dimitrijevic, E.; Zhang, L.; Gorelsky, S. I. *Angew. Chem. Int., Ed.* **2008**, *47*, 1410. (b) Moran, J.; Gorelsky, S. I.; Dimitrijevic, E.; Lebrun, M.-E.; Bédard, A.-C.; Séguin, C.; Beauchemin, A. M. *J. Am. Chem. Soc.* **2008**, *130*, 17893. For related work, see: (c) Bourgeois, J.; Dion, I.; Cebrowski, P. H.; Loiseau, F.; Bédard, A.-C.; Beauchemin, A. M. *J. Am. Chem. Soc.* **2009**, *131*, 874. (d) Moran, J.; Pfeiffer, J. Y.; Gorelsky, S. I.; Beauchemin, A. M. *Org. Lett.* **2009**, *11*, 1895. (e) MacDonald, M. J.; Schipper, D. J.; Ng, P. J.; Moran, J.; Beauchemin, A. M. *J. Am. Chem. Soc.* **2011**, *133*, 20100.

13. For related reactivity of hydrazines and hydrazides, see: (a) Roveda, J.-G.; Clavette, C.; Hunt, A. D.; Gorelsky, S. I.; Whipp, C. J.; Beauchemin, A. M. *J. Am. Chem. Soc.* **2009**, *131*, 8740. (b) Cebrowski, P. H.; Roveda, J.-G.; Moran, J.; Gorelsky, S. I.; Beauchemin, A. M. *Chem. Commun.* **2008**, 492. (c) Loiseau, F.; Clavette, C.; Raymond, M.; Roveda, J.-G.; Burrell, A.; Beauchemin, A. M. *Chem. Commun.* **2011**, *47*, 562.

14. Conditions primarily used for the reaction of *N,N*-dimethyl-propargylamine. The resulting oxime was used as a ligand in the synthesis of novel Ni(II) clusters: Brunet, G.; Habib, F.; Cook, C.; Pathmalingam, T.; Loiseau, F.; Korobkov, I.; Burchell, T. J.; Beauchemin, A. M.; Murugesu, M. *Chem. Commun.* **2012**, *48*, 1287.

94

Appendix
Chemical Abstracts Nomenclature; (Registry Number)

1-Ethynylcyclohexanol; (78-27-3)
Hydroxylamine; (7803-49-8)
E-1-(1-Hydroxycyclohexyl)ethanone oxime; (62114-93-6)

André M. Beauchemin grew up in Québec city and completed his B.Sc. in 1996 from Université Laval. He pursued his Ph.D. under the guidance of André B. Charette at the Université de Montréal and worked as a NSERC postdoctoral fellow with Prof. David A. Evans at Harvard University. He began his independent career at the University of Ottawa in August 2004. His research group is interested in novel methods and concepts for accessing nitrogen-containing molecules, and has developed a metal-free amination portfolio for the synthesis of hydroxylamines, hydrazines, oximes, saturated nitrogen heterocycles, pyridines, pyrazines, β-aminocarbonyl motifs and enantioenriched diamines.

Francis Loiseau was born and raised in Montreal, Canada, and received his B.Sc. degree in Chemistry from McGill University in 2005, where he performed undergraduate research in the laboratories of James L. Gleason and Mark P. Andrews. He went on to work for two years at Handy Chemicals in Candiac, Canada, where he worked at the synthesis of concrete and paint additives. He is currently a FQRNT Fellow at the University of Ottawa pursuing a Ph. D. degree under the direction of Professor André Beauchemin. His research focuses on the development of metal-free Cope-type hydroamination.

Nikolas Huwyler was born and raised in Zurich, Switzerland, and obtained a M.Sc. degree in Chemistry from ETH Zurich in 2009. During his undergraduate education he performed research in the laboratories of Prof. Erick M. Carreira and Prof. Antonio Togni at the ETH Zurich and completed his studies with a Master's thesis in the laboratories of Prof. Timothy F. Jamison at Massachusetts Institute of Technology, Cambridge MA, USA, where he had a 6 month stay as a visiting scientist. He is currently pursuing doctoral studies in synthetic organic chemistry with Prof. Carreira focusing on the total synthesis of natural products.

Preparation of 1,5-Disubstituted 1,2,3-Triazoles via Ruthenium-catalyzed Azide Alkyne Cycloaddition

Submitted by James S. Oakdale and Valery V. Fokin.[1]
Checked by Satoshi Umezaki and Tohru Fukuyama.

1. Procedure

1-Benzyl-5-phenyl-1H-1,2,3-triazole. Benzyl azide (10.0 g, 75 mmol, 1.0 equiv) (Note 1) is placed in a 500-mL, three-necked round-bottomed flask and equipped with an 8 × 30 mm, octagon-shaped Teflon coated-magnetic stirring bar. The center neck is equipped with a three-way stopcock connected to argon line and bubbler. The other two necks are equipped with a rubber septum. The reaction vessel is purged with argon supplied by an argon filled balloon, after which the three-way stopcock is reconnected. 1,2-Dichlorethane (150 mL) (Note 2) and phenylacetylene (Note 3) (8.06 g, 8.66 mL, 79 mmol, 1.05 equiv) are added sequentially to the flask and the stirred solution is placed in a 45 °C oil bath. After five min, a solution of chloro(1,5-cyclooctadiene)(pentamethylcyclopentadiene)-ruthenium (285 mg, 0.75 mmol, 1 mol %) (Note 4) in DCE (3 mL) is added to the reaction vessel via syringe. The orange solution becomes dark brown (Notes 5 and 6). After 30 min the solution is cooled to room temperature and silica gel (35 g) (Note 7) is added. The solvent is removed by rotary evaporation (35 °C, 40 mmHg). The resulting light brown powder is placed in a 5 cm diameter column and flushed with ethyl acetate (2 × 200 mL). The dark brown solution is concentrated again by rotary evaporation (35 °C, 40 mmHg) to give a dark brown solid. The solid material is transferred to a 250-mL one-necked round-bottomed flask, hexanes (200 mL) are added and the heterogeneous solution is triturated (Note 8) for 15 h. The mixture is filtered using a Büchner funnel, rinsed with hexanes (4 x 25 mL) and dried *in vacuo* (rt, 18 mmHg) to afford 15.9–16.2 g (90–92%) of the titled compound as a beige powder (Note 9).

Org Synth. **2013**, *90*, 96-104
Published on the Web 9/21/2012
© 2013 Organic Syntheses, Inc.

2. Notes

1. The submitters purchased benzyl azide, pract. (94%) from Frinton Laboratories, Inc., and used after flushing the compound through a short column of silica gel with diethyl ether. The submitters noted that three times as much catalyst was needed to complete the reaction without purification. The checkers purchased benzyl azide (97%) from Wako Pure Chemical Industries, Ltd. and used as received.

2. 1,2-Dichloroethane was purchased from Fisher Scientific (certified ACS, submitters) or Kanto Chemical Co., Inc. (>99.5%, conforms to ACS, checkers), and degassed prior to use by bubbling a stream of nitrogen (submitters) or argon (checkers) through the solvent for 1 hour. No precautions were taken to dry the solvent.

3. Phenylacetylene (98%) was purchased from Acros Organics and used as received.

4. Chloro(1,5-cyclooctadiene)(pentamethylcyclopentadienyl)ruthenium(II) (98%) was purchased from Stream Chemicals Inc. (submitters) or Aldrich Chemical Company, Inc. (checkers) and used as received.

5. The submitters monitored the reaction by GC-MS. Gas chromatography was performed using a Hewlett Packard 5890A Gas Chromatograph. GC-MS data was recorded on an Agilent 7890A GC system with an Agilent 5975C Inert MSD system operating in the electron impact (EI+) mode. Column; Agilent 19091S-433: 325 °C: 30 m × 320 μm × 0.25 μm, H-5MS 5% phenyl methyl silox. Split Ratio 20:1. Front Inlet; 200 °C. Front Detector FID; 300 °C. Oven program; 50 °C for 2.25 min then 60 °C/min to 300 °C for 4 min. Retention times: phenylacetylene (2.9 min), benzyl azide (4.03 min) and 1-benzyl-5-phenyl-1H-1,2,3-triazole (6.88 min).

6. The checkers monitored the reaction by ^{1}H NMR. Chemical shifts of the benzyl proton were used for monitoring: benzyl azide (4.35 ppm) and 1-benzyl-5-phenyl-1H-1,2,3-triazole (5.55 ppm).

7. The submitters purchased silica gel 60 from EMD Chemicals Inc.; particle size 40-63 μm. The checkers purchased silica gel (acidic) from Kanto Chemical Co., Inc., particle size 40-100 μm.

8. Three stir bars (8 × 30 mm, octagon-shaped) were added to the hexane suspension and the mixture was vigorously stirred in order to grind the solid into a fine powder. The submitters note that it is critical that the material be completely solid prior to the addition of hexanes.

9. 1-Benzyl-5-phenyl-1*H*-1,2,3-triazole has the following physical and spectroscopic data: mp 76–78 °C; ^{1}H NMR (CDCl$_3$, 400 MHz) δ: 5.55 (s, 2 H), 7.06–7.10 (m, 2 H), 7.24–7.32 (m, 5 H), 7.38–7.47 (m, 3 H), 7.75 (s, 1 H); ^{13}C NMR (CDCl$_3$, 100 MHz) δ: 51.7, 126.9, 127.1, 128.1, 128.8, 128.8, 128.9, 129.4, 133.2, 135.6, 138.1; IR (thin film) cm^{-1}: 1483, 1455, 1435, 1242, 1210, 1126; HRMS (ESI) *m/z* calcd. for C$_{15}$H$_{13}$N$_3$Na ([M+Na]$^+$) 258.1007; found 258.1006; Anal. calcd. for C$_{15}$H$_{13}$N$_3$: C, 76.57; H, 5.57; N, 17.86. found: C, 76.57; H, 5.72; N, 17.65.

Handling and Disposal of Hazardous Chemicals

The procedure in this article is intended for use only by persons with prior training in experimental organic chemistry. All hazardous materials should be handled using the standard procedures for work with chemicals described in references such as "Prudent Practices in the Laboratory" (The National Academies Press, Washington, D.C., 2011 www.nap.edu). All chemical waste should be disposed of in accordance with local regulations. For general guidelines for the management of chemical waste, see Chapter 8 of Prudent Practices.

The procedure must be conducted at one's own risk. *Organic Syntheses, Inc.*, its Editors, and its Board of Directors do not warrant or guarantee the safety of individuals using this procedure and hereby disclaim any liability for any injuries or damages claimed to have resulted from or related in any way to the procedures herein.

3. Discussion

Ruthenium-catalyzed azide-alkyne cycloaddition (RuAAC)[2] is a sister process of a widely utilized CuAAC reaction.[3] In contrast to the copper catalysis, which requires the formation of σ-copper(I) acetylides (when terminal alkynes are used), ruthenium catalysts activate alkynes via π-interactions, increasing their nucleophilicity and thereby promoting the addition of the alkyne's most nucleophilic carbon to the electrophilic N3 terminus of the azide. The currently accepted mechanism of the reaction is shown in **Scheme 1**.[4] In reactions with terminal alkynes, RuAAC produces 1,5-disubstituted 1,2,3-triazoles (instead of the 1,4-regiosiomers that are the exclusive products of CuAAC). Ruthenium catalysts are also active in the cycloaddition of internal alkynes with organic azides to provide fully substituted 1,2,3-triazoles.[2,4,6] Regioselectivity is often high for unsymmetrically substituted internal alkynes and is generally influenced by the electronic properties of the substituents, their steric demands, and the ability to engage in hydrogen bonding.[2,4-7] Thus, hydrogen bond donors (alcohols and amines) in the propargylic position of the alkyne invariably end up at the C-5 position of the product triazole.[4] This directing effect can be explained by the formation of a strong H-bond between the chloride ligand on the ruthenium and the H-bond donor group. The effect of the electronic properties of the alkyne on regioselectivity is exemplified by reactions of ynones and propiolic esters, which usually result in regioselective formation of substituted triazoles with those groups at the C-4 position of the heterocycle.

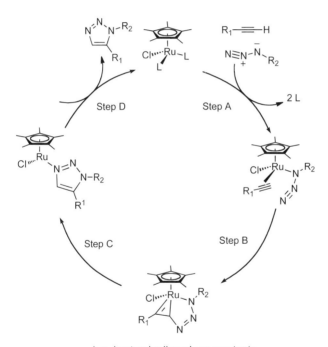

L = bystander ligands or reactants

Scheme 1. Key intermediates in the RuAAC catalytic cycle.

Among the many ruthenium catalysts that have been examined, only four complexes containing the [Cp*RuCl] fragment proved viable for this reaction: Cp*RuCl(PPh₃)₂, [Cp*RuCl]₄,[5] Cp*RuCl(COD) and Cp*RuCl(NBD). For a mechanistic study as well as reaction scope and limitations, see references 4,6.

The RuAAC reaction is compatible with a range of aprotic solvents including 1,2-dichloroethane, acetone, toluene, dioxane, dichloromethane, dimethylformamide, and chloroform, and it is important that the solution be completely homogeneous. Water, ethyl acetate, methanol, isopropyl alcohol, hexanes, and diethyl ether were detrimental to the catalysis. The procedures provided in this work call for moderate temperatures of 50 °C; however, for some substrates room temperature is sufficient to achieve high conversions with low catalyst loading. In general, ruthenium complexes containing labile ligands, such as cyclooctadiane, or the "ligand-free" tetramer [Cp*RuCl]₄ require lower temperature to achieve full conversion, as opposed to the bis(triphenylphosphine) complexes where it is advisable to perform the reactions at temperatures exceeding 60 °C.

At this stage of development, RuAAC is not as robust as the CuAAC with respect to functional group tolerance and reaction conditions compatibility, primarily due to potential catalyst deactivation, as shown in **Scheme 2**. For example, organic azides react with **1** to generate ruthenium tetraazadiene complex **2**. These complexes are exceedingly stable, can be isolated by column chromatography and are often easily identified by the presence of a green spot on TLC (or a green band on a silica gel column). Thus, the catalyst should not be mixed with the azide in the absence of alkyne.

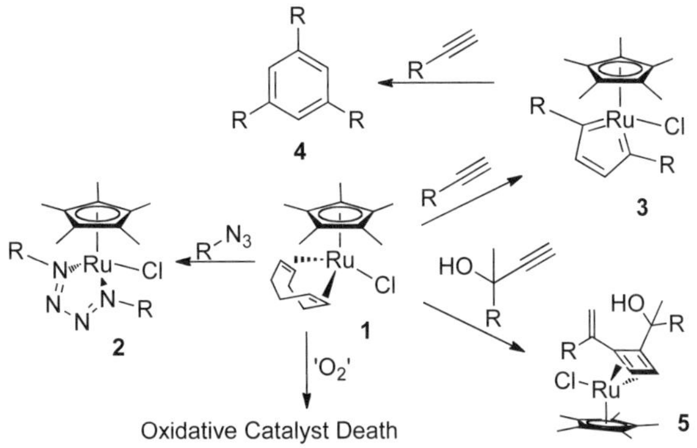

Scheme 2. Catalyst deactivation pathways in CuAAC.

Ruthenium-catalyzed reactions of alkynes are numerous and have evolved into an extraordinary number of synthetically useful transformations during the last decades. Among them is the cyclotrimerization of alkynes to generate benzene rings (**4**),[8] which was shown to proceed through a ruthenacyclopentatriene **3** intermediate.[8c] There is no direct evidence that ruthenacycle **3** results in deactivation of the catalyst or is even formed during standard RuAAC reactions. However, an isolated ruthenacycle similar to **3** was catalytically incompetent and also did not react with azides, even at the elevated temperature.[4]

Cyclobutadiene ruthenium complex **5** (characterized by X-ray crystallography) was recently isolated from a dimerization reaction of propargylic alcohols.[9] It is a very non-polar compound, stable to column chromatography, and reminiscent of ruthenium species prepared by Kirchner et. al.[10] As with **2** and **3**, complex **5** does not react with azides.

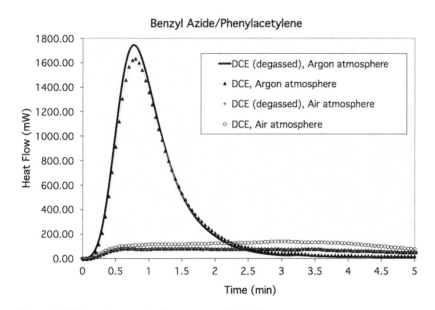

Figure 1. Effects of atmospheric oxygen on RuAAC.

Calorimetry trace of the reaction between benzyl azide and phenyl acetylene (0.1M in DCE). Several conditions were examined using 4 mol % catalyst loading (0.004M in DCE) including: degassed DCE under argon, degassed DCE under atmospheric oxygen, untreated DCE under atmospheric oxygen, and untreated DCE under argon. These results illustrate the necessity of placing the RuAAC reaction under an inert atmosphere.

RuAAC reaction is quite sensitive to atmospheric oxygen and must be run under an inert atmosphere. It is known that Cp*RuCl(PPh$_3$)$_2$ reacts with dioxygen to give Cp*RuCl(O$_2$)(PPh$_3$), where the dioxygen ligand is reportedly not tightly bound and can be replaced by a phosphite ligand, Cp*RuCl(L)(PPh$_3$).[11] However, when employing the Cp*RuCl(COD) catalyst the effects of molecular oxygen can be drastic in the RuAAC reaction. **Figure 1** shows the rate profile of the reaction of benzyl azide with phenylacetylene (0.1 M in DCE with 0.004 M Cp*RuCl(COD), at 45 °C) using reaction heat flow calorimetry. The area under the curve is directly correlated with the overall conversion. As seen in this example, the maximum rate and overall yield is significantly higher when the reaction is performed under argon. The conversion was quantitative under argon and only 20% under air.

The RuAAC process significantly expands the reach of catalytic dipolar cycloaddition reactions. Its applications are beginning to appear[12] and should continue to grow as the reaction matures and new and improved catalysts are developed.

References

1. The Scripps Research Institute, 10550 North Torrey Pines Road, La Jolla, California 92037. E-mail: fokin@scripps.edu. This work was supported by the National Science Foundation (CHE-0848982).
2. Zhang, L.; Chen, X.; Xue, P.; Sun, H. H. U.; Williams, I. D.; Sharpless, K. B.; Fokin, V. V.; Jia, G. *J. Am. Chem. Soc.* **2005**, *127*, 15998–15999.
3. (a) Rostovtsev, V. V.; Green, L. G.; Fokin, V. V.; Sharpless, K. B. *Angew. Chem., Int. Ed.* **2002**, *41*, 2596–2599. (b) Tornøe, C. W.; Christensen, C.; Meldal, M. *J. Org. Chem.* **2002**, *67*, 3057–3064.
4. Boren, B. C.; Narayan, S.; Rasmussen, L. K.; Zhang, L.; Zhao, H.; Liu, Z.; Jia, G.; Fokin, V. V. *J. Am. Chem. Soc.* **2008**, *130*, 8923–8930.
5. Rasmussen, L. K.; Boren, B. C.; Fokin, V. V. *Org. Lett.* **2007**, *9*, 5337–5339.
6. Majireck, M. M.; Weinreb, S. M. *J. Org. Chem.* **2006**, *71*, 8680–8683.
7. Oppilliart, S.; Mousseau, G.; Zhang, L.; Jia, G.; Thuery, P.; Rousseau, B.; Cintrat, J.-C. *Tetrahedron* **2007**, *63*, 8094.
8. For examples see (a) Yamamoto, Y.; Kinpara, K.; Saigoku, T.; Takagishi, H.; Okuda, S.; Nishiyama, H.; Itoh, K. *J. Am. Chem. Soc.*

2005, *127*, 605–613. (b) Yamamoto, Y.; Arakawa, T.; Ogawa, R. J.; Itoh, K. *J. Am. Chem. Soc.* **2003**, *125*, 12143–12160. (c) Kirchner, K.; Calhorda, M. J.; Schmid, R.; Veiros, L. F. *J. Am. Chem. Soc.* **2003**, *125*, 11721–11729.

9. Oakdale, J. S.; Fokin, V. V. Unpublished Results.
10. Paih, J. P.; Dérien, S.; Demerseman, B.; Bruneau, C.; Dixneuf, P. H.; Toupet, L.; Dazinger, G.; Kirchner, K. *Chem. Eur. J.* **2005**, *11*, 1312–1324.
11. Trost, B. M. *Acc. Chem. Res.* **2002**, *35*, 695–705.
12. (a) Poecke, S. V.; Negri, A. Gago, F.; Van Daele, I.; Solaroli, N.; Karlsson, A.; Balzarini, J.; Van Calenbergh, S. *J. Med. Chem.* **2010**, *53*, 2902–2912. (b) Horne, W. S.; Olsen, C. A.; Beierle, J. M.; Montero, A.; Ghadiri, M. R. Angew. *Chem. Int. Ed.* **2009**, *48*, 4718–4724. (c) Chemama, M.; Fonvielle, M.; Arthur, M.; Valéry, J.-M.; Etheve-Quelquejeu, M. *Chem. Eur. J.* **2009**, *15*, 1929–1938. (d) Hirose, T.; Sunazuka, T.; Sugawara, A.; Endo, A.; Iguchi, K.; Yamamoto, T.; Ui, H.; Shiomi, K.; Watanabe, T.; Sharpless, K. B.; Ōmura, S. *J. Antibiot.* **2009**, *62*, 277–282. (e) Nulwala, H.; Takizawa, K.; Odukale, A.; Khan, A.; Thibault, R. J.; Taft, B. R.; Lipshutz, B. H.; Hawker, C. J. *Macromolecules* **2009**, *42*, 6068–6074. (f) Kelly, A. R.; Wei, J.; Kesavan, S.; Marié, J.-C.; Windmon, N.; Young, D. W.; Marcaurelle, L. A. *Org. Lett.* **2009**, *11*, 2257–2260. (g) Tam, A.; Arnold, U.; Soellner, M. B.; Raines, R. T. *J. Am. Chem. Soc.* **2007**, *129*, 12670–12671.

Appendix
Chemical Abstracts Nomenclature; (Registry Number)

Benzyl Azide: Azidomethyl benzene; Triazotoluene; (622-79-7)

Chloro(1,5-cyclooctadiene)(pentamethylcyclopentadienyl)ruthenium(II): (92390-26-6)

Phenylacetylene: Ethynyl benzene; Acetylene benzene; (536-74-3)

Valery V. Fokin received his undergraduate education at the University of Nizhny Novgorod, Russia, and his Ph.D. degree at the University of Southern California under the tutelage of Professor Nicos A. Petasis. After a postdoctoral stint with Professor K. Barry Sharpless at The Scripps Research Institute in La Jolla, California, he joined the Scripps faculty, where he is currently Associate Professor in the Department of Chemistry. His research is focused on new reactivity of organometallic compounds and catalysis, and on applying them to the studies of macromolecular and biological phenomena. His research group is working on the development of new catalytic reactions, asymmetric synthesis, studies of organic and organometallic mechanisms, medicinal chemistry, and smart polymeric materials.

James Oakdale received a B.Sc. in chemistry in 2009 from the University of California, Davis, where he pursued undergraduate research in the laboratory of Prof. Mark Kurth. He is currently pursuing a Ph.D. as an NSF graduate research fellow at The Scripps Research Institute under the advisement of Professor Valery V. Fokin. His research efforts include the development and mechanistic investigation of ruthenium catalyzed reactions and the design and manipulation of functional materials.

Satoshi Umezaki was born in Kumamoto, Japan in 1984. He received his B.S. in 2007 from the University of Tokyo under the direction of Professor Koichi Narasaka. He then moved to the laboratories of Professor Tohru Fukuyama, the University of Tokyo. He received his M.S. in 2009 and he is pursuing a Ph.D. degree. His current research interest is total synthesis of natural products.

104

Discussion Addendum for:
Nickel-catalyzed Homoallylation of Aldehydes with 1,3-Dienes

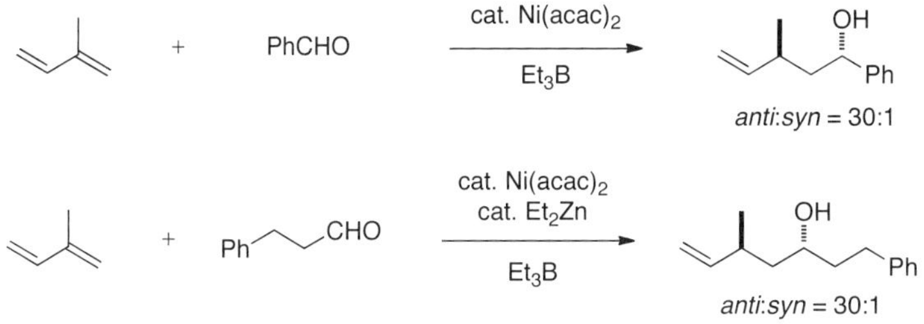

Prepared by Masanari Kimura.*[1]
Original article: Tamaru, Y.; Kimura, M. *Org. Synth.* **2006**, *83*, 88.

Reductive coupling reactions are among the most useful strategies for the efficient synthesis of complicated molecules and physiologically active molecules.[2] We reported for the first time that Ni-catalysts are able to promote the homoallylation of aldehydes with a wide variety of 1,3-dienes in the presence of triethylborane or diethylzinc to yield bishomoallyl alcohols with excellent regio- and stereoselectivities.[3] Based on these procedures, the scope of reductive coupling reactions have been significantly expanded in recent years. Herein, the scope of the homoallylation of aldehydes and ketones with conjugated dienes, asymmetric homoallylations of carbonyls, synthesis of natural products involving Ni-catalyzed reductive coupling, and bis-metallic homoallylations are summarized.

Scope of Homoallylation of Carbonyls with Various Conjugated Dienes

The results of the Ni-catalyzed homoallylation of benzaldehyde with conjugated dienes bearing electron-donating substituents, such as siloxy and alkoxy groups, were described (Table 1).[4] C1- and C2-siloxy-1,3-butadienes reacted with benzaldehyde highly regio- and stereoselectively and furnished homoallyation products as a single isomer in good to excellent yields. A siloxy group surpasses a methyl group in regiocontrolling effect; a substituted 1,3-butadiene possessing a siloxy group at C2 and a methyl group at C3 reacted with benzaldehyde via the C1 carbon atom to afford 4-methyl-3-siloxy-4-penten-1-ol exclusively (entry 2, Table 1). While a C2-siloxy group enhanced the reactivity of butadiene at the C1 position, a C1-siloxy group induced reactivity at the C4 position exclusively (entries 3-5, Table 1). The present homoallylation with siloxy and methoxy-substituted butadienes is of great synthetic value as synthetic equivalents of functionalized butenyl carbanion and not as silyl enol ethers.

Table 1. Ni-Catalyzed Homoallylation of Aldehydes with Conjugated Diene

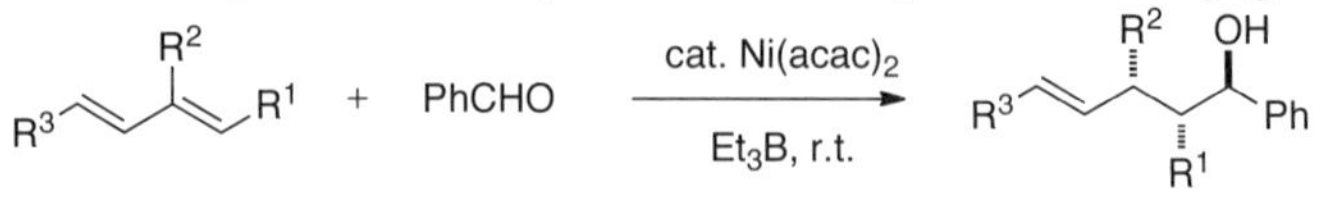

entry	diene	product (% yield) [1,3-*anti* :1,3-*syn*]
1	OTIPS	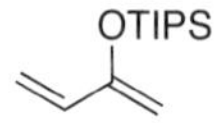 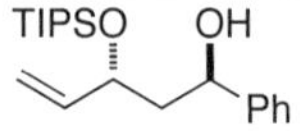 TIPSO OH (65%) [>30:1]
2	OTIPS	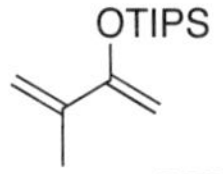 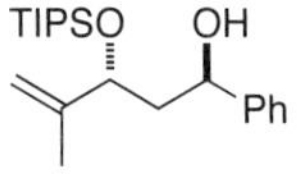 TIPSO OH (84%) [>30:1]
3	OTIPS, Ph	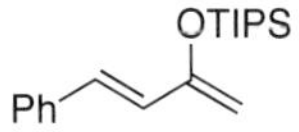 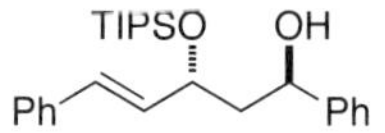 TIPSO OH (81%) [>30:1]
4	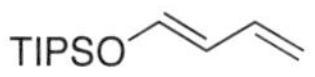 TIPSO	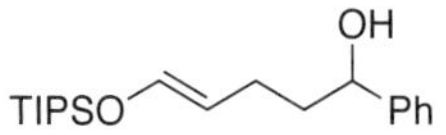 TIPSO OH (92%)
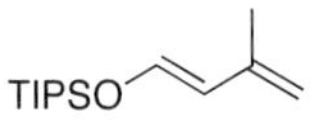 5	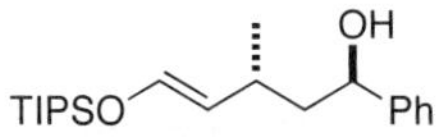 TIPSO	TIPSO OH (92%) [>30:1]
6	OSiMe₃, MeO	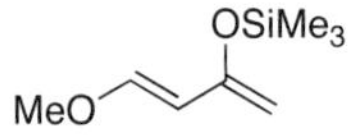 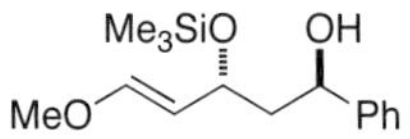 Me₃SiO OH, MeO (69%) [>30:1]

Cyclohexadiene failed to react with benzaldehyde using the Ni/Et₃B system, whereas the Ni/Et₂Zn system accelerated the reaction (Figure 1). However, the major product was *syn*-1-(2-cyclohexenyl)-1-phenylmethanol, which was obtained by nucleophilic allylation of benzaldehyde with 2-cyclohexenyl benzoate through umpolung of π-allylpalladium.[5]

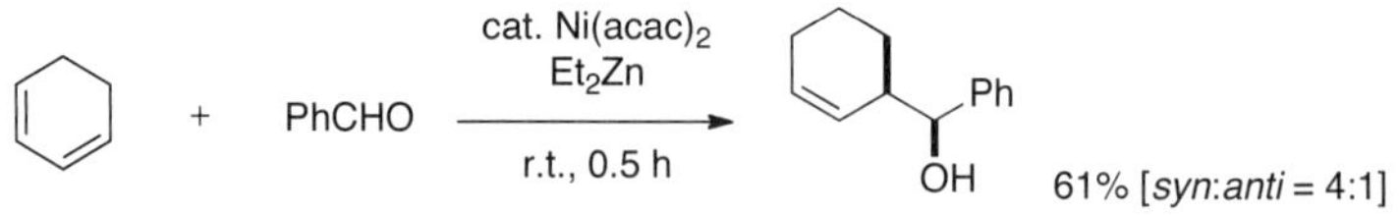

Figure 1. Ni-Catalyzed Allylation of PhCHO with 1,3-Cyclohexadiene

Org. Synth. **2013**, *90*, 105-111

Asymmetric reductive coupling of dienes and aldehydes was also reported. When Ni-catalyzed homoallylation of aldehydes with 1,3-dienes was performed with bulky spirobiindane phosphoramidite ligands and diethylzinc as a reducing agent, chiral bishomoallyl alcohols were produced in high yields with excellent diastereoselectivities and enantioselectivities (Figure 2).[6]

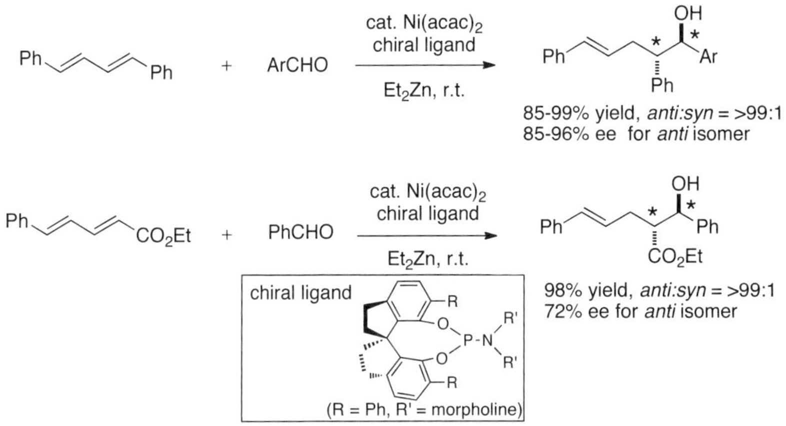

Figure 2. Asymmetric Ni-Catalyzed Homoallylation of Aldehydes

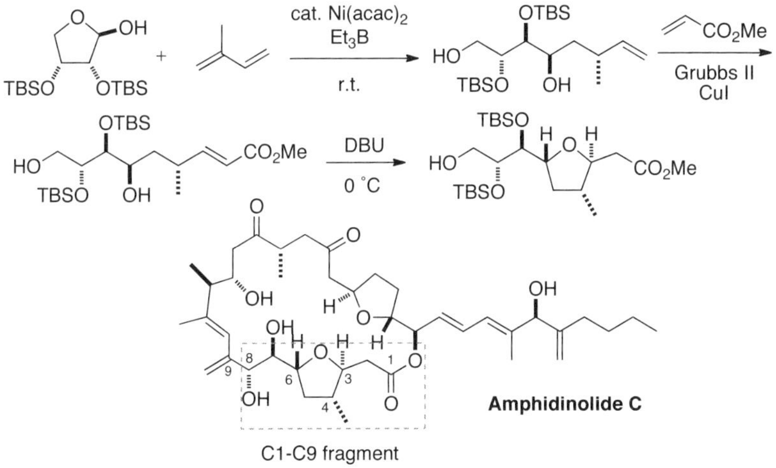

Figure 3. Formal Synthesis of C1-C9 Fragment of Amphidinolide C

Ni-catalyzed homoallylation of aldehydes with isoprene served as an important synthetic strategy for the construction of the tetrahydrofuran fragment from amphidinolide C (Figure 3).[7] Homoallylation of erythrolactol followed by cross metathesis of the hydroxyl alkene with methyl acrylate afforded ω-hydroxy α,β-unsaturated esters, which underwent cyclization upon treatment with DBU to produce tetrahydrofurans with the correct relative configuration for the C1-C9 fragment of amphidinolide C.

Bismetallative Homoallylation of Aldehydes with Dienes

Nickel-catalyzed borylative coupling reactions of diene and aldehyde were studied. In the presence of Ni catalyst, phosphine ligand, and bis(pinacolato)diboron, benzaldehyde reacted with 1,3-dienes, followed by oxidative hydrolysis with hydrogen peroxide, to provide 4-pentene-1,3-diols and 2-pentene-1,5-diols (Figure 4).[8] In this case, bis(pinacolato)diboron participated in the σ-bond metathesis providing access to allyl boronic esters. Their selectivity depends on the kind of phosphine ligands. Homoallylation products were observed exclusively using the $P(SiMe_3)_3$ ligand, whereas $P(t\text{-Bu})_3$ provided allylation products. This borylative carbonyl-diene coupling reaction could be extended to the reaction of ketones, which afforded bis-homoallyl alcohols using $Ni(cod)_2/P(t\text{-Bu})_3$ as catalyst, in contrast to the allylation product obtained with aldehydes.

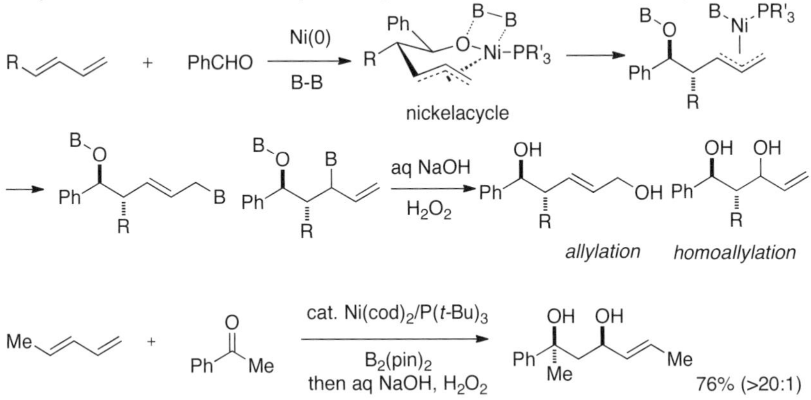

Figure 4. Borylative Coupling Reaction of Diene and Carbonyls

Ni-catalyzed silylborative coupling reactions with aldehydes and 1,3-dienes gave the corresponding bishomoallyl alcohols having an allylsilane unit in the side chain (Figure 5).[9] This procedure was applied to the synthetic approach to optically active α-chiral allylsilane derivatives.

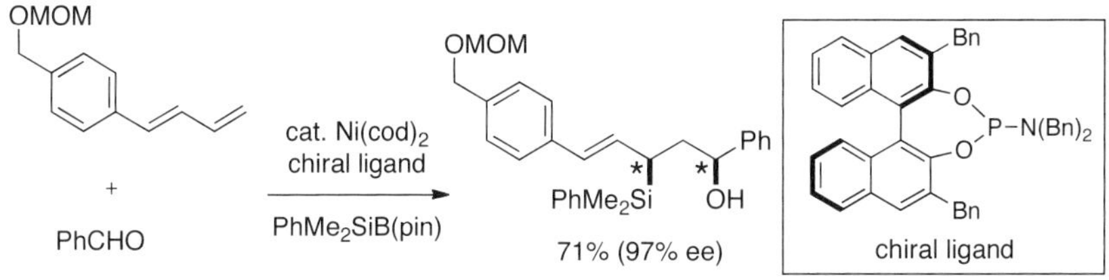

Figure 5. Borylative Coupling Reaction of Diene and Carbonyls

Transition Metal Catalyzed Coupling Reaction of Aldehydes with Dienes Promoted by Triethylborane

Transition metal catalyzed C-C bond formations involving conjugated dienes and carbonyls have been developed. When a mixture of 1,3-diene and aldehyde is used in the presence of triethylborane, the reaction changes dramatically depending on the kind of transition metal catalysts. Rh(I) catalysts promote the reductive coupling reaction of a wide variety of aldehydes with conjugated dienes in the presence of a stoichiometric amount of triethylborane to provide homoallyl alcohols in good yields (Figure 6).[10] Isoprene reacts with benzaldehyde at the C3- and C2-positions to give a regioisomeric mixture of 3-butenyl alcohols in a 6:1 ratio. Methyl sorbate reacts with aldehydes at the α-carbon position with excellent stereoselectivity producing the (Z)-3-hexen-1-ol framework as a single isomer. The similar coupling reactions of 1,3-dienes and aldehydes were observed by employment of Ru and Ir catalysts using isopropyl alcohol and formic acid as reducing agents (Figure 7).[11]

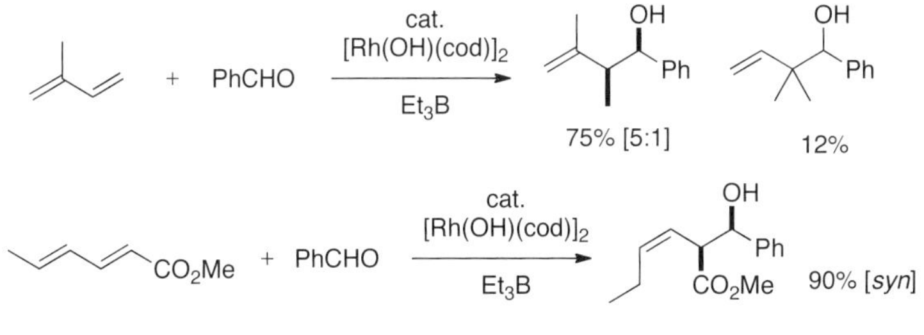

Figure 6. Rh(I) Catalyzed Reductive Coupling of Aldehydes and Dienes

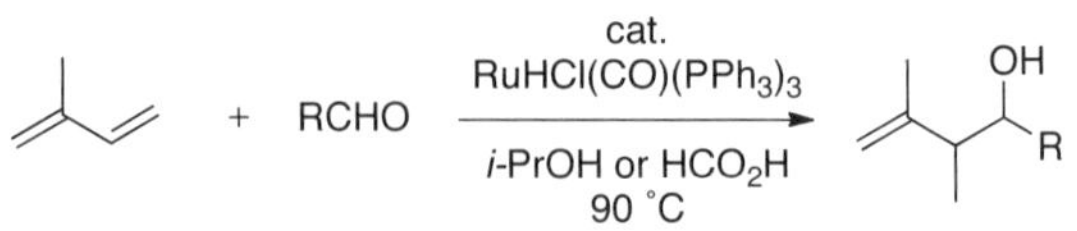

Figure 7. Ru-Catalyzed Reductive Coupling of Aldehydes and Dienes

A combination of Pd-catalyst/Xantphos and triethylborane promoted the ene-type reaction of aldehydes and conjugated dienes to afford dienyl homoallyl alcohols in excellent yields (Figure 8).[12] Methyl sorbate underwent formal Baylis-Hillman type C-C bond formation at the α-position to provide (2E,4E)-hexadien-1-ol with excellent stereoselectivity.

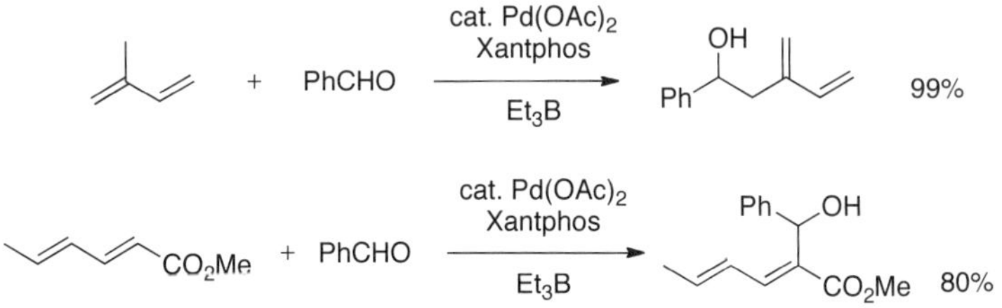

Figure 8. Pd/Xantphos Promoted Coupling Reaction of Aldehydes and Conjugated Dienes Promoted by Triethylborane

Thus, the combination of transition metal catalyst and triethylborane nicely promoted the coupling reaction of carbonyls and conjugated dienes to furnish the stereodefined bishomoallyl and homoallyl alcohols. The reductive coupling reactions offer great efficiency and a useful strategy for the synthesis of important physiologically active molecules, such as terpene alcohols and macrolides.

1. Graduate School of Engineering, Nagasaki University, 1-14 Bunkyo-machi, Nagasaki 852-8521, Japan. masanari@nagasaki-u.ac.jp
2. Kimura, M. and Tamaru, Y. *Topics in Current Chemistry.* **2007**, *279*, 173-207.
3. (a) Kimura, M.; Ezoe, A.; Shibata, K.; Tamaru, Y. *J. Am. Chem. Soc.* **1998**, *120*, 4033-4034. (b) Kimura, M.; Fujimatsu, H.; Akihiro, E.;

 Org. Synth. **2013**, *90*, 105-111

Shibata, K.; Shimizu, M.; Matsumoto, S.; Tamaru Y. *Angew. Chem. Int. Ed.* **1999**, *38*, 397-400.

4. Kimura, M.; Ezoe, A.; Mori, M.; Iwata, K.; Tamaru, Y. *J. Am. Chem. Soc.* **2006**, *128*, 8559-8568.

5. Tamaru, Y.; Tanaka, A.; Yasui, K.; Goto, S.; Tanaka, S. *Angew. Chem. Int. Ed.* **1995**, *34*, 787-789.

6. Yang, Y.; Zhu, S.-F.; Duan, H.-F.; Zhou, C.-Y.; Wang, L.-X.; Zhou, Q. L. *J. Am. Chem. Soc.* **2007**, *129*, 2248-2249.

7. (a) Paudyal, M. P.; Rath, N. P.; Spilling, C. D. *Org. Lett.* **2010**, *12*, 2954-2957. (b) Pei, W.; Krauss. I. J. *J. Am. Chem. Soc.* **2011**, *133*, 18514-18517.

8. (a) Cho, H. Y.; Morken, J. P. *J. Am. Chem. Soc.* **2008**, *130*, 16140-16141. (b) Cho, H. Y.; Morken, J. P. *J. Am. Chem. Soc.* **2010**, *132*, 7576-7577. (c) Cho, H. Y.; YU, Z.; Morken, J. P. *Org. Lett.* **2011**, *13*, 5267-5269.

9. (a) Saito, N.; Mori, M.; Sato, Y. *J. Organomet. Chem.* **2007**, *692*, 460-471. (b) Saito, N.; Kobayashi, A.; Sato, Y. *Angew. Chem. Int. Ed.* **2012**, *51*, 1228-1231.

10. Kimura, M.; Nojiri, D.; Fukushima, M.; Oi, S.; Sonoda, Y.; Inoue, Y. *Org. Lett.* **2009**, *11*, 3794-3797.

11. (a) Shibahara, F.; Bower, J. F.; Krische, M. J. *J. Am. Chem. Soc.* **2008**, *130*, 6338-6339. (b) Bower, J. F.; Patman, R. L.; Krische, M. J. *Org. Lett.* **2008**, *10*, 1033-1035. (c) Smejkal, T.; Han, H.; Breit, B.; Krische, M. J. *J. Am. Chem. Soc.* **2009**, *131*, 10366-10367.

12. Fukushima, M.; Takushima, D.; Kimura, M. *J. Am. Chem. Soc.* **2010**, *132*, 16346-16348.

Masanari Kimura was born in Nagasaki, Japan on 29 May 1967. He obtained his Ph.D. in 1995 under the direction of Professor Yoshinao Tamaru (Nagasaki University). He was Research Fellow of the Japan Society for the Promotion of Science in 1992-1995. He joined the faculty of Nagasaki University as assistant in 1995. He was a postdoctoral fellow MIT (S. L. Buchwald) in 1999-2000. He was promoted to associate professor in 2008, and to full professor in 2010. His current research interest includes the development of new reactions based on transition metals and heterocyclic chemistry.

Enantioselective Alkylation of *N*-(Diphenylmethylene)glycinate *tert*-Butyl Ester: Synthesis of (*R*)-2-(Benzhydrylidenamino)-3-Phenylpropanoic Acid *tert*-Butyl Ester

(*S*)-1 (Ar = 3,4,5-F$_3$-C$_6$H$_2$)

Submitted by Seiji Shirakawa, Kenichiro Yamamoto, Kun Liu, and Keiji Maruoka.[1]

Checked by Spencer Eggen and Scott E. Denmark.

1. Procedure

(R)-2-(Benzhydrylidenamino)-3-phenylpropanoic acid tert-butyl ester. An oven-dried, 300-mL, four-necked, round-bottomed flask, equipped with an overhead mechanical stirrer, a 125-mL pressure-equalizing dropping funnel capped with an argon inlet, and two rubber septa (Note 1) is charged with *N*-(diphenylmethylene)glycine *tert*-butyl ester (5.00 g, 16.9 mmol, 1.0 equiv) (Note 2), (*S*)-4,4-dibutyl-2,6-bis(3,4,5-trifluorophenyl)-4,5-dihydro-3*H*-dinaphtho[2,1-c:1',2'-e]azepinium bromide [(*S*)-1] (12.7 mg, 17.0 µmol, 0.001 equiv) (Note 3), benzyl bromide (2.41 mL, 3.47 g, 20.3 mmol, 1.20 equiv) (Note 4), and toluene (56.4 mL) (Note 5). The stirring mixture is cooled to ~0 °C in a 2-propanol bath (Note 6), and aqueous 48% potassium hydroxide solution (56.4 mL) (Note 7) is added dropwise over 33 min through the dropping funnel (Note 8). The mixture is stirred for 20 h at ~0 °C at a constant stirring speed (300 rpm) under an argon atmosphere (Note 9). The reaction mixture is diluted by the addition of water (113 mL) and the 2-propanol bath is removed. After being stirred for 15 min at ambient temperature, the reaction mixture is transferred to a 1-L separatory

Org. Synth. **2013**, *90*, 112-120
Published on the Web 11/1/2012
© 2013 Organic Syntheses, Inc.

funnel, and toluene (100 mL) is added. The aqueous layer is separated and extracted twice with toluene (100 mL). The combined organic layers are washed with brine (100 mL) and dried over anhydrous magnesium sulfate (15 g). The resulting suspension is filtered and concentrated on a rotary evaporator (Note 10). The residue is loaded onto a column (diameter: 6 cm) wet-packed with silica gel (150 g) and hexanes/*tert*-butyl methyl ether (*t*-BuOMe), 15:1 containing 1% of triethylamine (Note 11). The crude product is purified by gradient elution with hexanes/*t*-BuOMe (15:1 to 10:1) to afford product (7.87 g) (Note 12). After extensive drying, the product (6.40 g, 98% yield) is obtained as a colorless, viscous oil (Notes 13, 14, 15, 16, and 17).

2. Notes

1. The oven temperature is ~205 °C. The round-bottomed flask has three 24-40 necks and one 9-30 neck. The addition funnel, mechanical stirrer adapter, and septum are fitted on the 24-40 necks, while the 9-30 neck is fitted with a small rubber septum. The Teflon paddle of the mechanical stirrer is 4.65 cm in length and 1.9 cm in height. The internal temperature was monitored with an Omega Type K microprocessor digital thermometer with 0.2 cm-width probe.

2. *N*-(Diphenylmethylene)glycine *tert*-butyl ester (99%) was purchased from Alfa Aesar and used as received. The submitters used *N*-(diphenylmethylene)glycine *tert*-butyl ester (>98%) purchased from Tokyo Chemical Industry Co., Ltd. (TCI) and used as received.

3. (*S*)-4,4-Dibutyl-2,6-bis(3,4,5-trifluorophenyl)-4,5-dihydro-3*H*-dinaphtho-[2,1-c:1',2'-e]azepinium bromide [(*S*)-1] (99%) was purchased from Strem Chemicals, Inc. and used as received. The submitters used [(*S*)-1] (> 90%) from Kanto Chemical Co., Inc. as received. The checkers found that an increase in catalyst loading from 0.05 (recommended by the submitters) to 0.10% was essential to obtain enantiomeric ratios greater than 96:4 er.

4. Benzyl bromide (98%) was purchased from Sigma-Aldrich Co. and was freshly distilled over MgSO₄ (10 mmHg, 73–75 °C) under a nitrogen atmosphere. The submitters used benzyl bromide (98%) purchased from Sigma-Aldrich Co. as received.

5. Toluene (Certified ACS grade) was purchased from Thermo Fisher Scientific, Inc. and used as received.

6. The 2-propanol bath temperature (0 ± 2 °C) was maintained by a Thermo Scientific Neslab CC-100 Immersion Cooler with Cryotrol dial controller. As the solution cools, some of the starting material precipitates, forming a whitish, cloudy suspension.

7. The submitters used 48% potassium hydroxide solution from Kanto Chemical Co., Inc. without purification. The checkers prepared a 48% potassium hydroxide solution from solid pellets (Sigma-Aldrich Co., puriss. p.a., ≥86% (T) pellets) in water purified by a Millipore Milli-Q Integral Water Purification System.

8. The solution is stirred at 300 rpm until the internal temperature equilibrated to 1 °C. At this time, the addition of the base solution is begun. During the addition of the potassium hydroxide solution, the internal temperature is kept under 3 °C. Upon completion of the addition of base, the addition funnel is replaced by an argon inlet and the large septum is replaced with a glass stopper.

9. TLC analysis of the reaction mixture indicates the consumption of the starting materials after 10 h. The TLC analysis was performed on EMD Millipore Merck silica gel 60 F_{254} plates [*N*-(diphenylmethylene)glycine *tert*-butyl ester (starting material): R_f = 0.31, product: R_f = 0.41, 15:1 hexanes/*t*-BuOMe, visualized with 254 nm UV lamp].

10. The suspension is filtered through a Pyrex 150 mL course fritted glass funnel (packed with 7.0 g of Celite) into a 1-L suction filtration flask. The filtrate is then transferred with toluene to a 1-L teardrop-shape flask. Toluene is removed by rotary evaporation (16.5–38 °C, 18–20 mmHg) to yield 8.20 g (>100%) of crude product.

11. Neutralization of the silica gel with hexanes/*t*-BuOMe, 15:1 containing 1% triethylamine is necessary to minimize hydrolysis of the product on the column. The neutralized silica was rinsed with 20 column volumes of 15:1 hexanes/*t*-BuOMe (~5 L total) to achieve an eluate pH near 7. The checkers used silica gel purchased from Sigma-Aldrich Co. (Merck, grade 9385, 230-400 mesh, 60 Å). The submitters used silica gel purchased from Daiso Co., Ltd. (Daisogel IR-60–40/63). The hexanes used were Optima grade (Sigma-Aldrich Co.). The *t*-BuOMe used was Reagent Grade (98%) from Sigma-Aldrich Co. The triethylamine was 99% grade from Alfa-Aesar and freshly distilled over calcium hydride under a nitrogen atmosphere.

12. The column was eluted with hexanes/ *t*-BuOMe with a ratio of 15:1 (500 mL) followed by 10:1 (1000 mL) (172 fractions, 8 mL fraction size).

TLC indicates the presence of the desired alkylation product and benzophenone (R_f = 0.29, hexanes/t-BuOMe, 15:1, visualized with a 254 nm UV lamp). Fractions 47-164 were combined and concentrated (11.5–20 °C, 18–20 mmHg). ^{1}H NMR analysis indicates the presence of ~136 mole % of hexanes.

13. The checkers obtained a 92% yield on a ½- scale run (e.r. 98:2).

14. Hexanes were reduced to ~2.7 mole % by using a Büchi GKR-51 Kugelrohr distillation apparatus with continuous turning under vacuum (0.04–0.05 mmHg) at ambient temperature. The amount of hexanes present was determined by ^{1}H NMR spectroscopic analysis of aliquots at 0, 36, 96, and 186 hours. NMR analysis also indicates the presence of ~2.7 mole % benzophenone.

15. *(R)*-2-(Benzhydrylidenamino)-3-phenylpropanoic acid *t*-butyl ester has the following physical and spectroscopic data: e.r. = 98.5:1.5; $[\alpha]^{24}_D$ = 190.9 (c = 0.20, CHCl$_3$); ^{1}H NMR (500 MHz, CDCl$_3$) δ: 1.44 (s, 9 H), 3.16 (dd, J = 9.4, 13.4 Hz, 1 H), 3.23 (dd, J = 4.2, 13.4 Hz, 1 H), 4.11 (dd, J = 4.2, 9.3 Hz, 1 H), 6.60 (br d, J = 5.4 Hz, 2 H), 7.03–7.08 (m, 2 H), 7.13–7.21 (m, 3 H), 7.24–7.39 (m, 6 H), 7.55–7.59 (m, 2 H); ^{13}C NMR (500 MHz, CDCl$_3$) δ: 28.03, 39.58, 67.92, 81.09, 126.13, 127.64, 127.90, 128.01, 128.05, 128.16, 128.69, 128.85, 130.04, 136.35, 138.35, 139.55, 170.24, 170.80; IR 3060 (m), 3027 (m), 2977 (m), 2929 (m), 1731 (s), 1623 (m), 1446 (m), 1367 (s), 1285 (s), 1148 (s), 697 (s) cm^{-1}. MS (EI) *m/z* (relative intensity): 294.2 (12.7), 285.1 (11.0), 284.1 (47.7), 244.1 (32.8), 243.1 (17.9), 242.1 (100.0), 238.1 (29.9), 227.0 (16.2), 207.1 (14.6), 206.1 (11.8), 193.1 (27.1), 192.1 (70.7), 165.1 (21.6), 117.1 (10.5), 115.1 (25.3), 91.1 (14.0), 77.1 (11.1). HRMS (ESI-TOF) calcd for C$_{26}$H$_{28}$NO$_2$$^+$: 386.2120 ([M + H]$^+$). Found: 386.2112. The enantioselectivity was determined by chiral stationary phase HPLC analysis [Daicel Chiralcel OD-H column, 1% 2-propanol/hexane, 0.50 mL/min, 18 bar, λ = 251 nm, retention times: (*R*)-enantiomer (major): 14.19 min, (*S*)-enantiomer (minor): 25.91 min]. Chiral HPLC analysis was calibrated using a 77:23 ratio of the two enantiomers.

16. Some hydrolysis of the product to the free amine and benzophenone is observed during large-scale chromatographic separation. On the basis of ^{1}H NMR integration, there is 2.3 mole % benzophenone relative to product. A pure sample was prepared as follows. A small portion of the product (150 mg) contaminated with benzophenone from the first chromatographic separation can be further purified by a second chromatographic separation at reduced temperature. The product is loaded onto a jacketed column

(diameter: 2 cm) wet-packed with silica gel (29 g) and hexanes/t-BuOMe, 49:1. The product is purified by gradient elution (hexanes/t-BuOMe, 49:1 to 9:1) to yield 0.145 g of product (Note 16), which required extensive drying to remove solvent (Note 17).

17. A portion of silica gel (150 g) was neutralized by stirring with 5 mL of triethylamine, 10 mL of t-BuOMe, and 485 mL of hexanes in a 1-L round-bottom flask. The suspension was filtered through a 150-mL coarse fritted funnel and the silica gel was dried by rotary evaporation (17–22 °C, 18–20 mm Hg). The neutralized silica gel was rinsed with 750 mL t-BuOMe then with 1.0 L of hexanes/t-BuOMe, 49:1 prior to loading. Poly(vinyl) tubing is attached to the jacketed column is run through a water/ice slush. Gradient elution proceeds with 100 mL of each solution (49:1 through 9:1) in increments of 49:1, 24:1, etc. (146 fractions, 2.5 mL fraction size). The solutions were pre-chilled in a water/ice slush. Fractions 79-130 were combined and concentrated (15.5–22 °C, 18–20 mmHg). TLC conditions were the same as those in Note 12, except that the TLC plates were first neutralized with 4% triethylamine in dichloromethane.

18. The product was transferred to a one-piece medium Kugelrohr distillation bulb and attached to a diffusion pump apparatus. The apparatus consists of a Büchi GKR-50 heating chamber connected to an Edwards SI100 Diffusion Pump with 90 V AC pump heater electrical supply. The product was heated between 50–70 °C at a pressure range of 2.9–3.0×10^{-5} mm Hg. The viscosity of the product oil dramatically decreases as temperature is increased, but the oil does not distill under these conditions.

Handling and Disposal of Hazardous Chemicals

The procedures in this article are intended for use only by persons with prior training in experimental organic chemistry. All hazardous materials should be handled using the standard procedures for work with chemicals described in references such as "Prudent Practices in the Laboratory" (The National Academies Press, Washington, D.C., 2011 www.nap.edu). All chemical waste should be disposed of in accordance with local regulations. For general guidelines for the management of chemical waste, see Chapter 8 of Prudent Practices.

These procedures must be conducted at one's own risk. *Organic Syntheses, Inc.*, its Editors, and its Board of Directors do not warrant or guarantee the safety of individuals using these procedures and hereby

116

disclaim any liability for any injuries or damages claimed to have resulted from or related in any way to the procedures herein.

3. Discussion

α-Amino acids are a biologically significant class of compounds, since they are the basic building blocks for peptides and proteins, and they are the biopolymers responsible for both the structure and function of most living things. Accordingly, the development of truly efficient methods for the preparation of α-alkyl-α-amino acid, especially in an enantiomerically pure form, has become of great importance. Asymmetric synthesis of α-amino acids by phase-transfer alkylation using a chiral catalyst and a prochiral protected glycine derivative would provide a particularly attractive method for the preparation of optically active α-alkyl-α-amino acids.[2] First catalytic asymmetric alkylation of a glycine derivative under phase-transfer conditions was achieved by using a cinchona alkaloid derived chiral phase-transfer catalyst **2a** with moderate enantioselectivity.[3] The *N*-9-anthracenylmethyl group substituted catalyst **2b** was developed to improve the enantioselectivity.[4,5] After these reports, many types of cinchona-derived chiral phase-transfer catalysts were developed for the reaction.[2] Purely synthetic binaphthyl-modified new *N*-spiro-type chiral phase-transfer catalyst (*S,S*)-**3** was also developed for the reaction.[6] Compared to other phase-transfer catalysts, such as cinchona-derived catalysts, binaphthyl-modified *N*-spiro-type catalysts of type **3** generally require lower catalyst loading (1 mol%) and provide alkylated products in higher enantiomeric excesses. Based on the design of binaphthyl-modified *N*-spiro-type catalysts of type **3**, very reactive high-performance catalyst (*S*)-**1** was developed.[7] Most notably, the reaction proceeded smoothly under mild phase-transfer conditions in the presence of only 0.05~0.1 mol% of (*S*)-**1** to afford the corresponding alkylation products with excellent enantioselectivities (Table 1).

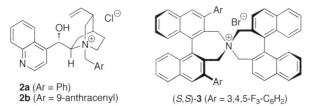

2a (Ar = Ph)
2b (Ar = 9-anthracenyl)

(*S,S*)-**3** (Ar = 3,4,5-F$_3$-C$_6$H$_2$)

Figure 2. Chiral Phase-Transfer Catalysts

Table 1. Enantioselective Alkylation of Glycine Derivative

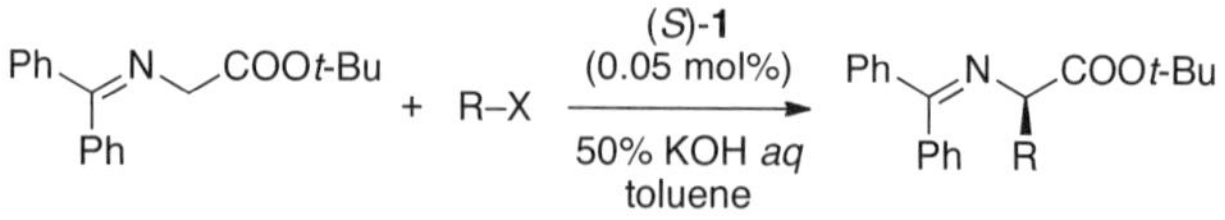

entry	R–X	conditions	yield (%)	ee (%)
1	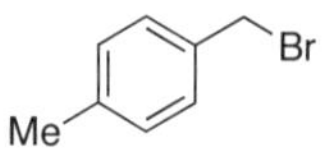	0 °C, 4 h	99	98
2	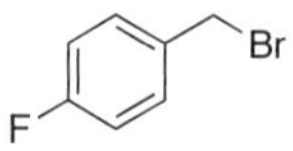	0 °C, 5 h	99	98
3	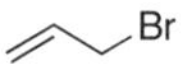	0 °C, 3 h	87	98
4	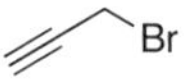	0 °C, 4 h	88	98
5ᵃ		–20 °C, 1 h	67	99

ᵃ Use of 0.1 mol% of (S)-**1** and CsOH•H₂O

1. Department of Chemistry, Graduate School of Science, Kyoto University, Sakyo, Kyoto, 606-8502, Japan. E-mail: maruoka@kuchem.kyoto-u.ac.jp

2. (a) Maruoka, K.; Ooi, T. *Chem. Rev.* **2003**. *103*, 3013. (b) O'Donnell, M. J. *Acc. Chem. Res.* **2004**, *37*, 506. (c) Lygo, B.; Andrews, B. I. *Acc. Chem. Res.* **2004**, *37*, 518. (d) Ooi, T.; Maruoka, K. *Angew. Chem., Int. Ed.* **2007**, *46*, 4222. (e) Ooi, T.; Maruoka, K. *Aldrichimica Acta* **2007**, *40*, 77. (f) Hashimoto, T.; Maruoka, K. *Chem. Rev.* **2007**, *107*, 5656. (g) Maruoka, K. *Org. Process Res. Dev.* **2008**, *12*, 679. (h) Maruoka, K., Ed. *Asymmetric Phase Transfer Catalysis*; Willy-VCH: Weinheim, 2008. (i) Shirakawa, S.; Maruoka, K. In *Catalytic Asymmetric Synthesis, 3rd Ed*; Ojima, I., Ed.; John Wiley & Sons: New Jersey, 2010; Chapter 2C, p 95.

3. (a) O'Donnell, M. J.; Bennett, W. D.; Wu, S. *J. Am. Chem. Soc.* **1989**, *111*, 2353. (b) Lipkowitz, K. B.; Cavanaugh, M. W.; Baker, B.;

Org. Synth. **2013**, *90*, 112-120

O'Donnell, M. J. *J. Org. Chem.* **1991**, *56*, 5181.

4. (a) Lygo, B.; Wainwright, P. G. *Tetrahedron Lett.* **1997**, *38*, 8595. (b) Lygo, B.; Crosby, J.; Lowdon, T. R.; Wainwright, P. G. *Tetrahedron* **2001**, *57*, 2391. (c) Lygo, B.; Crosby, J.; Lowdon, T. R.; Peterson, J. A.; Wainwright, P. G. *Tetrahedron* **2001**, *57*, 2403.

5. Corey, E. J.; Xu, F.; Noe, M. C. *J. Am. Chem. Soc.* **1997**, *119*, 12414.

6. (a) Ooi, T.; Kameda, M.; Maruoka, K. *J. Am. Chem. Soc.* **1999**, *121*, 6519. (b) Ooi, T.; Kameda, M.; Maruoka, K. *J. Am. Chem. Soc.* **2003**, *125*, 5139.

7. (a) Kitamura, M.; Shirakawa, S.; Maruoka, K. *Angew. Chem., Int. Ed.* **2005**, *44*, 1549. (b) Kitamura, M.; Arimura, Y.; Shirakawa, S.; Maruoka, K. *Tetrahedron Lett.* **2008**, *49*, 2026. (c) Kitamura, M.; Shirakawa, S.; Arimura, Y.; Wang, X.; Maruoka, K. *Chem. Asian J.* **2008**, *3*, 1702.

Appendix
Chemical Abstracts Nomenclature; (Registry Number)

(R)-2-(Benzhydrylidenamino)-3-phenylpropanoic acid tert-butyl ester; (119272-91-2)

N-(Diphenylmethylene)glycine *tert*-butyl ester; (81477-94-3)

(*S*)-4,4-Dibutyl-2,6-bis(3,4,5-trifluorophenyl)-4,5- dihydro-3*H*-dinaphtho [2,1-c:1',2'-e]azepinim bromide; (851942-89-7)

Benzyl bromide; (100-39-0)

Keiji Maruoka was born in 1953 in Mie, Japan. He graduated from Kyoto University (1976) and received his Ph.D. (1980) from University of Hawaii (Thesis Director: Professor Hisashi Yamamoto). He became an assistant professor of Nagoya University (1980) and was promoted to a lecturer (1985) and an associate professor (1990) there. He moved to Hokkaido University as a full professor (1995–2001), and he currently is a professor of chemistry in Kyoto University since 2000. His research interests are the design of high-performance organocatalysts and bidentate Lewis acid chemistry.

Seiji Shirakawa was born in 1974 in Ehime, Japan. He graduated from Nihon University (1997) and received his Ph.D. (2004) from Kyoto University (Thesis Director: Professor Keiji Maruoka). He worked with Professor James L. Leighton at Columbia University (2004–2005) and Professor Shu Kobayashi at University of Tokyo (2005–2007) as a postdoctoral fellow. He was appointed as an assistant professor at Nihon University in 2007, and he currently is an associate professor in Kyoto University since 2009.

Kenichiro Yamamoto was born in Madison, Wisconsin, USA, in 1978 and raised in Yamaguchi, Japan. He obtained his M.S. degree from Kyoto University in 2004 under the direction of Professor Keiji Maruoka. Since 2004, he has been working as process chemist in the Research & Development Center of NAGASE & CO., LTD. (Kobe, Japan). From 2009, he has been doing postgraduate studies in Kyoto University under the direction of Professor Keiji Maruoka.

Kun Liu was born in 1984 in Shandong Province, China. He graduated from Tianjin University (2006) and received his Master degree from the same university under supervision of Professor Jun-An Ma (2008). Currently he is a doctor-course student at Kyoto University under the supervision of Professor Keiji Maruoka.

Spencer Eggen received his B.S. degree in Chemistry with highest honors in 2011 from Texas Christian University. During his years as an undergraduate, he worked in the laboratories of Dr. Onofrio Annunziata, studying polymer-porphyrin interactions. In 2011, he joined Scott Denmark's research group at the University of Illinois, Urbana-Champaign. His research interests include the development of new Phase Transfer-catalyzed reactions.

Enantioselective Alkylation of 2-[(4-Chlorobenzyliden)Amino]Propanoic Acid *tert*-Butyl Ester: Synthesis of (*R*)-2-Amino-2-Methyl-3-Phenylpropanoic Acid *tert*-Butyl Ester

(*S*)-**1** (Ar = 3,4,5-F$_3$-C$_6$H$_2$)

Submitted by Seiji Shirakawa, Kenichiro Yamamoto, Kun Liu, and Keiji Maruoka.[1]
Checked by Lindsey R. Cullen and Scott E. Denmark.

1. Procedure

A. 2-[(4-Chlorobenzylidene)amino]propanoic acid tert-butyl ester. A 200-mL, three-necked, round-bottomed flask, equipped with an overhead mechanical stirrer (Note 1), a 10-mL pressure-equalizing dropping funnel (attached to an argon inlet), and a thermometer fitted with a thermometer adapter, is charged with L-alanine *tert*-butyl ester hydrochloride (5.00 g, 27.5 mmol, 1.05 equiv) (Notes 2 and 3), 4-chlorobenzaldehyde (3.69 g, 26.2 mmol, 1.00 equiv) (Note 4) and toluene (25.0 mL). The mixture is stirred at 400 rpm and triethylamine (4.20 mL, 30.1 mmol, 1.15 equiv) (Note 5) is added dropwise over 5 min through the dropping funnel at ambient temperature (Note 6). The addition funnel is removed and replaced with a

Org. Synth. **2013**, *90*, 121-129
Published on the Web 11/1/2012
© 2013 Organic Syntheses, Inc.

reflux condenser fitted with an argon inlet. The reaction mixture is heated to 70 °C in an oil bath and stirred at 400 rpm for 4 h (Note 7). After cooling to ambient temperature, the reaction mixture is diluted by the addition of water (50 mL). The reaction mixture is transferred to a 500-mL separatory funnel and toluene (50 mL) is added. The organic layer is separated and washed with brine (2 × 50 mL). The organic solution is dried over sodium sulfate (20 g), filtered, and concentrated on a rotary evaporator (15 mm Hg, 35 °C). The residue is dried by stirring under reduced pressure (0.2 mm Hg) at 23 °C for 2 h to afford the crude aldimine product (6.66 g, , 95% yield) (Note 8) as an amber oil (Note 9).

B. *(R)-2-Amino-2-methyl-3-phenylpropanoic acid tert-butyl ester.* A 300-mL, three-necked, round-bottomed flask, equipped with an overhead mechanical stirrer, 100-mL pressure-equalizing dropping funnel capped with an argon inlet, and a thermometer fitted with a thermometer adapter, is charged with crude 2-[(4-chlorobenzyliden)amino]propanoic acid *tert*-butyl ester (6.66 g, 24.9 mmol, 1.00 equiv) (Notes 8 and 10), (*S*)-4,4-dibutyl-2,6-bis(3,4,5-trifluorophenyl)-4,5-dihydro-3*H*-dinaphtho[2,1-c:1',2'-e]azepinium bromide (19.6 mg, 0.026 mmol, 0.001 equiv) (Note 11), benzyl bromide (3.74 mL, 31.4 mmol, 1.20 equiv) (Note 12) and toluene (52.4 mL). The mixture is stirred at 400 rpm and cooled to 0 °C in an ice-water bath. Aqueous cesium hydroxide solution (80 wt %, 24.5 g, 5.00 equiv) (Note 13) is added dropwise over 20 min through the dropping funnel (Note 14). The mixture is stirred for 18 h at 0 °C with constant stirring speed (400 rpm) under argon atmosphere (Note 15). The reaction mixture is diluted by the addition of water (100 mL) and the cooling bath is removed. After stirring for an additional 15 min at ambient temperature, the reaction mixture is transferred to a 500-mL separatory funnel and toluene (50 mL) is added. The organic layer is separated and washed with 10% aqueous sodium chloride solution (2 × 75 mL). The organic layer is dried over sodium sulfate (15 g), filtered, and concentrated on a rotary evaporator (15 mmHg, 35 °C). The pale yellow residue (10.4 g) is transferred to a 300-mL round-bottomed flask equipped with 3 cm egg-shaped stir bar and THF (50 mL) is added. Citric acid monohydrate (52.5 g, 250 mmol, 10 equiv) (Note 16) and water (150 mL) are added to the solution at ambient temperature. The cloudy, colorless mixture is stirred for 3 h at ambient temperature. The reaction mixture is transferred to a 500-mL separatory funnel and hexane (50 mL) is added. The water layer is separated and washed with hexane (2 × 50 mL). The aqueous solution is transferred to a 500-mL, three-necked, round-

122

bottomed flask equipped with a 3 cm egg-shaped stir bar, 100-mL pressure-equalizing dropping funnel, thermometer, septum, and is placed in a room temperature water bath. Aqueous sodium hydroxide solution (50 wt %, 39.3 g, 26.2 mL, 18.7 equiv) is added dropwise over 25 min (Note 17). The mixture is stirred for an additional 15 min at ambient temperature. The mixture is transferred to a 500-mL separatory funnel and *tert*-butyl methyl ether (150 mL) is added. The organic layer is separated and washed with 10% aqueous sodium chloride solution (2 × 75 mL). The organic layer is dried over sodium sulfate (15 g), filtered, and concentrated on a rotary evaporator (15 mm Hg, 23 °C). The residue is adsorbed onto Celite (8 g) and loaded as a solid onto a column (diameter: 5 cm, height: 10 cm) wet-packed with silica gel (100 g) (Note 18) in hexane/ethyl acetate, 4/1. Elution with a gradient of hexane/ethyl acetate (400 mL of each: 4:1 to 2:1 to 1:1 to 1:2) affords the product as a pale yellow oil [4.80 g, 78% yield (for 2 steps), 95–97% ee for the (*R*)-enantiomer] (Note 19 and 20).

2. Notes

1.   The Teflon paddle of the mechanical stirrer is 6.5 cm in length and 1.8 cm in height.

2.   L-Alanine *tert*-butyl ester hydrochloride (99%) was purchased from Acros Organics and used as received. This reagent is also available from Watanabe Chemical Industries, Ltd. and Sigma-Aldrich Co.

3.   D-Alanine and DL-alanine *tert*-butyl ester hydrochloride can be also used as starting materials instead of L-alanine *tert*-butyl ester hydrochloride.

4.   4-Chlorobenzaldehyde (>98.5%) was purchased from Acros Organics and used as received. This reagent is also available from Wako Pure Chemical Industries, Ltd.

5.   Triethylamine (>99%) was purchased from Sigma-Aldrich Co. and used as received. This reagent is also available from Kanto Chemical Co., Inc.

6.   The internal temperature was kept below 30 °C during the addition of triethylamine.

7. ^{1}H NMR analysis of the reaction mixture indicated nearly complete consumption of the starting materials. An aliquot of the reaction mixture (30 µL) was diluted with DMSO-d_6 (0.7 mL), and analyzed by ^{1}H NMR [4-chlorobenzaldehyde: δ 10.00 (Ar*CHO*, s, 1H), aldimine product: δ

8.39 (ArC*H*=N, s, 1H)]. A residual amount of 4-chlorobenzaldehyde (1%) was observed after 4 h.

8. The crude aldimine product includes 1mol % of 4-chlorobenzaldehyde (determined by ^{1}H NMR analysis).

9. The product gradually decomposes in air at room temperature. The compound should be kept in a refrigerator under an inert atmosphere where it is stable for at least 2 weeks. The product solidifies to a low-melting, white solid upon storage at –20 °C. ^{1}H NMR (500 MHz, DMSO-d_6) δ: 1.34 (d, J = 6.8 Hz, 3 H), 1.39 (s, 9 H), 4.09 (q, J = 6.6 Hz, 1 H), 7.53 (d, J = 8.3 Hz, 2 H), 7.78 (d, J = 8.6 Hz, 2 H), 8.39 (s, 1 H); ^{13}C NMR (125 MHz, DMSO-d_6) δ: 19.0, 27.6, 67.1, 80.3, 128.8, 129.7, 134.6, 135.5, 161.4, 170.9; ^{1}H NMR (500 MHz, CDCl$_3$) δ: 1.47 (s, 9 H), 1.48 (d, J = 7.1 Hz, 3 H), 4.04 (q, J = 6.9 Hz, 1 H), 7.38 (d, J = 8.6 Hz, 2 H), 7.71 (d, J = 8.6 Hz, 2 H), 8.25 (s, 1 H); ^{13}C NMR (125 MHz, CDCl$_3$) δ: 19.5, 28.2, 68.5, 81.4, 128.7, 129.8, 134.6, 137.0, 161.3, 171.8; IR (neat) 2975, 2926, 2871, 1732, 1642, 1365, 1153, 1125, 1087, 844, 824 cm^{-1}. HRMS (ESI-TOF) *m/z* calcd for C$_{14}$H$_{19}$ ClNO$_2^+$: 268.1104 ([M + H]$^+$). Found: 268.1108.

10. The full amount of crude aldimine product in procedure A was used for the next step (procedure B). Reagent equivalents in procedure B were based upon amount of 4-chlorobenzaldehyde (26.2 mmol).

11. (*S*)-4,4-Dibutyl-2,6-bis(3,4,5-trifluorophenyl)-4,5-dihydro-3*H*-dinaphtho[2,1-c:1',2'-e]azepinium bromide (Nagase purity) was purchased from Sigma-Aldrich Co. and used as received. This catalyst is also available from Kanto Chemical Co., Inc. and Strem Chemicals, Inc.

12. Benzyl bromide (98%) was purchased from Sigma-Aldrich Co. and used as received.

13. Cesium hydroxide solution (80 wt %) was prepared from cesium hydroxide monohydrate (96%) purchased from Alfa-Aesar Co. as follows: Cesium hydroxide monohydrate (44.81 g) was weighed into a neoprene bottle. The neoprene bottle was then cooled in an ice bath, followed by addition of distilled water (5.2 mL). The solution was shaken to incorporate solid into solution and allowed to stand for 2 h to ensure dissolution. The homogenous, colorless solution was stored in a neoprene bottle. [Caution: Cooling with an ice bath is necessary for preparation of 80 wt % cesium hydroxide solution. The dissolution of cesium hydroxide in water is very exothermic.] Cesium hydroxide monohydrate (>95%) is also available from Sigma-Aldrich Co.

14. During the addition of the cesium hydroxide solution, the internal temperature was kept under 5 °C.

15. ^{1}H NMR analysis of the reaction mixture indicates complete consumption of the starting materials. An aliquot of the reaction mixture (30 μL) was diluted with d_6-DMSO (0.7 mL), and analyzed by ^{1}H NMR [starting aldimine substrate: δ 8.39 (ArCH=N, s, 1H), alkylation product: δ 8.25 (ArCH=N, s, 1H)]. The checkers observed highly variable reaction times ranging from 18 h to 48 h to reach full conversion.

16. Citric acid monohydrate (>99.0%) was purchased from Sigma-Aldrich Co. and used as received. This reagent is also available from Kanto Chemical Co., Inc.

17. Sodium hydroxide solution (50 wt %) was purchased from Acros Organics and used as received. This reagent is also available from Kanto Chemical Co., Inc. During the addition of 50 wt % sodium hydroxide solution, the internal temperature was kept under 37 °C. A room temperature water bath was necessary to control the internal temperature.

18. Silica gel was purchased from SiliCycle., Inc. (SiliaFlash F-60–40/63). TLC: R_f 0.11 (hexane/EtOAc, 1:1, visualized using ninhydrin stain).

19. Column chromatography was performed without collecting a forecut and 20 mL fractions were collected using 16 mm × 150 mm test tubes. The desired product tailed significantly and was isolated from fractions 13 through 64 using rotary evaporation (15 mmHg, 35 °C).

20. The final product was obtained as a pale yellow oil but did not show the presence of an impurity by ^{1}H NMR. The colored impurity was removed from a portion of the final product (1.0 mL) by Kugelrohr distillation at 70 °C under reduced pressure (4.3×10^{-5} mmHg) to afford a colorless oil (956 mg) prior to elemental analysis and further characterization. Characterization data for the product: $[\alpha]^{24}_D$ = 15.3 (c = 1.0, CHCl$_3$); ^{1}H NMR (500 MHz, CDCl$_3$) δ: 1.34 (s, 3 H), 1.46 (s, 9 H), 1.51 (br s, 2 H), 2.77 (d, J = 13.2 Hz, 1 H), 3.11 (d, J = 13.2 Hz, 1 H), 7.21–7.29 (m, 5 H); ^{13}C NMR (125 MHz, CDCl$_3$) δ: 27.1, 28.1, 46.6, 58.9, 81.2, 126.9, 128.3, 130.3, 137.0, 176.4; IR 3372, 3310, 2975, 2926, 1722, 1365, 1153, 1105, 848, 737, 699 cm^{-1}. HRMS (ESI-TOF) m/z calcd for C$_{14}$H$_{22}$NO$_2{}^+$: 236.1651 ([M + H]$^+$); Found: 236.1657. Calcd. for C$_{14}$H$_{21}$NO$_2$: C; 71.46, H; 8.99, N; 5.95. Found: C; 71.50, H; 9.14, N; 6.17. The enantioselectivity was determined by chiral stationary phase HPLC analysis [Daicel Chiralcel AD-H column, 3.3% 2-propanol/hexane, 0.50 mL/min, λ = 220 nm, 5.0 mg/mL, retention times: (R)-enantiomer (major): 13.1 min, (S)-enantiomer (minor):

22.4 min]. The racemate was prepared for comparison following the above procedure by substituting tetrabutylammonium bromide as the catalyst.

Handling and Disposal of Hazardous Chemicals

The procedures in this article are intended for use only by persons with prior training in experimental organic chemistry. All hazardous materials should be handled using the standard procedures for work with chemicals described in references such as "Prudent Practices in the Laboratory" (The National Academies Press, Washington, D.C., 2011 www.nap.edu). All chemical waste should be disposed of in accordance with local regulations. For general guidelines for the management of chemical waste, see Chapter 8 of Prudent Practices.

These procedures must be conducted at one's own risk. *Organic Syntheses, Inc.*, its Editors, and its Board of Directors do not warrant or guarantee the safety of individuals using these procedures and hereby disclaim any liability for any injuries or damages claimed to have resulted from or related in any way to the procedures herein.

3. Discussion

Nonproteinogenic α,α-dialkyl-α-amino acids have played a special role in the design of peptides with enhanced properties. This is not only because they possess stereochemically stable quaternary carbon centers, but also their incorporation into peptides results in the significant influence on the conformational preferences, which eventually provides useful information for the elucidation of enzymatic mechanisms. Furthermore, α,α-dialkyl-α-amino acids themselves are often effective enzyme inhibitors and also constitute a series of interesting building blocks for the synthesis of various biologically active compounds. Accordingly, numerous studies have been conducted to develop truly efficient methods for their preparation,[2] and phase-transfer catalysis has made unique contributions.[3] In 1992, O'Donnell reported the first chiral phase-transfer-catalyzed alkylations of an alanine derivative by using an *N*-benzylcinchonium chloride with moderate enantioselectivity.[4] Lygo improved the enantioselectivity for the reaction by using an *N*-9-anthracenylmethyl substituted cinchona alkaloid-derived catalyst.[5] Highly enantioselective alkylation of an alanine derivative with broad generality of alkyl halides was achieved by using a binaphthyl-modified *N*-spiro-type chiral phase-transfer catalyst.[6] Based on the design of binaphthyl-modified *N*-spiro-type catalysts, the very reactive high-performance catalyst (*S*)-**1** was developed.[7] Most notably, the reaction proceeded smoothly under mild phase-transfer conditions in the presence of only 0.05–0.1 mol% of (*S*)-**1** to afford the corresponding alkylation products with excellent enantioselectivities (Scheme 1).

Scheme 1. Enantioselective Alkylation of Alanine Derivative

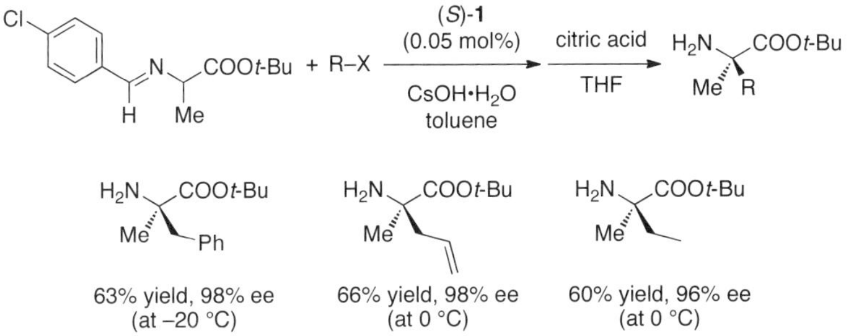

63% yield, 98% ee
(at –20 °C)

66% yield, 98% ee
(at 0 °C)

60% yield, 96% ee
(at 0 °C)

1. Department of Chemistry, Graduate School of Science, Kyoto University, Sakyo, Kyoto, 606-8502, Japan. E-mail: maruoka@kuchem.kyoto-u.ac.jp
2. (a) Schöllkopf, U. *Top. Curr. Chem.* **1983**, *109*, 65-84. (b) Cativiela, C.; Díaz-de-Villegas, M. D. *Tetrahedron: Asymmetry* **1998**, *9*, 3517. (c) Ohfune, Y.; Shinada, T. *Eur. J. Org. Chem.* **2005**, 5127. (d) Vogt, H.; Bräse, S. *Org. Biomol. Chem.* **2007**, *5*, 406.
3. (a) Maruoka, K.; Ooi, T. *Chem. Rev.* **2003**, *103*, 3013. (b) O'Donnell, M. J. *Acc. Chem. Res.* **2004**, *37*, 506. (c) Lygo, B.; Andrews, B. I. *Acc. Chem. Res.* **2004**, *37*, 518. (d) Ooi, T.; Maruoka, K. *Angew. Chem., Int. Ed.* **2007**, *46*, 4222. (e) Ooi, T.; Maruoka, K. *Aldrichimica Acta* **2007**, *40*, 77. (f) Hashimoto, T.; Maruoka, K. *Chem. Rev.* **2007**, *107*, 5656. (g) Maruoka, K. *Org. Process Res. Dev.* **2008**, *12*, 679. (h) Maruoka, K., Ed. *Asymmetric Phase Transfer Catalysis*; Willy-VCH: Weinheim, 2008. (i) Shirakawa, S.; Maruoka, K. In *Catalytic Asymmetric Synthesis, 3rd Ed*; Ojima, I., Ed.; John Wiley & Sons: New Jersey, 2010; Chapter 2C, p 95.
4. O'Donnell, M. J.; Wu, S. *Tetrahedron: Asymmetry* **1992**, *3*, 591.
5. Lygo, B.; Crosby, J.; Peterson, J. A. *Tetrahedron Lett.* **1999**, *40*, 8671.

6.   Ooi, T.; Takeuchi, M.; Kameda, M.; Maruoka, K. *J. Am. Chem. Soc.* **2000**, *122*, 5228.
7.   (a) Kitamura, M.; Shirakawa, S.; Maruoka, K. *Angew. Chem., Int. Ed.* **2005**, *44*, 1549. (b) Kitamura, M.; Shirakawa, S.; Arimura, Y.; Wang, X.; Maruoka, K. *Chem. Asian J.* **2008**, *3*, 1702.

Appendix
Chemical Abstracts Nomenclature; (Registry Number)

2-[(4-Chlorobenzyliden)amino]propanoic acid tert-butyl ester; (142274-97-3)
L-Alanine *tert*-butyl ester hydrochloride; (13404-22-3)
4-Chlorobenzaldehyde; (104-88-1)
Triethylamine; (121-44-8)
(R)-2-Amino-2-methyl-3-phenylpropanoic acid tert-butyl ester; (147714-90-7)
(*S*)-4,4-Dibutyl-2,6-bis(3,4,5-trifluorophenyl)-4,5-dihydro-3*H*-dinaphtho[2,1-c:1',2'-e]azepinium bromide; (851942-89-7)
Benzyl bromide; (100-39-0)
Cesium hydroxide monohydrate; (35103-79-8)

Keiji Maruoka was born in 1953 in Mie, Japan. He graduated from Kyoto University (1976) and received his Ph.D. (1980) from University of Hawaii (Thesis Director: Professor Hisashi Yamamoto). He became an assistant professor of Nagoya University (1980) and was promoted to a lecturer (1985) and an associate professor (1990) there. He moved to Hokkaido University as a full professor (1995–2001), and he currently is a professor of chemistry in Kyoto University since 2000. His research interests are the design of high-performance organocatalysts and bidentate Lewis acid chemistry.

Seiji Shirakawa was born in 1974 in Ehime, Japan. He graduated from Nihon University (1997) and received his Ph.D. (2004) from Kyoto University (Thesis Director: Professor Keiji Maruoka). He worked with Professor James L. Leighton at Columbia University (2004–2005) and Professor Shu Kobayashi at University of Tokyo (2005–2007) as a postdoctoral fellow. He was appointed as an assistant professor at Nihon University in 2007, and he currently is an associate professor in Kyoto University since 2009.

Kenichiro Yamamoto was born in Madison, Wisconsin, USA, in 1978 and raised in Yamaguchi, Japan. He obtained his M.S. degree from Kyoto University in 2004 under the direction of Professor Keiji Maruoka. Since 2004, he has been working as process chemist in the Research & Development Center of NAGASE & CO., LTD. (Kobe, Japan). From 2009, he has been doing postgraduate studies in Kyoto University under the direction of Professor Keiji Maruoka.

Kun Liu was born in 1984 in Shandong Province, China. He graduated from Tianjin University (2006) and received his Masters degree from the same university under supervision of Professor Jun-An Ma (2008). Currently he is a doctoral student at Kyoto University under the supervision of Professor Keiji Maruoka.

Lindsey R. Cullen received her bachelor's degree in Chemistry from the University of Detroit Mercy. During this time she performed research with Professor Matthew J. Mio on Cu(I)-catalyzed cross coupling reactions of aryl alkynes. In 2009 she began her graduate studies in Organic Chemistry at the University of Illinois, under the mentorship of Professor Scott E. Denmark. The focus of her research is the application of chiral quaternary ammonium salt catalysts in enantioselective reactions, specifically conjugate addition of nucleophiles via fluorodesilylation and phase transfer catalyzed rearrangements.

Preparation of α-Fluorobis(phenylsulfonyl)methane (FBSM)

A.

B.

Submitted by G. K. Surya Prakash, Nan Shao, Fang Wang, and Chuanfa Ni.[1]
Checked by Thomas D. Montgomery and Viresh H. Rawal.

Potassium hydride is a pyrophoric solid and must not be allowed to come into contact with the atmosphere. This reagent should only be handled by individuals trained in its proper and safe use.

1. Procedure

A. *Fluoromethyl phenyl sulfone* (**1**) (Note 1). An oven-dried (140 °C for 12 h) 100-mL round-bottomed flask equipped with a magnetic stir bar (25 x 8 mm, octagonal) is charged with spray-dried potassium fluoride (8.80 g, 152 mmol, 2.0 equiv) (Note 2) and 18-crown-6 (2.01 g, 7.6 mmol, 0.1 equiv) (Note 3). The flask is sealed with a rubber septum, into which is inserted a syringe needle attached to a nitrogen/vacuum line. The flask is evacuated and refilled with nitrogen 3 times. Anhydrous acetonitrile (50 mL) (Note 4) and chloromethyl phenyl sulfide (10.2 mL, 12.10 g, 76.0 mmol, 1.0 equiv) (Note 5) are added successively to the flask by syringe. A reflux condenser fitted with a nitrogen inlet adaptor is quickly attached and the apparatus is flushed with nitrogen three times. The stirred reaction mixture is heated to reflux in an oil bath (102–103 °C, bath temp) for 120 h (Note 6). The reaction mixture is then cooled in an ice bath (0 °C, bath temp) (Note 7), diluted with ice water (50 mL) (Note 8), and transferred to a 250 mL separatory funnel. The mixture is extracted with methylene chloride (4 x 25

130

ml, Note 9) (Note 10). The combined organic layer is washed with water (30 mL), dried over magnesium sulfate (ca. 10 g, 15 min) (Note 11), and filtered. The solvent is removed on a rotary evaporator (23 °C bath temp, 2 mmHg) to give crude fluoromethyl phenyl sulfide as a brownish oil (10.04–10.25 g, 93–95%) that is directly subjected to oxidation.

To a 1-L round-bottomed flask equipped with a magnetic stir bar (50 x 8 mm, octagonal) (Note 12) is added Oxone® (116.80 g, 190 mmol KHSO$_5$, 2.6 equiv) (Note 13) and distilled water (175 mL). The flask is capped loosely with a septum and placed in an ice bath. The septum is replaced with an addition funnel and a solution of crude fluoromethyl phenyl sulfide (10.20 g, 72 mmol, 1.0 equiv) in methanol (175 mL) (Note 14) is added dropwise over ca. 1 h (Note 15). The reaction mixture is allowed to slowly warm to room temperature and stirred for an additional 12 h. Methanol is removed via rotary evaporation (45 °C bath temp, 2 mmHg) (Note 16). The resulting residue contains a large amount of insoluble white precipitate, which is removed by filtration through a Büchner funnel (Note 17). The funnel is rinsed with methylene chloride (2 x 30 mL) and the filtrate is transferred to a 250-mL separatory funnel. After layer separation, the aqueous layer is further extracted with methylene chloride (Note 18) (5 × 30 mL, Note 9). The organic layers are combined, washed with water, dried over magnesium sulfate (ca. 10 g, 15 min) (Note 11), filtered, and concentrated to ca. 40 mL of a pale yellow solution. The solution is filtered through a plug of silica gel (230–400 mesh, 100 mL), which is further washed with methylene chloride (ca. 250 mL, Note 9) to give a clear solution (Note 19). The filtrate is concentrated via rotary evaporation (23 °C, 2 mmHg) and then placed under vacuum (room temperature, ca. 0.2–0.3 mmHg, 15–30 min) to result in a clear or slightly yellowish oil, which slowly solidifies at room temperature under vacuum. The solid is stirred with hot hexanes (ca. 50 mL, 60~65 °C) (Note 20) for 20 min, which forms two layers. Upon cooling to 0 °C in an ice bath, the bottom layer gradually crystallizes to yield colorless crystals over 15 min, which are collected by filtration on a Büchner funnel and washed with cold (0 °C) hexanes (2 x 10 mL) (Note 21) to afford fluoromethyl phenyl sulfone (1) (10.99–12.18 g, 83–92%) (Note 22).

B. *α-Fluorobis(phenylsulfonyl)methane (FBSM)* (2). An oven-dried (140 °C, 12 h) 250-mL round-bottomed flask, equipped with a magnetic stir bar (25 x 8 mm, octagonal), is charged with potassium hydride (21.80 g, 30% wt in oil, 163 mmol, 2.7 equiv) (Note 23), sealed with a rubber septum and connected through a syringe needle to a nitrogen/vacuum line. The flask is evacuated and purged with nitrogen three times and then placed in an ice bath. Excess oil is removed as follows. Anhydrous hexanes (20 mL) (Note 24) are added to the flask via syringe. The mixture is gently stirred for 10 min and allowed to stand unstirred for another 10 min before the removal of the hexanes-oil solution with a syringe. The hexanes-oil solution is added dropwise to an isopropyl alcohol solution. This washing procedure is repeated two more times. Anhydrous THF (130 mL) is then added (Note 25). Hexamethyldisilazane (40.9 mL, 31.5 g, 195 mmol, 3.2 equiv) (Note 26) is then added portion-wise to the stirred solution via syringe over a period of 20–30 min. The hydrogen evolution ceases within 15 min after the addition. The ice bath is removed and the reaction mixture is allowed to stand without stirring for 30 min at room temperature before use (Note 27).

An oven dried (140 °C, 12 h) 500-mL round-bottomed flask equipped with a magnetic stir bar (37.5 x 8 mm, octagonal) is charged with fluoromethyl phenyl sulfone (1) (10.66 g, 61.2 mmol, 1.0 equiv). The flask is sealed with a rubber septum and connected through a syringe needle to a nitrogen/vacuum line. The flask is evacuated and purged with nitrogen three times. Benzenesulfonyl fluoride (7.37 mL, 9.80 g, 61.2 mmol, 1.0 equiv) (Note 28) and anhydrous tetrahydrofuran (40 mL) (Note 25) are added successively via syringe. The flask is cooled in a dry ice-acetone bath (–78 °C) and the stirred contents are treated with the KHMDS solution in tetrahydrofuran prepared above (Note 29), which is added dropwise via cannula (Note 30) over 30 min. During the course of the addition, the reaction mixture becomes brownish, cloudy, and viscous. After 30 min at –78 °C, the reaction mixture is quenched by transfer via cannula over 30 min to another 500-mL round-bottomed flask maintained under a nitrogen atmosphere containing a stirred solution of 4M HCl (185 mL) (Note 31). The resultant mixture appears as a single opaque layer and is extracted with methylene chloride in a 500 mL separatory funnel (5 × 60 mL). The combined organic layer is washed with brine (50 mL), dried over magnesium sulfate (ca. 15 g, 15 min) (Note 11), and filtered. The filtrate is concentrated via rotary evaporation (23 °C bath temp, 2 mmHg) and further dried under vacuum (room temperature, ca. 0.2–0.3 mmHg) to afford crude α-fluorobis(phenylsulfonyl)methane (2) as a colorless solid (18.28 g, 95%).

Examination by ^{1}H NMR and ^{19}F NMR spectroscopy shows the crude product (2) to be satisfactory for most preparative purposes (>98% purity) (Note 32). Compound 2 can be further purified by recrystallization in

132

methylene chloride and hexanes as follows. The crude product is placed in a 250-mL round-bottomed flask equipped with a stir bar and a reflux condenser. Methylene chloride (35 mL) is added and the mixture is heated to reflux to dissolve the product. Hexanes (ca. 30 mL) are slowly added portion-wise through the top of the condenser, while maintaining the reflux. The solution is slowly cooled to room temperature. The solution is then transferred to a refrigerator set to 5 °C and held for 2 h. The resulting white precipitate is collected on a Büchner funnel, rinsed with 25 mL cold methylene chloride/hexanes (1:1, v/v; 0 °C) and allowed to air dry on the funnel for 15 min and then placed on a vacuum line for 15 min (rt, 0.2–0.3 mmHg) to render 14.04 g of **2**. The mother liquor is further concentrated to approximately one-half volume. An additional 20 mL of cold hexanes (0 °C) is then added causing the solution to become cloudy. After 15 minutes without stirring an additional 2.51 g of FBSM is isolated by the above mentioned procedure (combined yield 16.55 g, 85%) (Notes 32 and 33).

2. Notes

1. Fluoromethyl phenyl sulfone can be purchased from TCI America. Alternatively, the compound can be prepared according to a reported procedure.[2] Fluoromethyl phenyl sulfide was prepared according to a known protocol.[3] The oxidation of fluoromethyl phenyl sulfide is slightly modified from this procedure.

2. Spray-dried potassium fluoride (99%) was purchased from Sigma-Aldrich and used as received. It was stored in a desiccator between uses.

3. 18-Crown-6 (99%) was purchased (Sigma-Aldrich) and used as received.

4. Acetonitrile (Optima grade) was purchased from Fisher Scientific and was dried by passing through an alumina column, as part of an Innovative Technologies PureSolv system.

5. Chloromethyl phenyl sulfide is commercially available (Aldrich Chemical Company, Inc.). It can be prepared from thiophenol according to an *Organic Syntheses* procedure.[4a] The checkers prepared this compound from thioanisole using a simpler, alternate procedure that had been used in their lab. A very similar procedure has been described by Marko *et al.*, as follows:[4c] A 500-mL three-necked round-bottomed flask equipped with a magnetic stir bar (37.5 x 10 mm, octagonal) is charged with thioanisole

(35.2 mL, 0.30 mol) and anhydrous methylene chloride (230 mL), added sequentially via syringe. The flask is fitted with two rubber septa and a reflux condenser. To quench the HCl (g) generated through the chlorination, a Tygon® tube is affixed to the top of the condenser, and the end of the tubing is submerged in an Erlenmeyer flask containing 500 mL of 2M aqueous NaOH. The reaction is placed under a positive pressure of nitrogen via a needle connected to a nitrogen vacuum manifold, then heated to reflux in an oil bath while stirring (50 °C, bath temp). When the reaction reaches a steady reflux, a solution of sulfuryl chloride (24.1 mL, 0.33 mol, 1.1 equiv) in methylene chloride (70 mL) is added over 1 h via cannula. The reaction is refluxed for 2 h, removed from the oil bath and allowed to cool to room temperature. The reaction mixture is then carefully diluted with water (100 mL) and transferred to a 500-mL separatory funnel. The organic phase is separated and then washed with water (3 x 75 mL) and brine (50 mL) to give a pale pink translucent solution that is dried over magnesium sulfate (15 g, 15 min). The drying agent is removed by filtration and the filtrate is concentrated by rotary evaporation (23 °C, 2 mmHg) to give the crude product as an oil (44.30 g, 93%). The product is further purified using fractional vacuum distillation at (0.2–0.3 mmHg). A small amount of starting material is collected in the first fraction (60–63 °C at distillation head; 110 °C bath temp) and this is followed by the product (87–91 °C at the distillation head; 140 °C bath temp). The distilled product (25.37 g, 53.3%) was determined to be pure enough for the following reactions (>99.7% by ^{1}H NMR). Additional fractions from the distillation contained product (14.48 g, 30.4%) that was deemed insufficiently pure (95.0% by ^{1}H NMR) for use in the present sequence of reactions. It should be noted that impurities in the starting material are problematic in subsequent reactions.

6. The reaction mixture gradually turns brownish during the course of the reaction, and a large amount of white solid precipitates onto the wall of the flask.

7. The reaction may be monitored with the addition of α,α,α-trifluorotoluene (0.49 mL, 0.59 g, 4.0 mmol) as an internal standard. A small portion of the reaction mixture (the solution, ca. 0.5 mL) is withdrawn via syringe and monitored via ^{19}F NMR spectroscopy. This method provides an approximate assessment of the reactions progress (yield +/- 10%). α,α,α-Trifluorotoluene (≥99%) was purchased from Sigma-Aldrich and used as received.

8. Distilled water was used.

Org. Synth. **2013**, *90*, 130-144

9. Complete extraction was monitored by spotting a drop of the extract on a TLC plate and checking with a UV lamp to see the presence of the product.

10. Methylene chloride (Optima grade) was purchased from Fisher Scientific and was dried by passing through an alumina column, as part of an Innovative Technologies PureSolv system.

11. Anhydrous magnesium sulfate was purchased from Fisher Scientific and used as received.

12. The submitters carried this reaction out on a 5x scale and used an overhead mechanical stirrer.

13. Oxone® (potassium peroxymonosulfate, $2KHSO_5 \cdot KHSO_4 \cdot K_2SO_4$) was purchased from Sigma-Aldrich and used as received.

14. Methanol (HPLC grade) was purchased from Fisher Scientific and used as received.

15. The reaction is moderately exothermic. A fast addition of fluoromethyl phenyl sulfide leads to an increase in temperature of the reaction mixture.

16. The solution was placed on the rotary evaporator for 20 min (45 °C, 2 mmHg).

17. The presence of large quantities of insoluble solids (assumed to be from Oxone®) must be removed to avoid complications with the extraction of fluoromethyl phenyl sulfone.

18. The emulsion that forms during the extraction fully or partially resolves after 5-10 min. Subsequent extractions produce less emulsion.

19. The filtration through silica gel removes impurities such as residual 18-crown-6.

20. Hexanes (ACS reagent grade) were purchased from Fisher Scientific and used as received.

21. If pure fluoromethyl phenyl sulfone is not obtained at this point it is likely contaminated with methyl phenyl sulfone. TLC analysis: 2:1, Hex:EtOAc: $R_f = 0.40$ for fluoromethyl phenyl sulfone, $R_f = 0.18$ for methyl phenyl sulfone.

22. Fluoromethyl phenyl sulfone has the following physical and spectroscopic properties: mp 51–52 °C; ^1H NMR (500 MHz, CDCl$_3$) δ: 5.16 (d, $J = 47$ Hz, 2 H), 7.62 (t, $J = 8.3$ Hz, 2 H), 7.74 (tt, $J = 7.4$, 1.2 Hz, 1 H), 7.96 (d, $J = 7.3$ Hz, 2 H). ^{13}C NMR (125 MHz, CDCl$_3$) δ: 92.0 (d, $J = 217.5$ Hz), 129.0, 129.6, 134.9. ^{19}F NMR (500 MHz, CDCl$_3$) δ: –210.0 (td, $J = 50.0$, 2.25 Hz); IR (KBr) 3013 (w), 2950 (w), 1587 (w), 1447 (s), 1343 (s),

1314 (s), 1220 (m), 1155 (s), 1053 (s), 937 (m), 790 (s), 751 (s), 683 (s), 556 (s), 527 (s) cm^{-1}; Anal. Calcd for $C_7H_7FO_2S$: C, 48.27; H, 4.05. Found: C, 48.28; H, 3.89. The spectral data are in agreement with the reported values.[2]

23. Potassium hydride (30 wt% in oil) was purchased from Sigma-Aldrich Inc. Potassium hydride (50 wt% in paraffin) may also be used. Potassium hydride was used as received, the oil suspension was stirred prior to use with a dry glass rod to ensure the suspension was even. The suspension was then transferred to a tared receiving flask via a wide-tipped pipette. After the desired amount of potassium hydride had been transferred the flask was sealed with a rubber septum. All contaminated glassware was carefully quenched with isopropyl alcohol in a fume hood.

24. Hexanes (95%, anhydrous grade) was purchased from Sigma-Aldrich and used as received.

25. Tetrahydrofuran (Optima grade) was purchased from Fisher Scientific and dried on an alumina column as part of an Innovative Technologies PureSolv system. Dried THF was transferred from the solvent system to the reaction flask by syringe.

26. Hexamethyldisilazane (>96%) was purchased from Alfa Aesar and used as received.

27. More hydrogen evolution may occur on warming to room temperature and continue for several minutes. The solution should not be used until hydrogen evolution has stopped for at least 10 min.[5]

28. Benzenesulfonyl fluoride can be purchased from Aldrich Chemical Company, Inc. Access to high-purity benzenesulfonyl fluoride (>99%) is required for clean formation of the final product. As this compound is a liquid at room temperature, the known method for its synthesis, which calls for purification by recrystallization, was modified as follows:[6] A 500-mL round-bottomed flask equipped with a magnetic stirring bar (50 mm x 8 mm, octagonal) is charged with benzenesulfonyl chloride (39.0 mL, 53.90 g, 0.3 mol, 1.0 equiv), potassium fluoride (22.70 g, 0.39 mol, 1.3 equiv) and 18-crown-6 (3.96 g, 15 mmol, 0.05 equiv). The flask is then sealed with a rubber septum and connected via a syringe needle to nitrogen/vacuum line. The flask is evacuated/flushed with nitrogen three times, then acetonitrile (300 mL) is added via syringe and the mixture is stirred at room temperature for 24 h. The reaction mixture is then diluted with water (150 mL) and transferred to a 1L separatory funnel and extracted with diethyl ether (3 x 75 mL). The combined organic layer is washed with brine (30 mL) and dried over magnesium sulfate (15 g, 15 min). The drying agent is removed

136

by filtration and the solvent is removed by rotary evaporation (23 °C, 2 mmHg) to give a clear low-viscosity liquid. This product is dissolved in hexanes (50 mL) and washed with HCl solution (1 N, 5 x 20 mL) to remove residual 18-crown-6. The organic phase is dried over magnesium sulfate (15 g, 15 min), the drying agent is removed by filtration and the filtrate is concentrated by rotary evaporation (23 °C, 2 mmHg) and then further under high vacuum (room temp, 0.2–0.3 mmHg) to give benzenesulfonyl fluoride (28.2 mL, 37.50 g, 78 % yield).

29. Solid potassium bis(trimethylsilyl)amide (KHMDS, 95%) can also be purchased from Aldrich Chemical Company, Inc. and formulated into a 1M solution in THF, which can be employed to render similar results.

30. FBSM (**2**) is more acidic than fluoromethyl phenyl sulfone (**1**). The FBSM generated under the reaction conditions readily undergoes deprotonation, which consumes an extra equivalent of base. Thus, the reaction theoretically requires 2 equivalents of base. Employment of less than 2.5 equivalents of KHMDS resulted in an incomplete reaction.

31. HCl (12.1 N, ACS Plus) was purchased from Fisher Scientific.

32. Bis(phenylsulfonyl)methane was identified as the major impurity based on the characteristic signal of the methylene appearing in the ^{1}H NMR spectroscopy in $CDCl_3$ (δ = 4.73 ppm). According to the ^{1}H NMR integration, the amount of bis(phenylsulfonyl)methane was ca. 2 wt%.

33. α-Fluorobis(phenylsulfonyl)methane has the following physical and spectroscopic properties: mp 106.5–107.0 °C. ^{1}H NMR (500 MHz, $CDCl_3$) δ: 5.81 (d, J = 45.7 Hz, 1 H), 7.60 (t, J = 7.8 Hz, 4 H), 7.76 (t, J = 7.2 Hz, 2 H), 7.98 (d, J = 7.9 Hz, 4 H). ^{13}C NMR (125 MHz, $CDCl_3$) δ: 105.6 (d, J = 264 Hz), 129.6, 130.2, 135.3, 135.8. ^{19}F NMR (500 MHz, $CDCl_3$) δ: –167.4 (d, J = 48.6 Hz); IR (KBr) 3096 (w), 3071 (w), 2955 (m), 1581 (m), 1450 (s), 1358 (s), 1172 (s), 1077 (s), 797 (s), 683 (s), 533 (s), 520 (s) cm^{-1}; HRMS for $(C_{13}H_{11}FO_4S_2)Na^+$: Calcd 336.997500; Found 336.997129. Anal. Calcd for $C_{13}H_{11}FO_4S_2$: C, 49.67; H, 3.53. Found: C, 49.39; H, 3.45. The spectral data are in agreement with the reported values.[7,8]

Handling and Disposal of Hazardous Chemicals

The procedure in this article is intended for use only by persons with prior training in experimental organic chemistry. All hazardous materials should be handled using the standard procedures for work with chemicals described in references such as "Prudent Practices in the Laboratory" (The National Academies Press, Washington, D.C., 2011 www.nap.edu). All chemical waste should be disposed of in accordance with local regulations. For general guidelines for the management of chemical waste, see Chapter 8 of Prudent Practices.

The procedure must be conducted at one's own risk. *Organic Syntheses, Inc.*, its Editors, and its Board of Directors do not warrant or guarantee the safety of individuals using this procedure and hereby disclaim any liability for any injuries or damages claimed to have resulted from or related in any way to the procedures herein.

3. Discussion

Fluorinated organic compounds have received increasing interest in recent years due to their unique biological and physicochemical properties. Other than fluorinations of organic compounds via C-F bond forming reactions, the incorporation of fluorinated motifs using various fluoroalkylating reagents is of particular importance because of their synthetic advantages.[9] Among these reagents, the title compound, FBSM (2), has been developed as a versatile nucleophilic monofluoromethylating reagent. Notably, owing to the presence of the two phenylsulfonyl groups, FBSM possesses superior acidity than fluoromethyl phenyl sulfone (1) and can undergo feasible deprotonation to render a rather stable α-fluorocarbanion.[10] Thus, a variety of nucleophilic monofluoromethylation reactions have been achieved using FBSM, such as the ring-opening of epoxides and aziridines,[7] the allylic monofluoromethylation reaction,[8] the Mitsunobu reaction,[11] conjugate addition reactions,[12] the Mannich reaction,[13] the aldol reaction,[14] as well as many others (Scheme 1).[15] In particular, the facile reductive removal of the sulfonyl groups allows for the introduction of the unfunctionalized CH_2F motif using FBSM, thereby prevailing over many other monofluoromethylating reagents. In addition, FBSM can be further converted to fluoroiodobis(phenylsulfonyl)methane, which has been utilized as a viable radical monofluoromethylating reagent.[16]

Although extensively utilized in nucleophilic monofluoromethylation reactions, the synthetic approaches toward FBSM had been rather limited since the initial documentation of this compound. FBSM was originally synthesized in 49–60% yields through the reaction between Selectfluor® and bis(phenylsulfonyl)methide anion.[7,8] An alternative method was later

138

achieved via the electrochemical reaction of phenyl phenylsulfonylmethyl sulfide using tetraethylammonium fluoride-hydrogen fluoride mixture as the fluorine source.[17] The afforded product, α-fluoro-α-phenylthiomethyl phenyl sulfonyl, was further oxidized to generate FBSM in 44% yield in two steps. However, the major problems of the synthetic routes through the C-F bond formation are a) the fluorine sources are expensive and/or hazardous; b) the reactions can only afford the products in moderate yield (44–60%); c) the selectivity of the reactions is unsatisfactory due to the incomplete consumption of bis(phenylsulfonyl)methane and the formation of difluorinated product; d) the purification of the crude product necessitates the use of chromatography, which significantly limits the scalability of these methods.

Scheme 1. Synthetic Applications of FBSM (**2**)

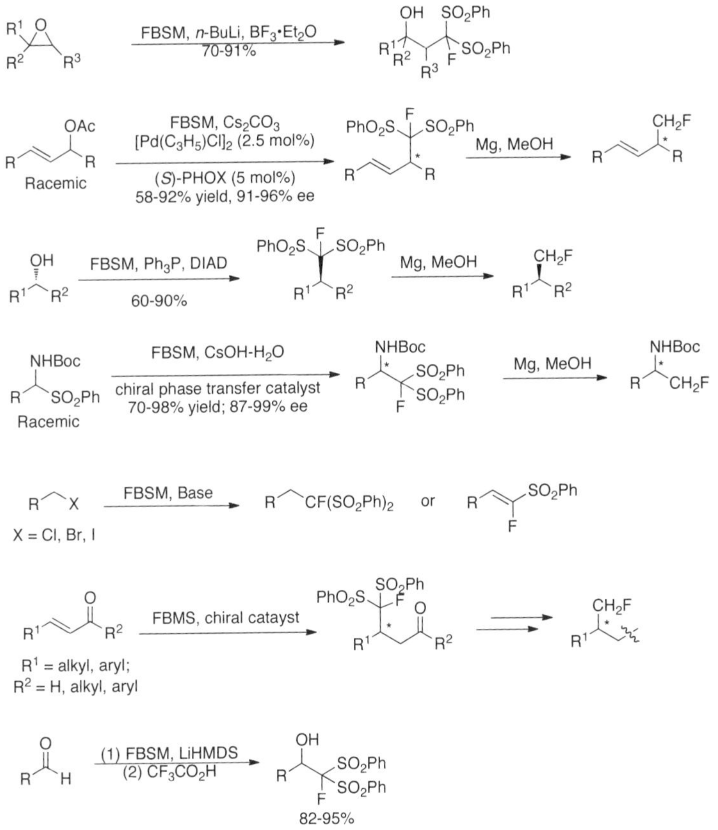

Lately, instead of applying the C-F bond forming strategy, an improved synthetic protocol has been reported by treating fluoromethyl phenyl sulfone with methyl benzenesulfinate followed by the oxidation of the intermediate. Facilitated by the C-S bond formation reaction, such a methodology has been demonstrated with remarkable selectivity, and affords FBSM with high purity and excellent yield without sophisticated purification processes. Although this method triumphs over the previous approaches, its practicality is somewhat diminished due to the employment of less available sulfinate, which also introduces an additional step in the preparation.

Inspired by the C-S bond forming strategy, our laboratory has achieved the aforementioned preparative-scale method[18] using readily available starting materials to yield FBSM with excellent yield and high purity.

Appendix
Chemical Abstracts Nomenclature; (Registry Number)

Chloromethyl phenyl sulfide: Benzene, [(chloromethyl)thio]-; (7205-91-6)

Potassium fluoride (KF); (7789-23-3)

Fluoromethyl phenyl sulfide: Benzene, [(fluoromethyl)thio]-; (60839-94-3)

α,α,α-Trifluorotoluene: Benzene, (trifluoromethyl)-; (98-08-8)

18-Crown-6: 1,4,7,10,13,16-Hexaoxacyclooctadecane; (17455-13-9)

Oxone: Potassium peroxymonosulfate sulfate ($K_5(HSO_5)_2(HSO_4)(SO_4)$); (70693-62-8)

Fluoromethyl phenyl sulfone: Benzene, [(fluoromethyl)sulfonyl]-; (20808-12-2)

Potassium hydride (KH); (7693-26-7)

Hexamethyldisilazane: Silanamine, 1,1,1-trimethyl-N-(trimethylsilyl)-; (999-97-3)

Potassium bis(trimethylsilyl)amide: Silanamine, 1,1,1-trimethyl-N-(trimethylsilyl)-, potassium salt (1:1); (40949-94-8)

Benzenesulfonyl fluoride; (368-43-4)

α-Fluorobis(phenylsulfonyl)methane (FBSM): Benzene, 1,1'-[(fluoromethylene)bis(sulfonyl)]bis-; (910650-82-7)

1. Loker Hydrocarbon Research Institute, University of Southern California, Los Angeles, CA 90089-1919, USA.
2. McCarthy J. R.; Matthews, D. P.; Paolini J. P. *Org. Synth.* **1995**, *72*, 209.
3. More, K. M.; Wemple, J. *Synthesis* **1977**, 791–792.
4. (a) Enders, D.; Berg, S.; Jandeleit B. *Org. Synth.* **2002**, *78*, 169. (b) Truce, W. E.; Birum, G. H.; McBee, E. T. *J. Am. Chem. Soc.* **1952**, *74*, 3594–3599. (c) Quinet, C.; Sampoux, L.; Markó, I. E. *Eur. J. Org. Chem.* **2009**, *11*, 1806–1811.
5. Brown, C. A. *J. Org. Chem.* **1974**, *39*, 3913–3918.
6. Lee, I; Shim, C. S.; Chung, S. Y.; Kim, H. Y.; Lee, H. W. *J. Chem. Soc., Perkin Trans. II* **1988**, 1919-1923.
7. Ni, C.; Li, Y.; Hu, J. *J. Org. Chem.* **2006**, *71*, 6829–6833.
8. Fukuzumi, T.; Shibata, N.; Sugiura, M.; Yasui, H.; Nakamura, S.; Toru, T. *Angew. Chem. Int. Ed.* **2006**, *45*, 4973–4977.
9. Selected reviews on fluoroalkylations: (a) Burton, D. J.; Yang, Z.-Y. *Tetrahedron* **1992**, *48*, 189–275. (b) McClinton, M. A.; McClinton, D. A. *Tetrahedron* **1992**, *48*, 6555–6666. (c) Dolbier Jr., W. R.; *Chem. Rev.* **1996**, *96*, 1557–1584. (d) Umemoto, T. *Chem. Rev.* **1996**, *96*, 1757–1778. (e) Prakash, G. K. S.; Yudin, A. K. *Chem. Rev.* **1997**, *97*, 757–786. (f) Singh, R. P.; Shreeve, J. M. *Tetrahedron* **2000**, *56*, 7613–7632; (g) Prakash, G. K. S.; Mandal, M. *J. Fluorine Chem.* **2001**, *112*, 123–131; (h) Langlois, B. R.; Billard, T. *Synthesis* **2003**, 185–194. (i) Prakash, G. K. S.; Beier, P. *Angew. Chem. Int. Ed.* **2006**, *45*, 2172–2174; (j) Prakash, G. K. S.; Hu, J. *Acc. Chem. Res.* **2007**, *40*, 921–930. (k) Brunet, V. A.; O'Hagan, D.; *Angew. Chem. Int. Ed.* **2008**, *47*, 1179–1182. (l) Ma, J.-A.; Cahard, D. *Chem. Rev.* **2008**, *108*, PR1–PR43. (m) Shibata, N.; Mizuta, S.; Kawai, H. *Tetrahedron: Asymmetry* **2008**, *19*, 2633–2644. (n) Hu, J. *J. Fluorine Chem.* **2009**, *130*, 1130–1139. (o) Cahard, D.; Xu, X. ; Couve-Bonnaire, S.; Pannecoucke, X. *Chem. Soc. Rev.* **2010**, *39*, 558–568. (p) Tomashenko O. A.; Grushin V. V. *Chem. Rev.* **2010**, *110*, 4475–4521.
10. Prakash, G. K. S.; Wang, F.; Shao, N.; Mathew, T.; Rasul, G.; Haiges, R.; Steward, T.; Olah, G. A. *Angew. Chem. Int. Ed.* **2009**, *48*, 5358–5362.
11. Prakash, G. K. S.; Chacko, S.; Alconcel, S.; Stewart, T.; Mathew, T.; Olah, G. A. *Angew. Chem. Int. Ed.* **2007**, *46*, 4933–4936.

12. (a) C. Ni; L. Zhang; J. Hu, *J. Org. Chem.* **2008**, *73*, 5699–5713. (b) Prakash, G. K. S.; Zhao, X. ; Chacko, S. ; Wang, F.; Vaghoo, H.; Olah, G. A. *Beilstein J. Org. Chem.* **2008**, *4*, No. 17. (c) Moon, H. W.; Cho, M. J.; Kim, D. Y. *Tetrahedron Lett.* **2009**, *50*, 4896–4898. (d) Alba, A.-N.; Companyó, X.; Moyano, A.; Rios, R. *Chem. Eur. J.* **2009**, *15*, 7035–7038; (e) Zhang, S.; Zhang, Y.; Ji, Y.; Li, H.; Wang, W. *Chem. Commun.* **2009**, 4886–4888.

13. Mizuta, S.; Shibata, N.; Goto, Y.; Furukawa, T.; Nakamura, S.; Toru, T. *J. Am. Chem. Soc.* **2007**, *129*, 6394–6395.

14. Shen, X.; Zhang, L.; Zhao, Y.; Zhu, L.; Li, G.; Hu, J. *Angew. Chem. Int. Ed.* **2011**, *50*, 2588–2592.

15. Excellent review articles on the synthetic applications of FBSM: (a) Prakash, G. K. S.; Chacko, S. *Curr. Opin. Drug Discov. Dev.* **2008**, *11*, 793–802. (b) Hu, J.; Zhang, W.; Wang, F. *Chem. Commun.* **2009**, 7465-7478. (c) Ni, C.; Hu, J. *Synlett* **2011**, 770–782. (d) Vallero, G.; Companyo, X.; Rios, R. *Chem. Eur. J.* **2011**, *17*, 2018–2037.

16. Prakash, G. K. S.; Ledneczki, I.; Chacko, S.; Ravi, S.; Olah, G. A. *J. Fluorine Chem.* **2008**, *129*, 1036 -1040.

17. Nagura, H.; Fuchigami, T. *Synlett* **2008**, 1714–1718;

18. Prakash, G. K. S.; Wang, F.; Ni, C.; Thomas, T.J.; Olah, G. A. *J. Fluorine Chem.* **2010**, *131*, 1007–1012.

G. K. Surya Prakash, born 1953 in India, earned a Bachelor's degree from Bangalore University and a Master's degree from the Indian Institute of Technology. Prakash joined George Olah's group at Case Western Reserve University to pursue graduate work in 1974. He moved with Olah to the University of Southern California (USC) in 1977 and obtained his Ph.D. in Physical Organic Chemistry in 1978. He joined the faculty of USC in 1981 and is currently a Professor and the holder of the Olah Nobel Laureate Chair in Hydrocarbon Chemistry and serves as the Director of the Loker Hydrocarbon Research Institute.

142

Nan Shao was born 1979 in Shanghai, China. He received his B.S. degree in biochemical engineering at the East China University of Science and Technology in 2001. After earning his M.S. degree in chemistry at the University of Minnesota Duluth in 2006, he started to pursue his Ph. D. in the laboratory of Professor G. K. Surya Prakash and Professor George A. Olah at the University of Southern California, where he has conducted research involving the development of novel fluoromethylating reagents and fluoromethylation protocols.

Fang Wang was born in 1983 in Shenyang, China. He received a B.S. degree in chemistry in 2006 from the Zhejiang University, China, where he began undergraduate research with Professor Ping Lu. While there, he also attended the Shanghai Institute of Organic Chemistry as a visiting student performing research under the guidance of Professor Jinbo Hu. Since 2006, he has been pursuing his doctoral degree in the laboratory of Professor G. K. Surya Prakash and Professor George A. Olah at the University of Southern California. His research focuses on asymmetric fluoromethylations and the related mechanistic studies.

Chuanfa Ni was born in Shandong, P. R. China in 1982. He obtained his B.S. degree in chemistry from Shandong Normal University in 2003. In the same year, he entered Shanghai Institute of Organic Chemistry, Chinese Academy of Sciences as a graduate student under the supervision of Professor Jinbo Hu. After receiving his Ph.D. degree in 2009, he moved to the University of Southern California. Currently, he is conducting post-doctoral research supervised by Professor G. K. Surya Prakash in the field of fluorine chemistry.

Thomas D. Montgomery was born in 1988 in Maryland, USA. He received a B.A. degree in chemistry and biology in 2010 from St. Mary's College of Maryland where he performed undergraduate research with Professor Andrew Koch. Since the fall of 2010 he has been pursuing his doctoral degree in Professor Viresh H. Rawal's lab at the University of Chicago. His research has focused on metal catalyzed functionalization of substituted indole cores and related compounds.

144

Direct Conversion of Benzylic and Allylic Alcohols to Diethyl Phosphonates

Submitted by Rebekah M. Richardson and David F. Wiemer.[1]
Checked by Laura S. Kocsis and Kay M. Brummond.

1. Procedure

A flame-dried 100-mL, two-necked, round-bottomed flask equipped with a cylindrical Teflon-coated magnetic stir bar (1.5 cm x 0.7 cm), a rubber septum, and a reflux condenser is maintained under an atmosphere of argon during the course of the reaction. The flask is charged with zinc iodide (12.2 g, 38.2 mmol, 1.2 equiv) (Note 1) and is evacuated and refilled with argon three times. Tetrahydrofuran (35 mL) (Note 2) is added by syringe to provide a white liquid with undissolved solid still present. Triethyl phosphite (8.2 mL, 48 mmol, 1.5 equiv) (Note 3) is added using a 5-mL disposable syringe, and the solution becomes colorless with a small quantity of solid still present. Benzyl alcohol (3.3 mL, 32 mmol, 1.0 equiv) (Note 4) is added using a disposable syringe. The reaction mixture is heated at reflux for 16 h (oil bath, 75 °C) (Note 5), at which time the reaction turns yellow and clear. The solution is allowed to cool for 10 min, slowly placed under vacuum, and concentrated at 2 mmHg for 5.5 h to remove volatiles (Note 6). The residue is then transferred to a 500-mL separatory funnel with diethyl ether (2 x 10 mL) and 2 N NaOH (2 x 10 mL), and diluted with additional diethyl ether (90 mL). The organic phase is washed with 2 N NaOH (350 mL) (Note 7), the aqueous phase is extracted with diethyl ether (3 x 125 mL), and the combined organic phase is dried over 12 g of MgSO$_4$. The solution is then filtered through a bed of Celite® (24 g) in a 350-mL coarse porosity sintered glass funnel, washed with diethyl ether (3 x 50 mL), and the filtrate is concentrated by rotary evaporation (23 °C, 10 mmHg) to afford a yellow oil. After transfer to a 25-mL round-bottomed flask equipped with a magnetic stir bar (Note 8), the product is purified via a short-path distillation with an 8-cm head under vacuum (0.5 mmHg) and collected with

Org. Synth. **2012**, *90*, 145-152
Published on the Web 11/6/2012

a fraction cutter or "cow" apparatus. A forerun of 1.2–1.6 g is collected, followed by product **2** distilling at 115–122 °C (0.5 mmHg) (Note 9) as a colorless oil (5.11 g, 70%) (Notes 10, 11, and 12).

2. Notes

1. ZnI_2 (≥98%) was obtained from Sigma-Aldrich and used as received. The bottle was purged with argon after each use due to the hygroscopic nature of the chemical.

2.   Anhydrous 99.9%, inhibitor free tetrahydrofuran was obtained from Sigma-Aldrich and purified with alumina using the Sol-Tek ST-002 solvent purification system directly prior to use. The submitters report that anhydrous tetrahydrofuran (≥99% containing 250 ppm BHT as inhibitor) was obtained from Sigma-Aldrich and distilled from sodium benzophenone ketyl immediately prior to use.

3.   Triethyl phosphite (98%) was obtained from Sigma-Aldrich and used as received. The bottle was purged with argon after each use to minimize air oxidation.

4.   Benzyl alcohol (≥99%) was obtained from Sigma-Aldrich and used as received.

5.   A longer reaction time is not detrimental to the reaction yield. The submitters performed the reaction for 12 h.

6.   The vacuum was attached at the head of the condenser in the round-bottomed flask. Occasional bumping occurs, and the condenser traps any material that bumps so that the residue can later be transferred to the separatory funnel for extraction, rather than being lost in the vacuum line. Alternatively, a rotary evaporator can be used to remove volatile materials, as long as care is taken to minimize oxidation of excess triethyl phosphite (bp ~156 °C) to the less volatile triethyl phosphate (bp ~215 °C).

7.   The initial washing with 2 N NaOH causes the aqueous phase to become cloudy with white precipitate. After dilution with 90 mL diethyl ether and 350 mL 2 N NaOH, the aqueous layer is colorless and slightly cloudy. After the first ether extraction, there is no visible precipitate in the aqueous layer.

8.   The submitters report that if left sitting overnight, the initial colorless oil sometimes turns yellow. The checkers only observed a crude yellow oil.

146

9. The temperature of the oil bath was increased by 25 °C increments from 50–150 °C every ten minutes to ensure collection of pure product. The forerun was collected from 100–105 °C. The submitters report that at pressures of 0.1 mmHg and 0.15 mmHg, the product was obtained distilling at 100–105 °C, and at pressures of 0.25 mmHg the product was obtained distilling at 108–115 °C.

10. Diethyl benzylphosphonate **2** has the following physical and spectroscopic properties: TLC analysis was performed on SiliCycle Inc. SiliaPlate silica gel plates (w/ UV254, glass-backed, 250 μm) and was visualized as a blue spot with cerium molybdate followed by heating. R_f = 0.23 in 70% EtOAc in hexane; IR (neat) 2983, 1247, 1027 cm^{-1}; LRMS (TOF MSMS ES+) m/z (%): 229 (28), 201 (38), 173 (100); HRMS (TOF MS ES+) [M+H]+ calcd for $C_{11}H_{18}O_3P$, 229.0994; found, 229.1004; ^{1}H NMR (400 MHz, CDCl$_3$) δ: 1.23 (t, J_{HH} = 7.2 Hz, 6 H), 3.14 (d, J_{HP} = 21.6 Hz, 2 H), 3.96–4.04 (m, 4 H), 7.24–7.30 (m, 5 H); ^{13}C NMR (100 MHz, CDCl$_3$) δ: 16.2 (d, J_{CP} = 6.0 Hz, 2C), 33.7 (d, J_{CP} = 137.0 Hz), 62.0 (d, J_{CP} = 7.0 Hz, 2C), 126.7 (d, J_{CP} = 3.0 Hz), 128.4 (d, J_{CP} = 3.0 Hz, 2C), 129.7 (d, J_{CP} = 6.0 Hz, 2C), 131.5 (d, J_{CP} = 8.0 Hz); ^{31}P (121 MHz, CDCl$_3$) δ: 26.4.

11. A yield of 70% was obtained on full scale, while a yield of 65% was obtained on half scale. The submitters report a 76% yield when reaction is performed on full scale. On small scale, products were typically purified through flash column chromatography on silica gel with yields of 80–90%.

12. The distillation gives material of ≥98% purity by ^{31}P NMR, which is sufficiently pure for most applications. To obtain an analytical sample, a portion of the distillate (0.2 mL) was purified by flash column chromatography [7.3 g silica gel, 2.5 x 45 cm glass column, diethyl ether-pentane (8:2) as eluent (Note 13)] to give 0.193 g (90% yield) of compound **2** as a colorless oil.

13. Standard grade silica gel (230 x 400 mesh) was obtained from Sorbent Technologies and used as received. In a typical purification, collecting 9-mL fractions, compound **2** would elute in fractions 11-27 as the pure product. The submitters reported that compound **2** elutes in fraction 4 as a slight mixture and fractions 5-11 as the pure product when ethyl acetate – hexanes (7:3) is used as eluent. The combined organic fractions were concentrated by rotary evaporation (23 °C, 10 mmHg) and then at 2 mmHg. Analysis by HPLC showed a single peak with a retention time of 17.6 min using 70% ethyl acetate-hexanes as the eluent and a flow rate of 4 mL/min.

HPLC experiments were conducted using a Varian Dynamax column (250 x 10.0 mm) and Varian UV/Vis detector operating at 254 nm.

Handling and Disposal of Hazardous Chemicals

The procedures in this article are intended for use only by persons with prior training in experimental organic chemistry. All hazardous materials should be handled using the standard procedures for work with chemicals described in references such as "Prudent Practices in the Laboratory" (The National Academies Press, Washington, D.C., 2011 www.nap.edu). All chemical waste should be disposed of in accordance with local regulations. For general guidelines for the management of chemical waste, see Chapter 8 of Prudent Practices.

These procedures must be conducted at one's own risk. *Organic Syntheses, Inc.*, its Editors, and its Board of Directors do not warrant or guarantee the safety of individuals using these procedures and hereby disclaim any liability for any injuries or damages claimed to have resulted from or related in any way to the procedures herein.

3. Discussion

Activated phosphonate esters serve as reagents in Horner-Wadsworth-Emmons reactions. Their synthesis often relies upon a classic three-step protocol involving formation of the mesylate from an alcohol, conversion to the corresponding halide, and a final Michaelis-Arbuzov reaction.[2-4] In some cases, this sequence has proven problematic[5] or difficult to conduct on a larger scale. This method allows for an inexpensive, one-flask procedure for conversion of benzylic and allylic alcohols to the corresponding phosphonates through treatment with triethyl phosphite and ZnI_2.[6,7]

Optimal reaction conditions can be slightly modified for each substrate with solvents including tetrahydrofuran, N,N-dimethylformamide, and toluene, and temperatures of 66–140 °C. The best yields often were obtained when the zinc-mediated reaction was conducted at 66 °C, an improvement upon the generally high temperatures required for the Michaelis-Arbuzov reaction.[3]

The reaction conditions are applicable to benzylic alcohols with varying electron donating or withdrawing substituents, and to furan systems (Table 1). Primary and secondary allylic alcohols (Table 2) also were

148

converted to phosphonates under parallel reaction conditions. As demonstrated in Table 2, tertiary allylic alcohols appear to undergo rearrangement to the primary allylic phosphonate. Application of this protocol to aliphatic systems led to mixed ester phosphonate and hydrogen phosphonate products, demonstrating the importance of a stabilized carbocation.

Table 1. Preparation of Benzyl Phosphonates[a]

Entry	Starting Alcohol	Rxn Temp °C	Solvent	Purified[b] Yield (%)
1[c]		66	THF	76
2		110	Toluene	69
3		140	Toluene	71
4		66	THF	85
5		66	THF	81
6		66	THF	57[d]
7		140	Toluene	64

[a]Reactions were carried out overnight using 3.0 equivalents of P(OEt)$_3$ and 1.5 equivalents of ZnI$_2$.[6b] [b]Products were purified through flash column chromatography using varying ratios of EtOAc in hexanes. [c]Reaction was carried out using 1.5 equivalents of P(OEt)$_3$ and 1.2 equivalents of ZnI$_2$. [d]70% based on recovered starting material.

Table 2. Preparation of Allylic Phosphonates[a]

Entry	Starting Alcohol	P(OEt)$_3$ equiv	Rxn Temp °C	Purified[b] Yield (%)
1	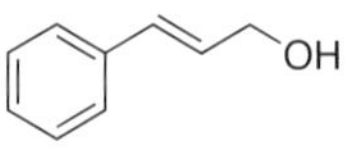	3.1	110	81
2[c]	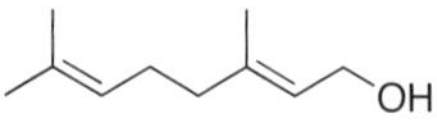	2.1	110	70
3	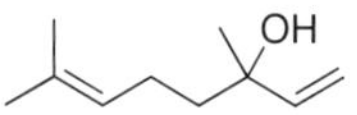	3.0	140	21[d]
4	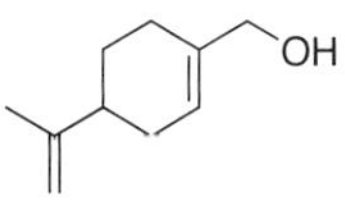	3.0	140	78

[a]Reactions were carried out overnight using 1.5 equivalents of ZnI$_2$ in toluene.[6] [b]Products were purified through flash column chromatography using varying ratios of EtOAc in hexanes. [c]Reaction was carried out using 1.1 equivalents of ZnI$_2$. [d]The tertiary alcohol rearranged to form diethyl geranylphosphonate as the product.

1. Department of Chemistry, University of Iowa, Iowa City, IA, 52242. E-mail: david-wiemer@uiowa.edu. Funding from the University of Iowa Graduate College, in the form of a Presidential Fellowship to R. M. R., and the Roy J. Carver Charitable Trust is gratefully acknowledged.
2. Maryanoff, B. E.; Reitz, A. B. *Chem. Rev.* **1989**, *89*, 863–927.
3. Bhattacharya, A. K.; Thyagarajan, G. *Chem. Rev.* **1981**, *81*, 415–430.
4. A number of other approaches to phosphonates and phosphonic acids are known. (a) For phosphonate synthesis from *o*-(hydroxymethyl)phenols: Böhmer, V.; Vogt, W.; Chafaa, S.;

Meullemeestre, J.; Schwing, M.-J.; Vierling, F. *Helv. Chim. Acta* **1993**, *76*, 139–149. (b) For phosphonate synthesis via Pd-couplings: Lavén, G.; Stawinski, J. *Synlett* **2009**, 225–228. (c) For synthesis of phosphonic acids from alcohols: Coudray, L.; Montchamp, J.-L. *Eur. J. Org. Chem.* **2008**, 4101–4103. (d) Bravo-Altamirano, K.; Montchamp, J.-L. *Tetrahedron Lett.* **2007**, *48*, 5755–5759.

5. (a) Ulrich, N. C.; Kodet, J. G.; Mente, N. R.; Kuder, C. H.; Beutler, J. A.; Hohl, R. J.; Wiemer, D. F. *Bioorg. Med. Chem.* **2010**, *18*, 1676–1683. (b) Kodet, J. G. PhD Thesis, University of Iowa, December 2010.

6. (a) Presented in part at the 240th ACS National Meeting, Boston, MA, August, 2010: Richardson, R. M.; Barney, R. J.; Wiemer, D. F. ORGN-1059. (b) Barney, R. J.; Richardson, R. M.; Wiemer, D. F. *J. Org. Chem.* **2011**, *76*, 2875–2879. (c) Richardson, R. M.; Barney, R. J.; Wiemer, D. F. *Tetrahedron Lett.* **2012**, *53*, 6682–6684.

7. Rajeshwaran, G. G.; Nandakumar, M.; Sureshbabu, R.; Mohanakrishnan, A. K. *Org. Lett.* **2011**, *13*, 1270–1273.

Appendix
Chemical Abstracts Nomenclature (Registry Number)

Zinc Iodide; (10139-47-6)
Triethyl Phosphite; (122-52-1)
Benzyl Alcohol; (100-51-6)\

David F. Wiemer was born and raised in southeastern Wisconsin. He received a B.S. degree in Chemistry from Marquette University, earned the Ph.D. degree at the University of Illinois, and was an NIH postdoctoral fellow at Cornell University. He joined the faculty in the Department of Chemistry at the University of Iowa as an assistant professor, and now holds the rank of F. Wendell Miller Professor of Chemistry. His research interests include synthetic methodology based on organophosphorus chemistry, synthesis of phosphonate analogues of isoprenoid phosphates as potential enzyme inhibitors, and synthesis of biologically active natural products, especially potential anti-cancer agents.

Rebekah Richardson was born in Peoria, IL. She graduated from Knox College with a B.A. in chemistry in 2007, where she performed undergraduate research at the USDA in Peoria. She then joined the group of Professor David F. Wiemer at the University of Iowa in 2007 as a recipient of the University of Iowa Presidential Graduate Fellowship. Her research focuses on C-P bond formation and natural product total synthesis.

Laura Kocsis was born in 1987 in Easton, Pennsylvania. In 2009, she received her B.S. degree in chemistry from Carnegie Mellon University in Pittsburgh, Pennsylvania. She is currently pursuing graduate studies at the University of Pittsburgh under the guidance of Prof. Kay Brummond. Her research currently focuses on exploring the novel synthesis and photophysical properties of functionalized cyclopenta[b]naphthalenes generated via a dehydrogenative Diels-Alder reaction.

152

Synthesis of 1-Naphthol via Oxidation of Potassium 1-Naphthyltrifluoroborate

Submitted by Gary A. Molander, Sabahat Zahra Siddiqui, and Nicolas Fleury-Brégeot.[1]
Checked by David Hughes.

1. Procedure

A. Potassium 1-Naphthyltrifluoroborate (2). A 1-L round-bottomed flask equipped with a 4-cm oval PTFE-coated magnetic stir bar and a thermocouple thermometer (Note 1) is charged with 1-naphthaleneboronic acid (22.0 g, 128 mmol, 1.0 equiv) and methanol (80 mL) (Notes 2 and 3). The solution is cooled to 5 °C using an ice-bath. A solution of KHF_2 (30.3 g dissolved in 100 mL water, 388 mmol, 3.0 equiv) is added in 5 portions over 10 min, resulting in a thick white slurry (Note 4). The ice bath is removed and the mixture is stirred for 20 min (Note 5). The stir bar is removed and the mixture is concentrated by rotary evaporation (45 °C bath temp, 100 to 20 mmHg) (Note 6). Acetonitrile (2 x 200 mL) flushes are used to azeotropically remove the remaining water, with the mixture concentrated by rotary evaporation after each flush (45 °C bath temp, 100 to 20 mmHg) (Notes 7 and 8). The flask is equipped with a 3-cm oval PTFE-coated magnetic stir bar and acetonitrile (300 mL) is added. The flask is placed in a heating mantle and the stirred mixture is heated at reflux for 90 min (Notes 9 and 10). After cooling to ambient temperature over 90 min, the resulting solids are removed by filtration through a 150-mL medium porosity sintered glass funnel and rinsed with acetonitrile (2 x 50 mL)

(Note 11). The filtrate is concentrated by rotary evaporation (55 °C bath temp, 70 mmHg) to afford potassium 1-naphthyltrifluoroborate (**2**) as a white solid (29.9 g) as a hemi-acetonitrile solvate (Note 12). The solid is heated under vacuum (70 mmHg) for 18–24 h at 90 °C to provide the de-solvated product (26.5 g, 89% yield) (Notes 13 and 14).

B. 1-Naphthol (3). A 1-L 3-necked round-bottom flask is equipped with a 3-cm oval PTFE-coated magnetic stir bar and a 500-mL addition funnel. The two side necks are sealed with septa, one of which is pierced with a thermocouple thermometer (Note 1). The flask is charged with potassium 1-naphthyltrifluoroborate (26.0 g, 111 mmol, 1.0 equiv) and acetone (300 mL) (Note 15). The mixture is stirred to dissolve all solids, then cooled to 3 °C using an ice bath. The addition funnel is charged with a solution of Oxone® (72 g dissolved in 300 mL water, 117 mmol, 1.05 equiv), which is added to the flask over 10 min (Note 16). The reaction mixture is stirred for an additional 5 min, then quenched with aq HCl (0.1 M, 360 mL, 0.3 equiv), added in one portion. The contents are transferred to a 2-L separatory funnel and extracted with dichloromethane (3 x 250 mL). The combined organic layers are washed with half-saturated aq. NaCl (200 mL), then dried by filtering through a bed of Na_2SO_4 (150 g) in a 350-mL sintered glass funnel. The filtrate is concentrated in portions in a 1-L flask by rotary evaporation (40 °C, 200 to 20 mmHg) to an orange oil, then flushed with toluene (75 mL) and concentrated (50 °C, 20 mmHg) to provide a brown wet solid (40 g). The product is recrystallized in the same flask by adding 15 mL of toluene and warming in a 60 °C water bath to dissolve all solids. A 3-cm oval PTFE-coated magnetic stir bar is added and the flask is equipped with a 150-mL addition funnel. The stirred mixture is allowed to cool to room temperature, upon which crystallization occurs. *n*-Heptane (140 mL) is added via the addition funnel over 1 h. After stirring for 4 h, the suspension is filtered through a 150-mL medium porosity sintered glass funnel, washed with 4:1 *n*-heptane:toluene (2 x 15 mL), and air-dried to afford 1-naphthol (12.0 g) as an off-white to slightly pink solid. A second crop of product is obtained as follows. The filtrate and washes from the initial crystallization are concentrated by rotary evaporation (50 °C, 20 mmHg) in a 1-L flask to afford 4 g of solids. Toluene (15 mL) is added along with a 3-cm oval PTFE-coated magnetic stir bar. The mixture is warmed in a 60 °C water bath to dissolve all solids, then allowed to cool to room temperature, upon which crystallization occurs. *n*-Heptane (60 mL) is added via a 150-mL addition funnel over 1 h, then the mixture is stirred for 14 h at room temperature. The suspension is filtered through a 150-mL medium porosity sintered glass funnel, washed with 4:1 n-heptane:toluene (2 x 10 mL), and air-dried to afford 1-naphthol as a pink solid (2.32 g). The two crops are combined (14.3 g, 89 % yield) (Notes 17 and 18).

2. Notes

1. The internal temperature was monitored using a J-Kem Gemini digital thermometer with a Teflon-coated T-Type thermocouple probe (12-inch length, 1/8 inch outer diameter, temperature range –200 to +250 °C).

2. The following reagents and solvents were used as received for Step A: 1-naphthaleneboronic acid (Oakwood), methanol (Fisher Optima), acetonitrile (Fisher Optima, water content 80 ppm based on Karl Fischer titration) and KHF_2 (Acros 99+%). Potassium hydrogen fluoride must be handled carefully as it is corrosive. In the presence of water, it slowly releases HF. As a consequence, when using normal glassware, etching of the flask may occur with time. The solution of KHF_2 was prepared fresh just prior to use. The submitters report the synthesis of organotrifluoroborates has also been carried out in Nalgene® bottles with a screw-cap lid fitted with a mechanical stirrer.

3. The submitters prepared 1-naphthaleneboronic acid as follows. An oven-dried 500-mL, 3-necked, round-bottomed flask, equipped with a stirring bar and a reflux condenser, is charged with oven-dried magnesium turnings (11.5 g, 459 mmol, 2.8 equiv, activated in an oven at 160 °C for 48 h) and flushed with N_2. The magnesium turnings are covered with dry tetrahydrofuran (50 mL) and stirred at 470 rpm. The reaction is initiated by the addition of iodine crystals (60 mg, 0.14 mol %) and a small portion (approximately 0.5 mL) of neat 1-bromonaphthalene. If necessary, the reaction flask can be further warmed using an oil bath. Once the reaction is started (as evident by the disappearance of the yellowish color of iodine) the flask is heated in a 75 °C oil bath, and 1-bromonaphthalene (34.0 g, 164 mmol, 1 equiv) in dry THF (155 mL) is added dropwise over 25 min. The dark green solution is refluxed for 17 h. The reaction is allowed to cool over 40 min, then the solution of 1-naphthalenemagnesium bromide is transferred by cannula in a dropwise fashion over 1 h to an N_2-flushed 1-L round-bottom flask containing a stirred solution of triethyl borate (42 mL, 246 mmol, 1.5 equiv) in dry THF (100 mL) placed in a dry-ice/acetone bath.

After the addition is complete, the cold bath is removed and the resulting brownish green solution is stirred for 5 h at room temperature. The color of the solution changes from dark brown to white and a precipitate is formed. The reaction flask is then placed in an ice bath and aqueous HCl (100 mL, 2 M, 1.2 equiv) is added dropwise over 10 min. The white precipitate dissolves to form a light yellow, clear solution. After 1 h stirring at room temperature, the reaction mixture is extracted with diethyl ether (2 x 150 mL) and the combined organic layers are washed with distilled water (3 x 300 mL). The organic layer is dried over anhydrous sodium sulfate (50 g) and the solvents are removed on a rotary evaporator at 35 °C at 75 mmHg. After drying under vacuum overnight, 1-naphthaleneboronic acid is obtained as a white solid (22.0–22.9 g, 78–81% yield). If necessary, the product may be recrystallized using 200 mL of distilled water and 20 mL of THF, sonicated for 5–7 min, refrigerated overnight, filtered, and dried under vacuum. Analytical data: mp: 194–196 °C. IR (cm^{-1}): 3261, 1575, 1508, 1348, 1318, 807, 778; ^{1}H NMR (500 MHz, DMSO-d^6) δ: 7.62–7.49 (m, 3 H), 7.78–7.76 (m, 1 H), 8.02–7.91 (m, 2 H), 8.35 (m, 3 H); ^{13}C NMR (125 MHz, acetone-d^6) δ: 125.6, 125.9, 126.1, 128.7, 129.3, 129.7, 132.6, 133.4, 136.1; ^{11}B NMR (128 MHz, acetone-d^6) δ: 30.6.

4.   The reaction is run open to air. The temperature rose to 15 °C at the end of the addition. If the slurry becomes too thick to stir, it should be broken up with a spatula.

5.   The submitters monitored completion of the reaction by ^{11}B-NMR. The disappearance of the boronic acid peak at 30.7 ppm indicated the completion of the reaction.

6.   The pressure is reduced gradually to minimize bumping. About 120 mL of solvent was removed, leaving a residue of 97 g of wet solids in the flask.

7.   The pressure is reduced gradually to minimize bumping. After each acetonitrile flush, the mixture was concentrated to 50–52 g of solids.

8.   The submitters removed water by lyophilization for 24 h.

9.   The mixture remains heterogeneous as the product dissolves, but the residual KHF$_2$ remains undissolved.

10.   The submitters subjected the crude material to Soxhlet extraction using 700 mL of acetonitrile in a 1-L round-bottomed flask placed in an oil bath (102 °C and 280 rpm) for 6 h. The filtrate was concentrated on a rotary evaporator under reduced pressure at (40 °C, 150 mmHg) until dry material was obtained.

11. The water content of the acetonitrile solution was determined to be 570 ppm based on Karl Fischer titration. Higher levels of water can result in some dissolution of KHF_2.

12. The hemi-solvate was determined based on ^1H NMR analysis in acetone-d^6. The solvate remained intact upon vacuum drying 2 days at 40 °C.

13. The submitters report the crude product can be purified by dissolving the material in acetonitrile (50 mL) and precipitating with diethyl ether (300 mL).

14. Potassium 1-naphthyltrifluoroborate has the following physical and spectroscopic properties: mp: > 300 °C. IR (cm^{-1}): 2980, 2884, 2360, 1382, 940, 668, 651; ^1H NMR (400 MHz, acetone-d^6) δ: 7.30–7.36 (m, 3 H), 7.64 (d, J = 8.2 Hz, 1 H), 7.74–7.77 (m, 2 H), 8.57–8.59 (m, 1 H); ^{13}C NMR (100 MHz, acetone-d^6) δ: 124.5, 124.8, 125.9, 126.7, 128.4, 130.0 (q, J_{C-F} = 3.2 Hz), 131.4, 134.5, 138.1, 147 (very broad); ^{11}B NMR (128 MHz, acetone-d^6) δ: 4.06; ^{19}F NMR (377 MHz, acetone-d^6) δ: –138.6; HRMS (ESI) m/z calcd. for $C_{10}H_7BF_3$ (M-K) 195.0593, found 195.0593. Anal calcd for $C_{10}H_7BF_3K$: C, 51.31; H, 3.01; Found: C, 51.50; H, 2.81.

15. The following reagents and solvents were used as received for Step B: acetone (Fisher certified ACS), Oxone® (Sigma-Aldrich), toluene (Fisher Optima), n-heptane (99%, ReagentPlus, Sigma-Aldrich), dichloromethane (Fisher, certified ACS reagent, stabilized) and 6 N HCl (Fisher). The solution of Oxone was prepared fresh just prior to use.

16. The reaction was carried out open to air. The temperature rose to 32 °C during the addition. The reaction mixture becomes milky white during the addition.

17. 1-Naphthol has the following physical and spectroscopic properties: mp: 95–96 °C (lit. mp 95 °C[2a], 94 °C[2b]); IR (cm^{-1}): 3300, 1597, 1358, 1269, 1083, 789, 765; R_f: 0.4 (hexanes/EtOAc 4:1); ^1H NMR (400 MHz, CDCl$_3$): δ: 5.17 (s, 1 H), 6.83 (dd, J = 7.4, 0.7 Hz, 1 H), 7.30–7.34 (m, 1 H), 7.43 (d, J = 8.3 Hz, 1 H), 7.48–7.53 (m, 2 H), 7.81–7.85 (m, 1 H), 8.17–8.20 (m, 1 H); ^{13}C NMR (100 MHz, CDCl$_3$): δ: 108.8, 121.0, 121.7, 124.6, 125.5, 126.0, 126.7, 127.9, 135.0, 151.5; MS (EI, 70eV) 144 (M+, 100), 116 (62), 115 (81), 89 (22), 63 (18); Purity by GC: 99% (t_R = 11.2 min, conditions: Agilent DB35MS column; 30 m x 0.25 mm; initial temp 60 °C, ramp at 20 °C/min to 280 °C, hold 15 min)

18. A yield of 87% was obtained at half scale.

Safety and Waste Disposal Information

The procedures in this article are intended for use only by persons with prior training in experimental organic chemistry. All hazardous materials should be handled using the standard procedures for work with chemicals described in references such as "Prudent Practices in the Laboratory" (The National Academies Press, Washington, D.C., 2011 www.nap.edu). All chemical waste should be disposed of in accordance with local regulations. For general guidelines for the management of chemical waste, see Chapter 8 of Prudent Practices.

These procedures must be conducted at one's own risk. *Organic Syntheses, Inc.*, its Editors, and its Board of Directors do not warrant or guarantee the safety of individuals using these procedures and hereby disclaim any liability for any injuries or damages claimed to have resulted from or related in any way to the procedures herein.

3. Discussion

Boron containing reagents are among the most versatile building blocks used in chemistry. Historically, boranes and boronic acids are the most encountered, but over the past decades, modified boron compounds, which overcome some of the limitations associated with the earlier boron reagents, have emerged.[3]

Potassium organotrifluoroborates, quaternary boron salts, represent a relatively new class of reagents that are air and moisture stable and are efficient partners in boron-based chemical transformations.[4] They are easily prepared by addition of KHF_2 to various organoboron reagents,[5] and once synthesized they can be stored indefinitely without particular precaution. The organotrifluoroborate group acts as a boronic acid surrogate that can be carried through several chemical transformations[6] before being reacted or deprotected to yield the boronic acid.[7] These features make them very attractive reagents, and they are the subject of many studies. They are mainly used as coupling partners in the Suzuki-Miyaura reaction[8] but are also very useful reagents in other transformations.[9]

Treatment of organotrifluoroborate salts with oxone[10] represents a very efficient and practical method to oxidize these boron reagents. Oxone is an inexpensive reagent that readily oxidizes a large range of organotrifluoroborates in high yields within minutes. The rapidity, ease of execution, and non-toxicity of the oxidizing agent, associated with the advantages inherent to the potassium organotrifluoroborates, make the procedure described above a very convenient, environmentally sound, and attractive transformation. The process is exceedingly general. Aromatic trifluoroborates are transformed to phenols, alkyltrifluoroborates are converted to the corresponding alcohols in a stereospecific manner (complete retention of configuration), and alkenyltrifluoroborates can be used to access aldehydes (Table 1). These transformations occur with great chemoselectivity, as other oxidizable groups such as alkenes, aldehydes, as well as sulfides and amines remain untouched under the mild reaction conditions.

Table 1. Oxidation of Various Organotrifluoroborates with Oxone

$$R\text{-}BF_3K \xrightarrow[\substack{\text{acetone : } H_2O \\ \text{rt, 2 min}}]{\text{Oxone}} R\text{-}OH$$

entry	starting material	product	isolated yield (%)
1	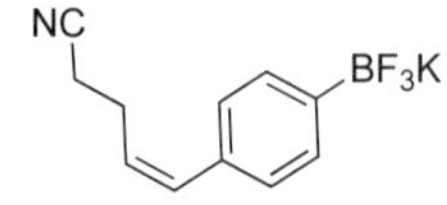	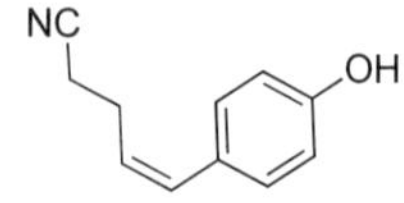	98
2	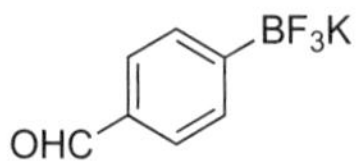	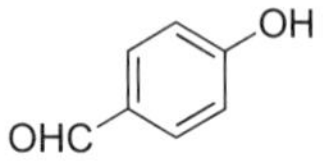	99
3	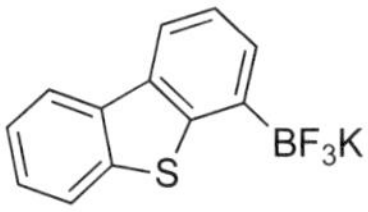	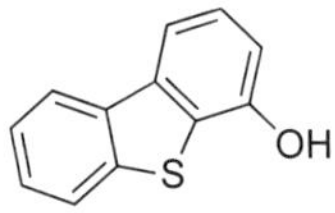	95
4	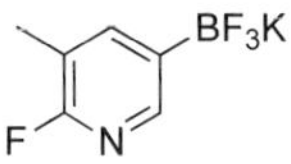	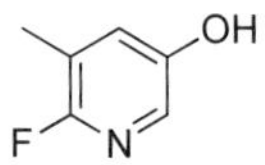	94
5	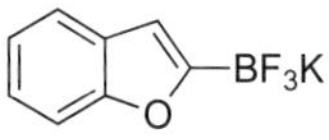	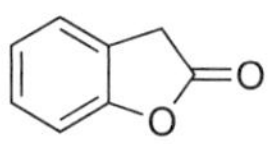	97
6	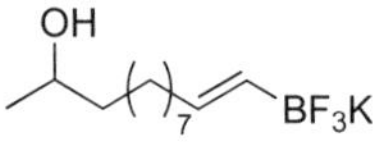	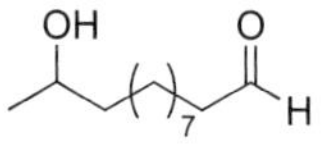	99
7	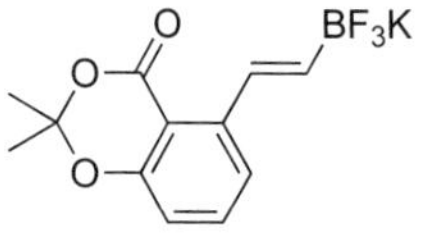	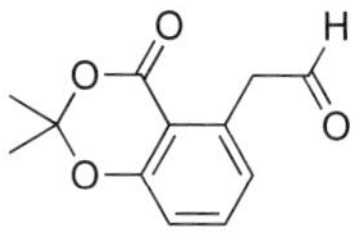	99
8	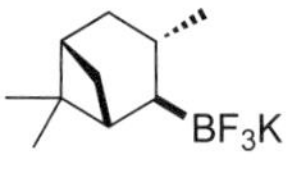	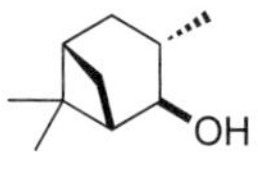	98
9	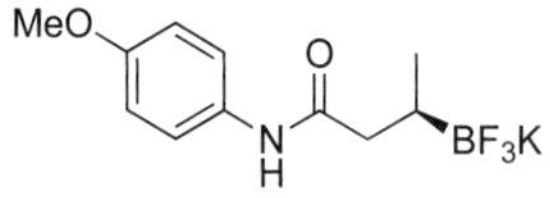	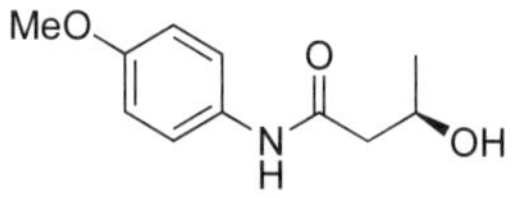	97

1. Roy and Diana Vagelos Laboratories, Department of Chemistry, University of Pennsylvania, 231 S. 34th Street, Philadelphia, PA 19104-6323

2. (a) Paul, R.; Ali, M. A.; Punniyamurthy, T. *Synthesis* **2010**, 4268–4272. (b) Hoffmann, R. W.; Ditrich, K. *Synthesis* **1983**, 107–109.

3. (a) Hall, D. G., Ed. Boronic Acids; Wiley-VCH: Weinheim, 2005; (b) Onak, T. Organoborane Chemistry; Academic Press: New York, 1975.

4. (a) Molander, G. A., Figueroa, R. *Aldrichim. Acta* **2005**, *38*, 49–56; (b) Molander, G. A., Ellis, N. *Acc. Chem. Res.* **2007**, *40*, 275–286; (c) Stefani, H. A., Cella, R., Adriano, S. *Tetrahedron* **2007**, *63*, 3623–3658; (d) Darses, S., Genêt, J.-P. *Chem. Rev.* **2008**, *108*, 288–325; for mechanistic details on the advantages of organotrifluoroborates over other boron reagents in the Suzuki reaction see: (e) Butters, M., Harvey, J. N., Jover, J., Lennox, A. J. J., Lloyd-Jones, G. C., Murray, P. M. *Angew. Chem. Int. Ed.* **2010**, *49*, 5156–5160.

5. (a) Vedejs, E.; Chapman, R. W.; Fields, S. C.; Lin, S.; Schrimpf, M. R. *J. Org. Chem.* **1995**, *60*, 3020: (b) Molander, G. A.; Cooper, D. J. *Encyclopedia of Reagents for Organic Synthesis*; John Wiley & Sons: New York, **2006**; DOI 10.1002/047084289X.rn00628.

6. For some examples see: (a) dihydroxylation: Molander, G. A., Figueroa, R. *Org. Lett.* **2006**, *8*, 75–78; (b) nucleophilic substitution: Molander, G. A., Ham, J. *Org. Lett.* **2006**, *8*, 2031–2034; (c) dipolar cycloaddition: Molander, G. A., Ham, J. *Org. Lett.* **2006**, *8*, 2767–2770; (d) oxidation: Molander, G. A., Petrillo, D. E. *J. Am. Chem. Soc.* **2006**, *128*, 9634–9635; (e) metalation: Molander, G. A., Ellis, N. M. *J. Org. Chem.* **2006**, *71*, 7491–7493; (f) ozonolysis: Molander, G. A., Cooper, D. J. *J. Org. Chem.* **2007**, *72*, 3558–3560; (g) olefination: Molander, G. A., Figueroa, R. *J. Org. Chem.* **2006**, *71*, 6135–6140; Molander, G. A., Ham, J., Canturk, B. *Org. Lett.* **2007**, *9*, 821–824; (h) reductive amination: Molander, G. A., Cooper, D. J. *J. Org. Chem.* **2008**, *73*, 3885–3891; (i) cross-coupling: Molander, G. A., Sandrock, D. L. *J. Am. Chem. Soc.* **2008**, *130*, 15792–15794; Molander, G. A., Sandrock, D. L. *Org. Lett.* **2009**, *11*, 2369–2372; (j) condensation: Molander, G. A., Febo-Ayala, W., Jean-Gerard, L. *Org. Lett.* **2009**, *11*, 3830–3833.

7. Molander, G.A., Cavalcanti, L. N., Canturk, B., Pan, P.S., Kennedy, L. E. *J. Org. Chem.* **2009**, *74*, 7364–7369.

8. Molander, G.A., Canturk, B. *Angew. Chem. Int. Ed.* **2009**, *48*, 9240–9261.

9.   For bromodeboronation see: (a) Kabalka, G. W., Mereddy, A. R.,
     Organometallics, **2004**, *23*, 4519–4521; (b) for iododeboronation, see:
     Kabalka, G. W., Mereddy, A. R. *Tetrahedron Lett.*, **2004**, *45*, 343–345;
     Kabalka, G. W., Mereddy, A. R. *Tetrahedron Lett.*, **2004**, *45*, 1417–
     1419; (c) for chlorodeboronation see: Molander, G. A., Cavalcanti, L.
     N. *J. Org. Chem.* **2011**, *76*, 7195–7203.
10. Molander, G. A., Cavalcanti, L.N. *J. Org. Chem.* **2011**, *76*, 623–630.

Appendix
Chemical Abstracts Nomenclature; (Registry Number)
1-Bromonaphthalene; (90-11-9)
Triethyl borate; (150-46-9)
Magnesium; (7439-95-4)
Iodine; (7553-56-2)
1-Naphthaleneboronic acid; (13922-41-3)
Potassium hydrogen fluoride; (7789-29-9)
Potassium 1-Naphthyltrifluoroborate: Borate(1-), trifluoro-1-naphthalenyl-,
     potassium (1:1), (*T*-4)-; (166328-07-0)
Oxone; (70693-62-8)
1-Naphthol; (90-15-3)

Professor Gary Molander received his B.S. degree at Iowa State
University in 1975, working with Professor Richard C. Larock.
He obtained his Ph.D. degree in 1979 under the direction of
Professor Herbert C. Brown. He joined Professor Barry Trost's
group at the University of Wisconsin, Madison as a National
Institutes of Health postdoctoral fellow in 1980, and in 1981 he
accepted an appointment at the University of Colorado, Boulder.
In 1999 he joined the faculty at the University of Pennsylvania.
His research interests focus on the development of new synthetic
methods for organic synthesis and natural product synthesis.

Sabahat Zahra Siddiqui was born in Lahore, Pakistan. She received her B.Sc degree from Lahore College for Women University (Roll of Honor), M.Sc degree in Organic Chemistry from Government College University, Lahore (Certificate of Merit) and M.Phil in Organic Chemistry from Punjab University, Lahore (Certificate of Merit). Currently, she is a Lecturer and a PhD scholar at Government College University, Lahore. She won an IRSIP scholarship 2011-2012 from the Higher Education Commission of Pakistan to visit the University of Pennsylvania, USA as a visiting research scholar where her research focused on synthetic methods using organotrifluoroborates under the guidance of Prof. Gary Molander.

Dr. Nicolas Fleury-Brégeot studied chemistry in France at the ESCOM (Ecole Supérieure de Chimie Organique et Minérale), where he received a Master of Science degree in 2005. He then joined the ICSN (Institut de Chimie des Substances Naturelles) in Gif-sur-Yvette where he obtained his Ph.D. in 2008 under the supervision of Dr. Angela Marinetti. After working for one year as a postdoctoral research associate in Tarragona (Spain) with Prof. Carmen Claver on asymmetric hydrogenation, he moved to Philadelphia. He is currently pursuing research in the group of Prof. Gary Molander as a postdoctoral research associate at the University of Pennsylvania. His research focuses on the development of new synthetic methods involving organotrifluoroborates.

Potassium *tert*-Butoxide Mediated Synthesis of Phenanthridinone

A.

B.

Submitted by Bhagat Singh Bhakuni, Kaustubh Shrimali, Amit Kumar and Sangit Kumar.[1]*
Checked by David Hughes.

1. Procedure

A. *2-Iodo-N-phenylbenzamide.* A 500-mL 3-necked round-bottomed flask, equipped with a 3-cm oval PTFE-coated magnetic stirring bar, is charged with 2-iodobenzoyl chloride (16.3 g, 61.1 mmol, 1.0 equiv) and dichloromethane (80 mL) (Note 1). The center joint is fitted with a 125-mL pressure-equalizing addition funnel equipped with a gas inlet adapter connected to a nitrogen line and a gas bubbler. The other two necks are capped with rubber septa; a thermocouple probe is inserted through one of the septa (Note 2). The flask is immersed in an ice-water bath and the contents cooled to 2 °C. Separately, aniline (7.0 g, 75 mmol, 1.2 equiv) and triethylamine (7.6 g, 75 mmol, 1.2 equiv) are added to a 50-mL Erlenmeyer flask and dissolved in dichloromethane (20 mL). This solution is transferred to the addition funnel then added to the cooled, stirred solution of the acid chloride over 20 min, keeping the internal temperature below 10 °C. The ice-bath is removed and the stirred mixture is allowed to warm to 22 °C and maintained at this temperature for 1.5 h (Note 3). The reaction is worked up by addition of dichloromethane (50 mL) and water (150 mL) and transferred to a 1-L separatory funnel. The layers are separated and the organic layer washed with half-saturated brine, then filtered through a bed of anhydrous sodium sulfate (50 g) in a 150-mL sintered glass funnel into a 500-mL

164

Published on the Web 1/24/2013
© 2013 Organic Syntheses, Inc.

round-bottomed flask. The sodium sulfate cake is rinsed with dichloromethane (2 x 30 mL). The filtrate is concentrated by rotary evaporation (40 °C water bath, 200 to 10 mmHg) to afford a tan solid (19.1 g). A 4-cm PTFE-coated oval magnetic stir bar and water (150 mL) are added to the flask and the slurry is stirred for 1 h at room temperature. The slurry is filtered through a 150-mL sintered glass funnel, using water (2 x 40 mL) to rinse the flask and wash the filter cake. The product is dried to constant weight in a vacuum oven (50 °C, 70 mmHg, 14 h) to afford 2-iodo-*N*-phenylbenzamide as an off-white crystalline solid (17.6 g, 89% yield) (Notes 4 and 5).

B. *Phenanthridin-6(5H)-one (1)*. A 1-L 3-necked round-bottomed flask, equipped with a 4-cm oval PTFE-coated magnetic stirring bar, is charged with 2-iodo-*N*-phenylbenzamide (7.0 g, 21.6 mmol, 1.0 equiv), 2,2'-azoisobutyryonitrile (AIBN) (0.77 g, 4.7 mmol, 0.2 equiv), potassium *t*-butoxide (12.0 g, 107 mmol, 5 equiv) and benzene (250 mL) (Notes 6 and 7). The center neck is fitted with a reflux condenser equipped with a gas inlet adapter connected to a nitrogen line and gas bubbler. One outer neck is sealed with a glass stopper; the other neck is capped with a rubber septum through which a thermocouple probe is inserted (Note 2). Using a heating mantle, the slurry is warmed to a gentle reflux (internal temp 81 °C) for 20 h (Notes 8 and 9). An additional portion of AIBN (50 mg, 0.3 mmol, 1.5 %) is added and the mixture refluxed for 5 h (Note 10). The stir bar, reflux condenser and thermocouple probe are removed and the mixture is concentrated by rotary evaporation (40 °C bath temp, 70 to 10 mmHg) to a tan powder (20 g). A 4-cm oval PTFE-coated magnetic stirring bar and water (150 mL) are added to the flask. The suspension is stirred for 1 h at ambient temperature then filtered through a 60-mL sintered glass funnel. Water (2 x 50 mL) is used to rinse the flask and wash the cake. The resulting solid is dried under vacuum (70 mmHg, 40 °C) for 20 h to afford a tan powder (4.04 g) (Note 11). The solid is transferred to a 200-mL round-bottomed flask equipped with a 3-cm oval PTFE-coated magnetic stirring bar. Methyl *t*-butyl ether (MTBE) (40 mL) is added and the slurry is stirred for 1 h at ambient temperature then filtered through a 60-mL sintered glass funnel. The cake is washed with MTBE (2 x 8 mL) then dried under vacuum (70 mmHg, 40 °C) for 4 h to afford phenanthridin-6(5*H*)-one (**1**) as a tan powder (3.59 g, 85% yield) (Notes 12 and 13).

2. Notes

1.   The following reagents and solvents were used as received for Step A: 2-iodobenzoyl chloride (Alfa Aesar, 98% purity label, but approx. 95% based on ^{1}H NMR analysis), aniline (Sigma Aldrich, ACS reagent, 99.5%), triethylamine (Sigma Aldrich, >99.5%), dichloromethane (Fisher, ACS reagent, stabilized).

2.   The internal temperature was monitored using a J-Kem Gemini digital thermometer with a Teflon-coated T-Type thermocouple probe (12-inch length, 1/8 inch outer diameter, temperature range –200 to +250 °C).

3.   The reaction was monitored by ^{1}H NMR. A 0.1 mL aliquot of the reaction mixture was evaporated then dissolved in 1 mL CDCl$_3$. No starting material (< 2%) was detected - diagnostic peaks, starting material δ 8.04-8.10 (m, 2 H); product 7.90 (d, 1 H).

4.   2-Iodo-*N*-phenylbenzamide has the following spectroscopic and physical properties: mp 143–145 °C (lit^2 mp 143–145°C); ^{1}H NMR (400 MHz, CDCl$_3$) δ: 7.12–7.20 (m, 2 H), 7.36–7.43 (m, 3 H), 7.61 (br s, 1 H), 7.50 (dd, J = 7.6, 1.2 Hz, 1 H), 7.64 (d, J = 7.9 Hz, 2 H), 7.90 (d, J = 7.9 Hz, 1 H); ^{13}C NMR (100 MHz, CDCl$_3$) δ: 92.6, 120.3, 125.1, 128.5, 128.7, 129.3, 131.7, 137.7, 140.2, 142.3, 167.4; MS (EI, 70 eV) *m/z* (relative intensity): 323 (M$^+$, 52), 231 (M$^+$- PhNH, 100), 203 (27), 76 (34); IR (plate) cm^{-1}: 3310, 1652, 1523, 1441, 1323, 1254, 1013, 739. GC purity: 97% (t$_R$ = 16.5 min; conditions: Agilent DB35MS column; 30 m x 0.25 mm; initial temp 60 °C, ramp at 20 °C/min to 280 °C, hold 15 min).

5.   An 87% yield was obtained at half scale.

6.   The following reagents and solvents were used as received for Step B: potassium *t*-butoxide (Acros, 98%), AIBN (Sigma Aldrich, 98%), benzene (Fisher, certified ACS, thiophene-free, 99.7%), and methyl *t*-butyl ether (Sigma-Aldrich, ACS reagent, >99%).

7.   Five equiv of KO^{t}Bu are essential for complete conversion.

8.   Efficient stirring is required since the reaction mixture becomes thick during the early phase of the reaction. If the mixture sets up and stirring stops, the solids should be broken up with a spatula.

9.   The reaction was monitored by TLC using 3:7 EtOAc-heptane, R$_f$ = 0.32 for starting material and 0.18 for product. The major by-product, benzanilide, R$_f$ = 0.35 (typical level 4-5%) nearly co-elutes with the starting material. To verify complete conversion, a 0.1 mL aliquot was evaporated then dissolved in DMSO-d$_6$ and analyzed by ^{1}H NMR. The N-H protons are

166

diagnostic (δ 11.7 for phenanthridinone, 10.25 for benzanilide, and 10.4 for 2-iodo-*N*-phenylbenzamide). Other distinguishing peaks include 7.76 (d, 2H) for benzanilide and 7.70 (d, 2H) for 2-iodo-*N*-phenylbenzamide.

10. Additional AIBN is charged if the reaction contains more than 5% unreacted starting material.

11. Based on ^{1}H NMR analysis, the crude material typically contains 1–2% unreacted starting material, 4–5% benzanilide, and low levels of unknown impurities. The product is nearly insoluble in MTBE and can be readily purified with high recovery by this slurry procedure.

12. Phenanthridin-6(5H)-one (**1**) has the following spectroscopic and physical properties: mp 294–296 °C (lit[3] mp 291–293 °C); ^{1}H NMR (400 MHz, DMSO-d$_6$) δ: 7.26 (dt, J = 1.2, 7.6 Hz, 1 H) 7.37 (dd, J = 1.1, 8.1 Hz, 1 H), 7.49 (dt, J = 1.1, 7.3 Hz, 1 H), 7.65 (dt, J = 1.0, 8.0 Hz, 1 H), 7.87 (dt, J = 1.5, 7.2 Hz, 1 H), 8.32 (dd, J = 1.1, 8.0 Hz, 1 H), 8.39 (dd, J = 1.0, 8.2 Hz, 1 H), 8.51 (d, J = 8.2 Hz, 1 H), 11.70 (bs, 1 H). ^{13}C NMR (100 MHz, DMSO-d$_6$) δ: 116.1, 117.5, 122.2, 122.6, 123.2, 125.7, 127.4, 127.9, 129.5, 132.7, 134.2, 136.6, 160.8; HRMS (ESI) *m/z* 196.0858, calcd for $C_{13}H_9ON + H^+$: 196.0757; MS (EI, 70 eV), *m/z* (relative intensity): 195 (M$^+$, 100), 167 (29), 166 (19), 140 (10), 139 (12), 83 (12); IR (plate), cm^{-1}: 3416, 1631, 1360, 1151, 748, 726. GC purity: 99% (t$_R$ = 16.1 min; conditions: Agilent DB35MS column; 30 m x 0.25 mm; initial temp 60 °C, ramp at 20 °C/min to 280 °C, hold 15 min). A wt % purity of 96% was determined by ^{1}H NMR (with a 10 s pulse delay) using 1,4-dimethoxybenzene as internal standard. Sample preparation: ca. 20 mg of 1,4-dimethoxybenzene and ca 30 mg of product were accurately weighed, then dissolved in 5 mL DMSO-d$_6$. Each of the 8 protons of the product were integrated and compared to the integration of the 6.8 ppm singlet of 1,4-dimethoxybenzene.

13. To avoid the use of benzene, the checker carried out reactions in toluene and in tetrahydropyran (based on a report that THP was an effective solvent for radical reactions, Yasuda, H.; Uenoyama, Y.; Nobuta, O.; Kobayashi, S.; Ryu, I. *Tetrahedron Lett.* **2008**, *49*, 367-370). Neither of these solvents was as effective as benzene, as summarized in the table below. The reactions in toluene were sluggish at 80 °C, with 10–12% unreacted starting material after 50–57 h even with an additional AIBN charge at the 24 h time point, and generated 12 % benzanilide from proto-deiodination. The reactions in THP went to full completion in 20 h, with or without AIBN, but generated 18–23% benzanilide.

Solvent	Scale (g)	Temp (°C)	Time (h)	AIBN (equiv)	SM %	Benzanilide %	Isolated yield[a]
Benzene	7.0	81	25	0.2	1.5	4	85
Benzene	3.5	81	25	0.2	2	4	83
Toluene	4.1	80	57	0.3	12	12	62
Toluene	3.5	80	50	0.3	10	12	68
THP	3.5	85	20	0.2	0	18	71
THP	3.5	85	20	0	0	23	61

[a]Reactions were purified by MTBE slurry as described in Step B. Products from reactions in THP required two slurry procedures to reduce benzanilide levels to <2%.

Safety and Waste Disposal Information

The procedures in this article are intended for use only by persons with prior training in experimental organic chemistry. All hazardous materials should be handled using the standard procedures for work with chemicals described in references such as "Prudent Practices in the Laboratory" (The National Academies Press, Washington, D.C., 2011 www.nap.edu). All chemical waste should be disposed of in accordance with local regulations. For general guidelines for the management of chemical waste, see Chapter 8 of Prudent Practices.

These procedures must be conducted at one's own risk. *Organic Syntheses, Inc.*, its Editors, and its Board of Directors do not warrant or guarantee the safety of individuals using these procedures and hereby disclaim any liability for any injuries or damages claimed to have resulted from or related in any way to the procedures herein.

3. Discussion

Phenanthridinones and related biaryl lactams are privileged cores present in many alkaloids and pharmaceutically relevant organic molecules (Figure 1).[4] As a result, numerous methods for their syntheses have been described, many involving multiple steps.[5] Palladium-catalyzed biaryl C-C coupling is a key step in most of the reported methodologies, most of which require protection of the N-H bond. A method that avoids a protection/ deprotection strategy and expensive palladium catalyst is highly desirable.

168

Figure 1. Biaryl Lactams

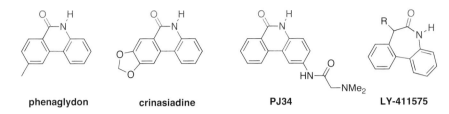

| phenaglydon | crinasiadine | PJ34 | LY-411575 |

Potassium *t*-butoxide-mediated C-C bond formation for the synthesis of biaryls has been explored by a number of groups.[6] Recently, we reported a carbon-carbon biaryl coupling reaction via C-H activation of anilines for the construction of six-membered phenanthridinones and seven-membered dibenzoazepinone biaryl lactams.[7] Treatment of 2-iodo-*N*-arylbenzamide substrates with potassium *t*-butoxide using catalytic AIBN or 1,10-phenanthroline in benzene solvent gave 81–91% yields of biaryl lactams (Table 1).

KO*t*Bu alone is not fully effective for the reaction, as only 35% of product was formed in 24 h with 60% unreacted substrate recovered. When 0.2 equiv of the radical initiator AIBN was employed, complete conversion occurred within 24 h and the coupled product was obtained in excellent yield.

Substrates with electron-withdrawing groups, such as fluoride, difluoride, chloride, and bromide, and the electron-donating groups, such as methyl, methoxy, and dimethoxy, on either aryl ring were tolerated under our reaction conditions. Furthermore, the scope of the reaction was extended to other aryl substrates, including naphthalene and thiophene-based 2-halo-arylamide substrates.[7] Regarding the aryl halide, a brief study indicated best yields were obtained with the iodo and bromo substrates (Table 1, entries 1 and 2). The chloride also reacted but gave a moderate yield (49%) of cyclized product.[7]

This metal-free synthesis of phenanthridinones offers several advantages over traditional Pd-catalyzed biaryl couplings with regard to cost, sustainability and lack of toxicity. In addition, the required 2-iodo-*N*-phenylbenzamide substrates are readily accessible by coupling of 2-iodobenzoyl chloride with an arylamine.

Table 1. Synthesis of Phenanthridinones and Dibenzoazepinones[a]

Entry	Amide	Lactam	Additive, Base	Time (h)	Yield (%)
1	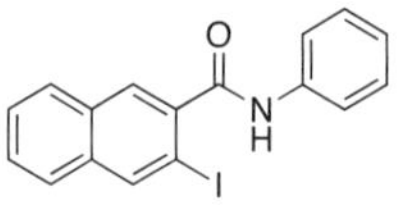	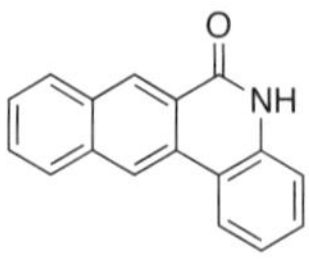	AIBN (0.2 equiv) KO^{t}Bu (5.0 equiv)	12	**2** (90)
2	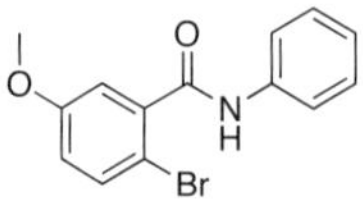	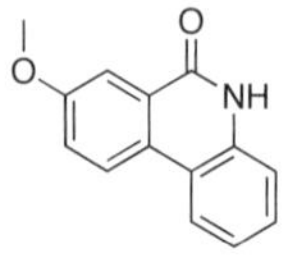	AIBN (0.2 equiv) KO^{t}Bu (5.0 equiv)	12	**3** (82)
3	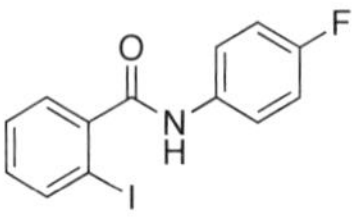	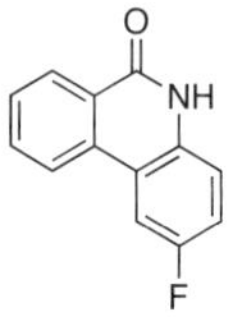	AIBN (0.2 equiv) KO^{t}Bu (5.0 equiv)	12	**4** (81)
4 5	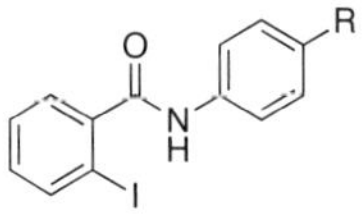	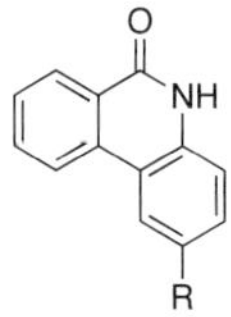	AIBN (0.2 equiv) KO^{t}Bu (5.0 equiv)	24 24	**5** R=Me (89)[b] **6** R=OMe (83)[b]
6	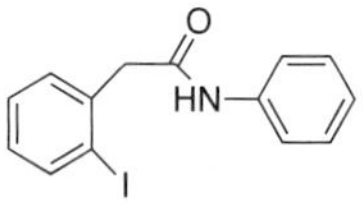	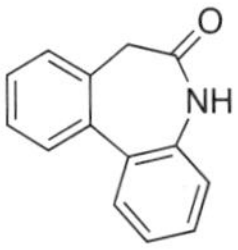	L (0.2 equiv) KO^{t}Bu (8.0 equiv)	24	**7** (91)[c]
7	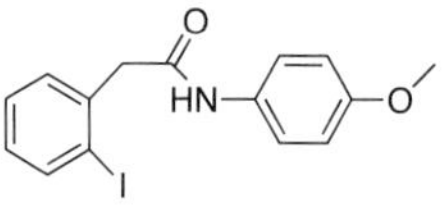	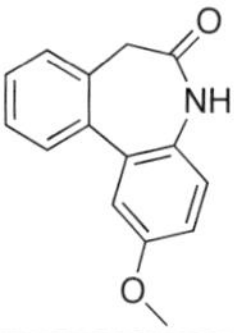	L (0.2 equiv) KO^{t}Bu (8.0 equiv)	24	**8** (82)

[a]Preparation of these compounds is described in reference 7. [b]Two isomers formed in 1:1 ratio, isolated yield for both isomers. [c]Reaction was carried out using 3 g of 2-iodobenzamide substrate following the procedure described herein for **1**. L = 1,10-phenanthroline.

1. IISER Bhopal and DST-New Delhi for financial support. Department of Chemistry, Indian Institute of Science Education and Research Bhopal, (IISER), Bhopal, MP 462 023, India. Department of Chemistry, Indian Institute of Science Education and Research Bhopal, MP, India, 462

023, Phone: (+91) 755-409-2325, Fax: (+91) 755-409-2392, E-mail: sangitkumar@iiserb.ac.in

2. Gabbutt, C. D.; Heron, B. M.; Instone, A. C. R. *Heterocycles* **2003**, *60*, 843–856.
3. Banwell, M. G.; Lupton, D. W.; Ma, X.; Renner, J.; Sydnes, M. O. *Org. Lett.* **2004**, *6*, 2741–2744.
4. (a) Simanek, V. In The Alkaloids; Brossi, A., Ed.; Academic Press: NewYork, 1985; Vol. 26, pp 185–229. (b) Lee, S.; Hwang, S.; Yu, S.; Jang, W.; Lee, Y. M.; Kim, S. *Arch. Pharm. Res.* **2011**, *34*, 1065. (c) Fang, S. D.; Wang, L. K.; Hecht, S. M. *J. Org. Chem.* **1993**, *58*, 5025. (d) Hegan, D. C.; Lu, Y.; Stachelek, G. C.; Crosby, M. E.; Bindra, R. S.; Glazer, P. M. *Proc. Natl. Acad. Sci. U.S.A.* **2010**, *107*, 2201.
5. (a) Furuta, T.; Kitamura, Y.; Hashimoto, A.; Fujii, S.; Tanaka, K.; Kan, T. *Org. Lett.* **2007**, *9*, 183. (b) Karthikeyan, J.; Cheng, C.-H. *Angew. Chem., Int. Ed.* **2011**, *50*, 9880. (c) Dubost, E.; Magnelli, R.; Cailly, T.; Legay, R.; Fabis, F.; Rault, S. *Tetrahedron* **2010**, *66*, 5008. (d) Baudoin, O.; Cesario, M.; Guenard, D.; Gueritte, F. *J. Org. Chem.* **2002**, *67*, 1199.
6. (a) Yanagisawa, S.; Ueda, K.; Taniguchi, T.; Itami, K. *Org. Lett.* **2008**, *10*, 4673. (b) Sun, C.-L.; Li, H.; Yu, D.-G.; Yu, M.; Zhou, X.; Lu, X.-Y.; Huang, K.; Zheng, S.-F.; Li, B.-J.; Shi, Z.-J. *Nat. Chem.* **2010**, *2*, 1044. (c) Liu, W.; Cao, H.; Zhang, H.; Zhang, H.; Chung, K. H.; He, C.; Wang, H.; Kwong, F. Y.; Lei, A. *J. Am. Chem. Soc.* **2010**, *132*, 16737. (d) Shirakawa, E.; Itoh, K.-i.; Higashino, T.; Hayashi, T. *J. Am. Chem. Soc.* **2010**, *132*, 15537. (e) Sun, C.-L.; Gu, Y.-F.; Wang, B.; Shi, Z.-J. *Chem. Eur. J.* **2011**, *17*, 10844. (f) Sun, C.-L.; Gu, Y.-F.; Huang, W.-P.; Shi, Z.-J. *Chem. Commun.* **2011**, *47*, 9813.
7. Bhakuni, B. S.; Kumar, A.; Balkrishna, S. J.; Sheikh, J. A.; Konar, S.; Kumar, S. *Org. Lett.*, **2012**, *14*, 2838.

Appendix
Chemical Abstracts Nomenclature (Registry Number)

2-Iodobenzoyl chloride (2042672)
2-Iodo-*N*-phenylbenzamide (15310-01-7)
Phenanthridin-6(5*H*)-one (2413-02-7)
2,2'-Azoisobutyryonitrile (AIBN) (78-67-1)

Sangit Kumar was born in Uttar Pradesh, India in 1978, received his PhD degree from Indian Institute of Technology, Bombay in 2004 under the supervision of Professor Harkesh B. Singh. He had first post-doctoral studies with Professor Lars Engman at Uppsala University, Sweden and second post-doctoral experience with Professor Michael R. Detty at University at Buffalo (SUNY), NY, USA. After completing his post-doctoral studies, he returned to India and joined as an assistant professor in the department of Chemistry, Indian Institute of Science, Education, and Research Bhopal. His research interest includes synthesis and biological activity of organochalcogen compounds and transition metal catalyzed carbon-heteroatom coupling reactions

Bhagat Singh Bhakuni was born in Almora, India in 1987. He had completed his graduation in bachelor of science (honors) Chemistry from Kirorimal College of University of Delhi in 2007. Subsequently, he obtained Master Degree in Chemistry from University of Delhi in 2009. He joined the PhD program at Indian Institute Science Education and Research Bhopal in 2009 under the supervision of Dr. Sangit Kumar. Currently he is studying radical chain breaking antioxidant activity of organochalcogen compounds.

Kaustubh Shrimali was born in Maharastra, India in 1990. He cleared 10+2 in year 2007 from S.I.C.A School, Indore, and qualified national level joint entrance exam (IIT JEE) in the year 2008 and was admitted to the BS-MS dual degree program in Indian Institute of Science Education and Research, Bhopal. He is currently pursuing his final year project work under the supervision of Dr. Sangit Kumar. His research interests are organometallics and organic synthesis.

172

Amit Kumar was born in Uttar Pradesh in 1986, and received his Bachelors of Science in 2006 and Master in Chemistry in 2008 from Maharaj Singh College Saharanpur affiliated with Ch. Charan Singh University Meerut. He joined PhD program in the Department of Chemistry, IISER Bhopal under the supervision of Dr. Sangit Kumar in 2010. He is working on the synthesis of organochalcogen compounds.

Preparation of N¹-Phenylacetamidine 4-Bromobenzoate Using 2,2,2-Trichloroethyl Acetimidate Hydrochloride

Submitted by Lulin Wei and Stéphane Caron.[1]
Checked by Margaret Faul, Rob Milburn, and John Colyer.

1. Procedure

A. *2,2,2-Trichloroethyl acetimidate hydrochloride.* To a 50-mL two-necked round-bottomed flask equipped with a thermocouple inserted into the septum and a magnetic stir bar is added acetonitrile (13.0 mL, 248 mmol, 1.19 equiv) (Note 1) and trichloroethanol (20.0 mL, 208 mmol, 1.00 equiv) (Note 2). The flask is capped with a vacuum receiving tube that has the central inlet extended with an Eppendorf pipet tip secured with teflon tape such that the pipet tip extends well into the solution inside the flask (Note 3). The central inlet of the vacuum receiving tube is connected to a hydrogen chloride gas cylinder using Tygon tubing, and the outlet of the vacuum receiving tube is connected via Tygon tubing to an empty 2-necked 150-mL round-bottomed flask (Note 4) that is in turn connected via Tygon tubing to a scrubber (Figure 1) (Note 5). The acetonitrile and trichloroethanol solution is cooled to 0 – 5 °C using an ice bath, and HCl gas is bubbled into the solution for 5 h while maintaining the internal

Org, Synth. **2013**, *90*, 174-181
Published on the Web 2/7/2013
© 2013 Organic Syntheses, Inc.

temperature at 0 – 5 °C (Notes 6, 7, and 8). A rate of 10-15 bubbles/min at the scrubber is maintained throughout the 5 h reaction time (Note 9). After complete reaction (Note 10), the flow of HCl gas is stopped; the reaction flask is removed from the ice bath and the stopper removed for approximately 5 min (Note 11). Seed is added to the flask (1 wt%), which is then sealed with a glass stopper and placed in a refrigerator kept at 0 °C for 66 h (Note 12). The product crystallizes as a single solid white mass that is broken up using a spatula and then filtered. The large crystals are further crushed, washed with cold acetonitrile (approx. 0 °C, 3 x 15 mL), and dried in a vacuum oven (125–250 mmHg) at 45 °C to a constant weight (Note 13) to afford **3** as a white solid (42.8–46.0 g, 91–97%) (Note 14).

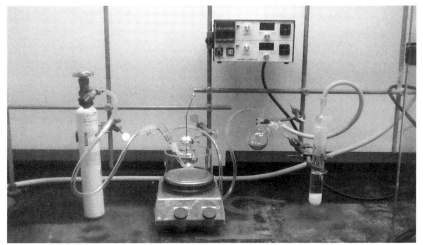

Figure 1. Apparatus assembly for Step A

B.   *N^I-Phenylacetamidine 4-bromobenzoate*. To a 100-mL round-bottomed flask equipped with a magnetic stir bar is added potassium carbonate (12.4 g, 89.9 mmol, 3 equiv) (Note 15), water (30 mL) (Note 16), and 2-methyltetrahydrofuran (20 mL) (Note 17). 2,2,2-Trichloroethyl acetimidate hydrochloride (8.60 g, 37.9 mmol, 1.25 equiv) is added, the biphasic mixture is stirred for 30 min, transferred to a 150 mL separatory funnel and the bottom aqueous layer is discarded. The remaining organic layer is transferred to a 250-mL round-bottomed flask, equipped with a magnetic stir bar, containing aniline (2.76 mL, 30.2 mmol, 1.00 equiv) (Note 18) in 2-methyltetrahydrofuran (100 mL) and the mixture is stirred at room temperature for 3 h. A solution of 4-bromobenzoic acid (6.06 g, 30.1 mmol,

1.00 equiv) (Note 19) in 2-methyltetrahydrofuran (100 mL) is then added and the mixture is stirred at room temperature overnight (Note 20). The resulting solids are filtered, washed with 2-methyltetrahydrofuran (3 x 30 mL), and dried in a vacuum oven (125–250 mmHg) at 45 °C (Note 21) to afford **6** as a white solid (8.88–8.91 g, 87–88%) (Note 22).

2. Notes

1.   Acetonitrile was purchased from Aldrich (271004-100ML, Lot # SHBB9999V) and used as received.
2.   2,2,2-Trichloroethanol was purchased from Aldrich (T54801-100G, lot # BCBG1806V) and used as received.
3.   A 1 mL disposable Eppendorf pipet tip attached to the inlet using teflon tape was used.

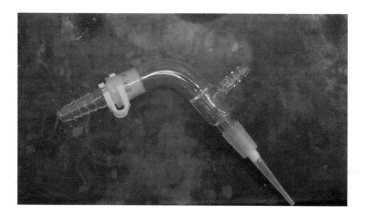

4.   The 150-mL 2-necked round-bottomed flask is required to prevent NaOH solution in the scrubber from being sucked back into the reaction during the initial stages of HCl addition.
5.   A Chemglass 29/42 "cold finger" charged with 300 mL of aqueous 5 N NaOH solution was used.
6.   Hydrogen chloride gas was purchased from Aldrich (295426-227G, Lot # MKBJ4524V) and used as received.
7.   An exotherm to ~8 °C is observed within the first 20 min, during which time the rate of HCl addition had to be adjusted to prevent NaOH from being sucked into the trap.
8.   The temperature is maintained by adding ice to the bath.

Org. Synth. **2013**, *90*, 174-181

9.   The rate of bubbling from the scrubber typically remained constant over the first 2 h of the reaction, at which time the flow of HCl was reduced to maintain the 10-15 bubbles/min at the scrubber.

10.   The reaction was typically complete within 5 h as determined by ^{1}H NMR using CH_2 resonance of trichlorethanol (4.86 ppm) and the acetimidate (5.35 ppm) Sample preparation: dilute 100 uL of reaction mixture was with 0.5 mL of DMSO.

11   This is done to ensure that there is no pressure build up. Temperature should be monitored due to increased off-gassing as the solution warms.

12.   The temperature of the refrigerator was measured at 0 °C. Colder temperature resulted in incomplete crystallization. The yield of the crystallization was increased 5-7% by addition of seed (1% w/w).

13.   Drying was conducted overnight.

14.   2,2,2-Trichloroethyl acetimidate hydrochloride has the following spectroscopic and physical properties: mp 187 °C; ^{1}H NMR (400 MHz, DMSO-d_6) δ: 2.55 (s, 3 H), 5.35 (s, 2 H), 12.29 (s, 2 H). 1 – 3% of acetamide and trichloroethanol from hydrolysis of 3 are typically observed in the ^{1}H NMR spectrum at δ 2.17 (s, 3 H, CH_3CO), 4.86 (s, 2 H, CCl_3CH_2), 7.31 (br s, 1 H, NH), 7.44 (br s, 1 H, OH), 7.57 (br s, 1 H, NH).

15.   Potassium carbonate (99%) was purchased from Sigma-Aldrich (347825-250G, Lot # MKBJ7723V) and used as received.

16.   Deionized water prepared via a reverse osmosis system was used.

17.   Anhydrous 2-methyltetrahydrofuran was purchased from Sigma-Aldrich (186562-1L, lot # SHBC3613V), and used as received.

18.   Aniline (99.5%) was purchased from Sigma-Aldrich (242284-100ML, Lot # 68396APV) and used as received.

19.   4-Bromobenzoic acid (99%) was purchased from Sigma-Aldrich (108510-100G, Lot # 08231KHV) and used as received.

20.   Solids start to appear after a few minutes

21.   Drying was conducted until all the 2-methyltetrahydrofuran was removed (typically 2–3 days).

22. N^1-Phenylacetamidine 4-bromobenzoate has the following spectroscopic and physical properties: mp 179 °C. ^{1}H NMR (400 MHz, DMSO-d_6) δ: 2.16 (br s, 3 H), 7.23 (d, J = 7.6 Hz, 2 H), 7.28 (t, J = 7.4 Hz, 1 H), 7.44 (t, J = 8.0 Hz, 2 H), 7.55 (d, J = 8.6 Hz, 2 H), 7.83 (d, J = 8.6 Hz, 2 H), 8.83 (br s, 1 H), 11.37 (br s, 2 H);^{13}C NMR (100 MHz, DMSO-d_6) δ: 123.8, 124.8, 126.5, 129.4, 130.6, 131.2, 136.5,137.8, 169.9. HRMS: [M +

H]$^+$ calcd for $C_8H_{10}N_2$: 135.0917. Found: 135.0920. Anal. Calcd for $C_{15}H_{15}BrN_2O_2$: C, 53.75; H, 4.51; N, 8.36. Found: C, 53.44; H, 4.25; N, 8.28.

Handling and Disposal of Hazardous Chemicals

The procedures in this article are intended for use only by persons with prior training in experimental organic chemistry. All hazardous materials should be handled using the standard procedures for work with chemicals described in references such as "Prudent Practices in the Laboratory" (The National Academies Press, Washington, D.C., 2011 www.nap.edu). All chemical waste should be disposed of in accordance with local regulations. For general guidelines for the management of chemical waste, see Chapter 8 of Prudent Practices.

These procedures must be conducted at one's own risk. *Organic Syntheses, Inc.*, its Editors, and its Board of Directors do not warrant or guarantee the safety of individuals using these procedures and hereby disclaim any liability for any injuries or damages claimed to have resulted from or related in any way to the procedures herein.

3. Discussion

Amidines have been a privileged functional group for the preparation of a wide range of compounds[2-5] and as synthetic intermediates for the elaboration of more complex structures such as heterocycles.[6-8] While the preparations of amidines from amide or nitrile starting materials are well precedented, new synthetic methods to access this functional group have continued to emerge.[9-13] The most common reagent for the preparation of acetamidines is ethyl acetimidate hydrochloride, a stable and relatively inexpensive reagent. Unfortunately, in the case of poor nucleophiles, the reaction can require extensive reaction time and lead to low yields. For instance, the reaction of a 2-amino-1,3,4-thiadiazole resulted in a 68% yield after 5 days.[14] We believed that modification of the ethoxy leaving group into a more electron withdrawing moiety would enhance the reactivity of the imidate and expand the scope of amine nucleophiles. Indeed it was demonstrated that the use of trifluoro- and trichloroethyl imidates allows for reactions with a number of different amines to generate amidines.[9] The hydrochloride salts of 2,2,2-trichloro imidates proved to be especially

178

attractive because of their crystallinity and stability. Indeed, we have found that samples of compound **3** stored on the benchtop for a year have not lost potency.

The conversion of imidate **3** to an amidine is conducted by neutralization of the hydrochloride salt and direct use of the 2-MeTHF solution from the extraction. Reaction with an amine generally proceeds rapidly to generate the amidine. We identified the 4-bromobenzoate as a suitable salt for direct crystallization of the amidine. This renders the reaction and isolation of the desired products operationally simple. In the case of aniline (**5**) this reaction proceeds in >90% yield. Interestingly, the same reaction protocol when conducted with ethyl acetimidate provided a modest ~60% yield.

This procedure has been exemplified using several different amines, as shown in Table 1.

Table 1. Preparation of amidines as *p*-bromobenzoate salts

Entry	R	Yield (%)
1	PhNH-	88
2	4-MePhNH-	93
3	4-ClPhNH-	96
4	4-MeOPhNH-	98
5	2,4-Me$_2$PhNH-	74
6	PhCH$_2$NH-	91
7	PhCH$_2$CH$_2$NH-	98
8	Ph—⟨ ⟩N-ξ	92*

* Prepared and isolated as the HCl salt without free basing

1. Chemical R&D, Pfizer Worldwide R&D, MS-8118D/4002, Eastern point Rd, Groton CT 06240. E-mail stephane.caron@pfizer.com

2. Coles, M. P. *Dalton Trans.* **2006**, 985–1001.

3. Ikeda, M.; Tanaka, Y.; Hasegawa, T.; Furusho, Y.; Yashima, E. *J. Am. Chem. Soc.* **2006**, *128*, 6806–6807.

4. Edwards, P. D.; Albert, J. S.; Sylvester, M.; Aharony, D.; Andisik, D.; Callaghan, O.; Campbell, J. B.; Carr, R. A.; Chessari, G.; Congreve, M.; Frederickson, M.; Folmer, R. H. A.; Geschwindner, S.; Koether, G.; Kolmodin, K.; Krumrine, J.; Mauger, R. C.; Murray, C. W.; Olsson, L.-L.; Patel, S.; Spear, N.; Tian, G. *J. Med. Chem.* **2007**, *50*, 5912–5925.

5. Peterlin-Masic, L.; Kikelj, D. *Tetrahedron* **2001**, *57*, 7073–7105.

6. Brain, C. T.; Brunton, S. A. *Tetrahedron Lett.* **2002**, *43*, 1893–1895.

7. Ito, K.; Kizuka, Y.; Hirano, Y. *J. Heterocycl. Chem.* **2005**, *42*, 583–588.

8. Langlois, M.; Guilloneau, C.; Tri Vo, V.; Jolly, R.; Maillard, J. *J. Heterocycl. Chem.* **1983**, *20*, 393–398.

9. Caron, S.; Wei, L.; Douville, J.; Ghosh, A. *J. Org. Chem.* **2010**, *75*, 945–947.

10. Wang, Y.-F.; Zhu, X.; Chiba, S. *J. Am. Chem. Soc.* **2012**, *134*, 3679–3682.

11. DeKorver, K. A.; Johnson, W. L.; Zhang, Y.; Hsung, R. P.; Dai, H.; Deng, J.; Lohse, A. G.; Zhang, Y.-S. *J. Org. Chem.* **2011**, *76*, 5092–5103.

12. Harjani, J. R.; Liang, C.; Jessop, P. G. *J. Org. Chem.* **2011**, *76*, 1683–1691.

13. Cortes-Salva, M.; Garvin, C.; Antilla, J. C. *J. Org. Chem.* **2011**, *76*, 1456–1459.

14. Chapleo, C. B.; Myers, P. L.; Smith, A. C. B.; Stillings, M. R.; Tulloch, I. F.; Walter, D. S. *J. Med. Chem.* **1988**, *31*, 7–11.

Appendix
Chemical Abstracts Nomenclature; (Registry Number)

2,2,2-Trichloroethanol; (115-20-8)

Aniline; (62-53-3)

4-Bromobenzoic acid; (586-76-5)

2-Methyltetrahydrofuran; (96-47-9)

2,2,2-Trichloroethyl acetimidate hydrochloride; (16507-47-4)

N^1-Phenylacetamidine 4-bromobenzoate; (1207066-26-9)

Stéphane Caron received his B.Sc. and M.Sc. degrees from Université Laval in Canada under the direction of Robert Burnell. He joined the group of Clayton H. Heathcock at the University of California-Berkeley where he received his Ph.D. in 1995 upon completion of the total synthesis of Zaragozic Acid A. He started his industrial career at Pfizer in Groton Connecticut where he is now a Senior Director in Chemical R&D. He is the editor of "Practical Synthetic Organic Chemistry; Reactions, Principles, and Techniques."

Lulin Wei received his B.S. degree in analytic chemistry from Lanzhou University in China and M.S. degree in organic chemistry from Southern Illinois University. After several years studying palladium catalysis in the labs of Professor Richard Larock at Iowa State University, he joined Pfizer in Groton, CT in 1996, where he is currently a senior scientist in Chemical R&D in the Post-POC group.

Robert R. Milburn studied chemistry at the University of Waterloo, Canada and completed an Honors B.Sc in 1996. He then pursued a Ph.D (1997-2002) at Queen's University, Canada under the guidance of Professor Victor Snieckus, exploring organolithiation chemistry and methodology development. After a postdoctoral appointment (2003-2005) at TSRI with Professor K.C. Nicolaou, Robert joined the Chemical Process Research & Development department at Amgen in Thousand Oaks, where he has been involved in the development of synthetic processes of numerous clinical candidates.

John T. Colyer was born in Columbus, Indiana, in 1977. In 2000 he earned his B. S. in chemistry from Indiana University-Purdue University Indianapolis working under the guidance of Professor William H. Moser. He completed his M.S. in 2004 at the University of Arizona under the guidance of Professor Michael P. Doyle, where he studied dirhodium (II) carboxamidate catalysis. In 2004 he joined the Chemical Process Research & Development group at Amgen in Thousand Oaks, CA.

Discussion Addendum for:
Stereoselective Synthesis of *anti* α-Methyl-β-Methoxy Carboxylic Compounds

Submitted by Pedro Romea* and Fèlix Urpí.*[1]
Original article: Gálvez E.; Romea P.; Urpí F. *Org. Synth.* **2009**, *86*, 81–91.

Prior and subsequent to our original report in *Organic Syntheses*, we have described a number of highly diastereoselective Lewis acid-mediated additions of titanium enolates from chiral *N*-acyl thiazolidinethiones to acetals.[2] These involve the addition of *N*-propanoyl and *N*-acetyl-4-isopropyl-1,3-thiazolidine-2-thione to dialkyl acetals (eq 1 and 2 in Scheme 1)[3-5] and dimethyl ketals from methyl ketones (eq 3 in Scheme 1).[6] Furthermore, the scope of the acyl groups has been recently expanded to *N*-glycolyl derivatives. Especially, *O*-pivaloyl protected *N*-glycolyl 4-isopropyl-1,3-thiazolidine-2-thione undergoes highly diastereoselective additions to a wide array of dimethyl and dibenzyl acetals (eq 4 in Scheme 1).[7]

Org. Synth. **2013**, *90*, 182-189
Published on the Web 3/4/13
© 2013 Organic Syntheses, Inc.

Scheme 1. Stereoselective additions of titanium enolates from N-acyl-4-isopropyl-1,3-thiazolidine-2-thione to dialkyl acetals and ketals

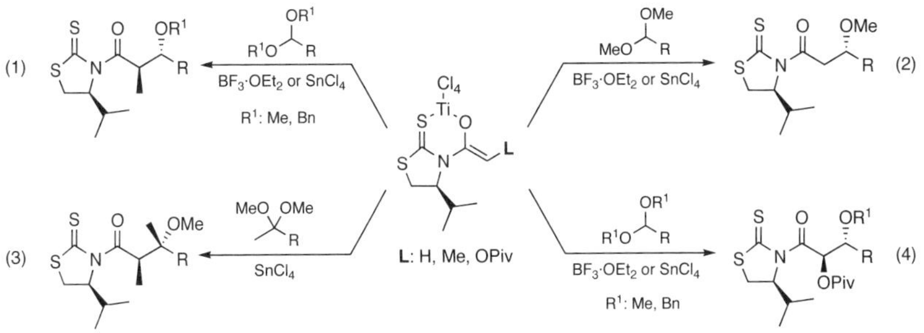

These reactions likely proceed through a mechanism in which an oxocarbenium intermediate approaches to the less hindered face of the titanium enolate. Taking advantage of such a picture, this chemistry has been successfully applied to glycals. Thereby, activation of glycals by Lewis acids triggers the formation of cyclic and conjugated oxocarbenium intermediates that may participate in highly diastereoselective α- and β-C-glycosidation processes with the abovementioned titanium enolates. Importantly, the appropriate choice of the chiral auxiliary and the C6-protecting group determines the stereochemical outcome of these carbon-carbon bond forming reactions and permits the modular preparation of three of the four possible diastereomers of α- or β-1'-methyl-substituted C-glycosides (Scheme 2).[8-10]

Scheme 2. Stereoselective C-glycosidation reactions

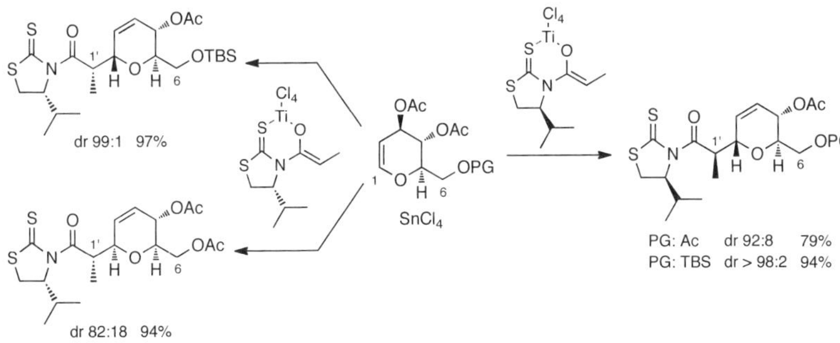

Synthetic applications

These methods have been already applied to the synthesis of natural products. For instance, our group took advantage of the Lewis acid-mediated addition of the titanium enolates from (*S*) *N*-acetyl and *N*-propanoyl 4-isopropyl-1,3-thiazolidine-2-thiones (**1** and **2** respectively in Scheme 3) to dialkyl acetals for the construction of the C9–C21 fragment of debromoaplysiatoxin.[11] As shown in Scheme 3, chromatographic purification of the product formed from the reaction between **1** and the dimethyl acetal of an aromatic aldehyde afforded diastereomerically pure adduct **3** in 82% yield. Removal of the chiral auxiliary gave alcohol **4**, which was further elaborated to chiral dibenzyl acetal **5**. Then, the stage was set for the introduction of two new stereocenters. Indeed, the asymmetric induction imparted by the titanium enolate of **2** and the *Felkin* bias of the oxocarbenium cation from **5** provided adduct **6** as a single diastereomer (dr > 98:2) in 74% yield. Finally, treatment of **6** with MeNHOMe furnished Weinreb amide **7** in 90% yield.[12]

Scheme 3. Stereoselective synthesis of C9–C21 fragment of debromoaplysiatoxin

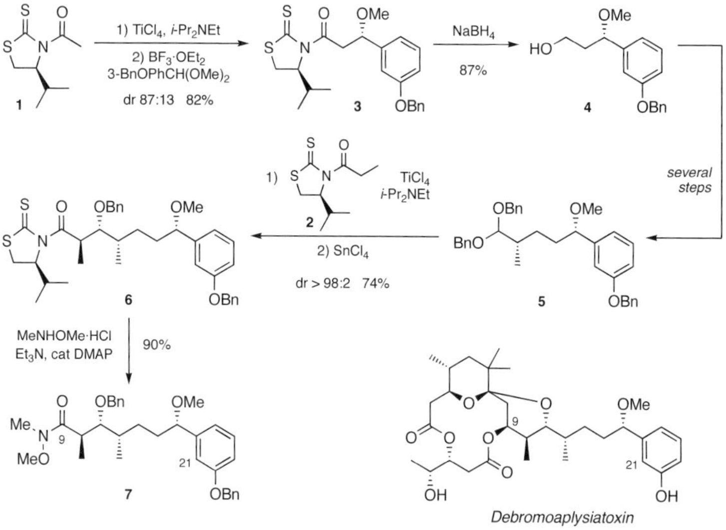

Crimmins and Chakraborty have also used these transformations in the total syntheses of pironetin and a protected form of the monomeric unit

184

of rhizopodin respectively (Scheme 4).[13,14] Thereby, Crimmins reported that the addition of **2** to a chiral dimethyl acetal provided adduct **8** as a single diastereomer in 64% yield, which was further converted into aldehyde **9** by treatment with *i*-Bu$_2$AlH (eq 1 in Scheme 4).[13] In turn, Chakraborty described a similar process involving an achiral dimethyl acetal in which adduct **10** was obtained in an excellent diastereomeric ratio (dr 93:7) and 86% yield (eq 2 in Scheme 4).[14] Interestingly, removal of the chiral auxiliary by the lithium salt of the dimethyl methylphosphonate furnished ketophosphonate **11** ready to participate in an HWE olefination.

Scheme 4. Total syntheses of pironetin and a monomeric unit of rhizopodin

Protected monomeric unit of rhizopodin

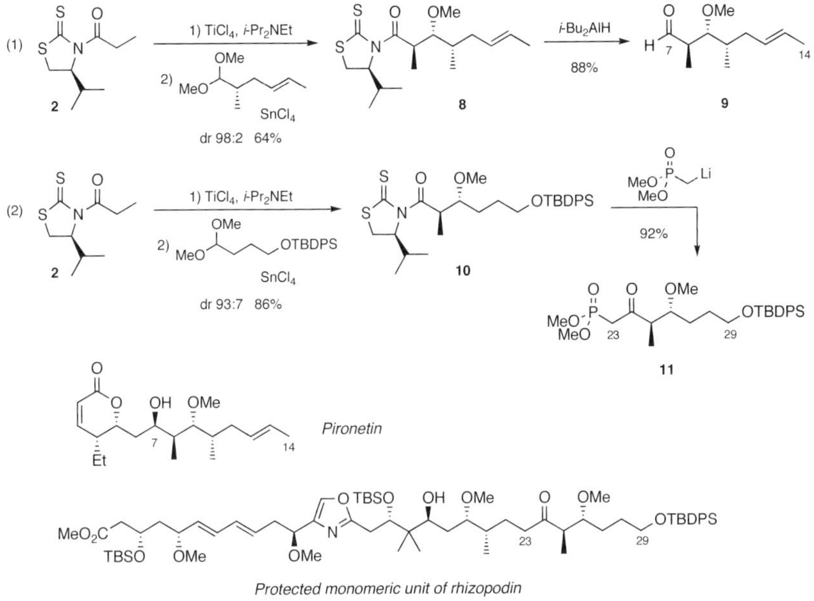

Other reports have also established that these Lewis acid acid-mediated couplings of *N*-acyl thiazolidinethiones and acetals are flexible enough to be successfully adapted to the synthesis of a broad array of natural products. For instance, Crimmins took advantage of the highly diastereoselective reaction of the titanium enolate from *N*-(4-pentenoyl) thiazolidinethione **12** and a dibenzyl acetal, the mild removal of the chiral auxiliary from adduct **13**, and the RCM of the resultant alcohol **14** to obtain in a few steps the cyclic core **15** of aldigenin B (eq 1 in Scheme 5).[15] In turn, Hodgson reported that the addition of the titanium enolate of (*R*)-

phenylglycine-derived *N*-acetyl thiazolidinethione **16** to the benzaldehyde diallyl acetal provided diastereomerically pure adduct **17** in 74% yield after chromatography. Further treatment of this adduct with magnesium acetoacetate and imidazole led to β-keto ester **18**, a key intermediate in the total synthesis of hyperolactone C (eq 2 in Scheme 5).[16] Finally, the synthesis of hennoxazole A by Smith includes a comprehensive study on the Lewis acid-mediated additions of *N*-acetyl thiazolidinethiones to the dimethyl acetal of bisoxazole **19**.[17] This revealed that the ability of an oxazole nitrogen atom to coordinate with the titanium center altered the stereochemical outcome of such additions and the undesired diastereomer was obtained in 80:20 diastereomeric ratio and 67% yield. Looking for alternative conditions, Smith found that the enolization of *N*-acetyl thiazolidinethione **20** using Sammakia's conditions (PhBCl$_2$/sparteine) followed by the BF$_3$·OEt$_2$-mediated addition of the resultant boron enolate to **19** provided the desired adduct **21** in 86:14 diastereomeric ratio and 53% yield (eq 3 in Scheme 5). Eventually, reduction of **21** with *i*-Bu$_2$AlH gave aldehyde **22**, which was immediately used in the next step.

Scheme 5. Total syntheses of aldigenin B, hyperolactone C, and hennoxazole

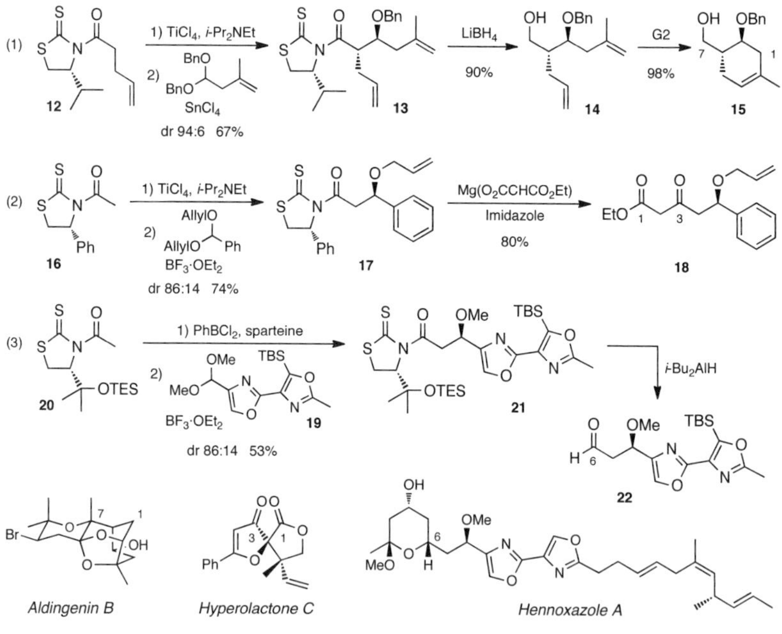

Dialkyl acetals are not the only suitable substrates for these transformations. Certainly, oxocarbenium cations from cyclic hemiacetals and glycals (see Scheme 2) can also undergo highly diastereoselective additions to titanium enolates from *N*-acyl thiazolidinethiones, as was used by our group in the stereoselective synthesis of the western hemisphere of salinomycin. Indeed, the SnCl$_4$-mediated addition of the titanium enolate from *N*-butanoyl thiazolidinethione **23** to hemiacetal **24** afforded α-*C*-glycoside **25** as a single diastereomer (dr > 97:3), which was treated with methanol to obtain methyl ester **26** in 75% overall yield (eq 1 in Scheme 6).[18] More recently, Leighton has reported that the coupling of structurally complex *O*-acetyl hemiacetal **27** with the titanium enolate of ***ent*-2** produced α-*C*-glycoside **28** as a single diastereomer in an outstanding 91% yield (eq 2 in Scheme 6).[19] Methanolysis proceeded exceptionally smoothly to give **29** in quantitative yield, which was finally converted into zincophorin methyl ester in a few steps.

Scheme 6. Syntheses of antibiotic polyethers

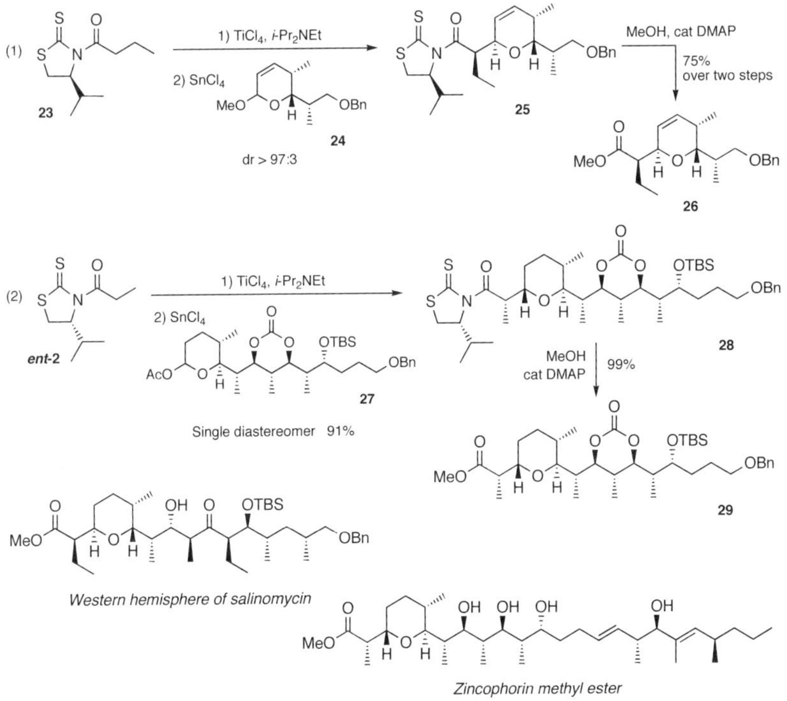

In summary, the Lewis acid-mediated addition of titanium enolates from *N*-acyl thiazolidinethiones to dialkyl acetals, hemiacetals, or glycals represents a powerful synthetic tool for the stereoselective construction of carbon-carbon bonds leading to *anti* α-alkyl-β-alkoxy carboxylic derivatives. Importantly, the chiral auxiliary can be easily removed using a number of mild conditions, which confers to these methodologies a remarkable appeal for the synthesis of natural products.

1. Departament de Química Orgànica, Universitat de Barcelona, Martí i Franqués 1-11, 08028 Barcelona, Catalonia, Spain. E-mail: pedro.romea@ub.edu; felix.urpi@ub.edu. Financial support from the Spanish Ministerio de Ciencia e Innovación (MICINN), Fondos FEDER (Grant No. CTQ2009-09692), and the Generalitat de Catalunya (2009SGR825) are acknowledged.

2. For the influence of the chiral auxiliary on the stereochemical outcome of these transformations, see: Baiget, J.; Cosp, A.; Gálvez, E.; Gómez-Pinal, L.; Romea, P.; Urpí, F. *Tetrahedron* **2008**, *64*, 5637–5644.

3. Cosp, A.; Romea, P.; Talavera, P.; Urpí, F.; Vilarrasa, J.; Font-Bardia, M.; Solans, X. *Org. Lett.* **2001**, *3*, 615–617.

4. Cosp, A.; Romea, P.; Urpí, F.; Vilarrasa, J. *Tetrahedron Lett.* **2001**, *42*, 4629–4631.

5. Cosp, A.; Larrosa, I.; Vilasís, I.; Romea, P.; Urpí, F.; Vilarrasa, J. *Synlett* **2003**, 1109–1112.

6. Checa, B.; Gálvez, E.; Parelló, R.; Sau, M.; Romea, P.; Urpí, F.; Font-Bardia, M.; Solans, X. *Org. Lett.* **2009**, *11*, 2193–2196.

7. Baiget, J.; Caba, M.; Gálvez, E.; Romea, P.; Urpí, F.; Font-Bardia, M. *J. Org. Chem.* **2012**, *77*, 8809–8814.

8. Larrosa, I.; Romea, P.; Urpí, F.; Balsells, D.; Vilarrasa, J.; Font-Bardia, M.; Solans, X. *Org. Lett.* **2002**, *4*, 4651–4654.

9. Gálvez, E.; Larrosa, I.; Romea, P.; Urpí, F. *Synlett* **2009**, 2982–2986.

10. Gálvez, E.; Sau, M.; Romea, P.; Urpí, F.; Font-Bardia, M. *Tetrahedron Lett.* **2013**, *54*, 1467–1470.

11. Cosp, A.; Llàcer, E.; Romea, P.; Urpí, F. *Tetrahedron Lett.* **2006**, *47*, 5819–5823.

12. For an approach to the synthesis of the γ-amino acid embedded in the structure of bistramides based on this chemistry, see: Gálvez, E.; Parelló, R.; Romea, P.; Urpí, F. *Synlett* **2008**, 2951–2954.

13. Crimmins, M. T.; Dechert, A.-M. R. *Org. Lett.* **2009**, *11*, 1635–1638.
14. Pulukuri, K. K.; Chakraborty, T. K. *Org. Lett.* **2012**, *14*, 2858–2861.
15. Crimmins, M. T.; Hughes, C. O. *Org. Lett.* **2012**, *14*, 2168–2171.
16. Hodgson, D. M.; Man, S. *Chem. Eur. J.* **2011**, *17*, 9731–9737.
17. Smith, T. E.; Kuo, W.-H.; Balskus, E. P.; Bock, V. D.; Roizen, J. L.; Theberge, A. B.; Carroll, K. A.; Kurihara, T.; Wessler, J. D. *J. Org. Chem.* **2008**, *73*, 142–150.
18. Larrosa, I.; Romea, P.; Urpí, F. *Org. Lett.* **2006**, *8*, 527–530.
19. Harrison, T. J.; Ho, S.; Leighton, J. L. *J. Am. Chem. Soc.* **2011**, *133*, 7308–7311.

Pedro Romea completed his BSc in Chemistry in 1984 at the University of Barcelona. That year, he joined the group of Professor Jaume Vilarrasa, at the University of Barcelona, receiving his Master Degree in 1985, and he followed PhD studies in the same group from 1987 to 1991. Then, he joined the group of Professor Ian Paterson at the University of Cambridge (UK), where he participated in the total synthesis of oleandolide. Back to the University of Barcelona, he became Associate Professor in 1993. His research interests have focused on the development of new synthetic methodologies and their application to the stereoselective synthesis of naturally occurring molecular structures

Fèlix Urpí received his BSc in Chemistry in 1980 at the University of Barcelona. In 1981, he joined the group of Professor Jaume Vilarrasa, at the University of Barcelona, receiving his Master Degree in 1981 and PhD in 1988, where he was an Assistant Professor. He then worked as a NATO postdoctoral research associate in titanium enolate chemistry with Professor David A. Evans, at Harvard University (USA). He moved back to the University of Barcelona and he became Associate Professor in 1991. His research interests have focused on the development of new synthetic methodologies and their application to the stereoselective synthesis of naturally occurring molecular structures.

Bimolecular Oxidative Amidation of Phenols: 1-(Acetylamino)-4-oxo-2,5-cyclohexadiene-1-acetic acid, Methyl Ester

1 → **2**

Phl(OAc)₂, TFA / CH₃CN, 3 h

Submitted by Jaclyn Chau[1] and Marco A. Ciufolini.*[1]
Checked by Aaron Bedermann and John L. Wood.

1. Procedure

A. *Methyl 2-(1-acetamido-4-oxocyclohexa-2,5-dien-1-yl)acetate (2).* An oven-dried (110 °C), three-necked, 1-L round-bottomed flask is equipped with a 1.25-inch egg-shaped magnetic stirring bar and three rubber septa, one containing a temperature probe and another a nitrogen inlet. Solid (diacetoxyiodo)benzene ("DIB," 23.9 g, 74.2 mmol, 1.35 equiv) (Note 1) is added under purging N₂ followed by CH₃CN (480 mL) (Note 2). The flask is purged for 5 min and the contents are maintained under inert atmosphere during all subsequent operations. The stirrer is set to 560 rpm and neat trifluoroacetic acid (TFA, 6.4 mL, 83.6 mmol, 1.5 equiv) (Note 3) is carefully added via syringe over 3 min at room temperature, whereupon the mixture becomes homogeneous.

A solution of methyl 4-hydroxyphenylacetate (**1**) (9.1 g, 54.8 mmol, 1.0 equiv) (Note 4) in CH₃CN (20 mL), prepared in a flame-dried 50-mL round-bottomed flask, is added to the DIB solution over 3 h via syringe pump (Note 5). During the addition, the tip of the needle is immersed into the solution to a depth of 0.5 inch and the progress of the reaction is followed by TLC (Note 6). At the end of the addition (Note 7), the solution turns light brown. The mixture is transferred to a one-necked 1-L round-bottomed flask and concentrated under reduced pressure (mechanical pump vacuum, 175 mmHg) at 32 °C (rotary evaporator bath temperature). Toluene (10 mL) (Note 8) is then added and the resulting suspension is concentrated under reduced pressure to azeotropically remove residual TFA. Addition of

190

Org. Synth. **2013**, *90*, 190-199
Published on the Web 3/6/2013
© 2013 Organic Syntheses, Inc.

toluene and evaporation are repeated twice again to ensure removal of essentially all of the TFA.

The resultant residue is dissolved in CH_2Cl_2 (100 mL) (Note 9) and silica gel (40.0 g) (Note 10) is added to the solution. The CH_2Cl_2 is then removed under reduced pressure (rotary evaporator, 175 mmHg, 32 °C). Residual solvent is further removed under high vacuum (0.5 mmHg). The resultant free-flowing powder is loaded onto a chromatography column (3.0 inches in diameter) that had been dry-packed with silica (160.0 g). A layer of sand (1.0 inch) was placed on top of the dry silica and a mobile phase (gradient 50%, 75%, 100% EtOAc/Hexanes, 25%, 50%, 75% MeCN/CH_2Cl_2, 200 mL each) (Notes 11 and 12) is passed through the silica, followed by MeCN (400 mL) to flush out the remaining product. The fractions determined to contain product via TLC analysis (Note 6) are combined and concentrated via rotary evaporator. (Note 13) Further solvent removal under high vacuum provides the pure product (**2**) as a brown solid (7.7 g, 34.5 mmol, 63% yield) (Notes 14 and 15).

2. Notes

1.   (Diacetoxyiodo)benzene, also described as iodobenzene diacetate, +98%, was purchased from Alfa Aesar and was used as received.

2.   Acetonitrile, Spectroscopic grade, UltimAr®, was purchased from Macron Chemicals and was used as received.

3.   Trifluoroacetic acid, ReagentPlus 99%, was purchased from Sigma Aldrich, Inc. and was used as received.

4.   Methyl 4-hydroxyphenylacetate, 99%, was purchased from Sigma Aldrich, Inc. and was used as received.

5.   A 30 mL syringe with a diameter of 1 inch containing a 12 inch, 18-gauge needle, was used to contain a 20 mL CH_3CN solution of **1**.

6. R_f of **1** in 4:6 EtOAc/hexanes = 0.56; R_f of **2** in the same solvent system = 0; R_f of **2** in 50% MeCN/CH_2Cl_2 = 0.30.

7.   Normally, the phenol is undetectable by TLC at this point.

8.   Toluene, HPLC grade, was purchased from Fisher Scientific and was used as received.

9. CH_2Cl_2, certified ACS, was purchased from Fisher Scientific and was used as received.

10.   Silica gel SilicaFlash® F60 (40-63 µm/230-500 mesh) was purchased from Silicycle.

11. EtOAc, hexanes, and MeCN, certified ACS, which were used in the chromatography column, were purchased from Fisher Scientific and used as received.

12. Fractions were collected by filling 16 x 150 mm culture tubes and the fractions containing product, R_f = 0.30, determined by TLC in 50% MeCN/CH$_2$Cl$_2$, were combined.

13. The chromatographic procedure detailed herein was developed by the checkers. The submitters described the following alternative purification method: The brown residue obtained upon azeotropic removal of TFA with toluene was dissolved in a mixture of Et$_2$O (5 mL) and CH$_2$Cl$_2$ (5 mL) and the solution was adsorbed onto a pad of silica gel (55 g), covered with a layer of sand (0.25-inch), that had been prepared inside a fritted glass filter funnel (2.5-inch diameter). Water aspirator vacuum was used to draw the solution through the silica. The residue remaining in the reaction flask was dissolved with more CH$_2$Cl$_2$ (10 mL) and the solution was also applied to the silica gel pad using water aspirator vacuum. The silica gel pad was first eluted (water aspirator vacuum) using 300 mL of Et$_2$O to remove PhI and some brown tar. The filtrate was analyzed by TLC (2.5:1, Et$_2$O/CH$_3$CN; R_f of **2** = 0.54) to ascertain that it contained little or no product; then it was discarded. Further elution (water aspirator vacuum) with a 2.5:1 (vol/vol) mixture of Et$_2$O/CH$_3$CN (700 mL) retrieved the product. Vacuum concentration of the eluate afforded a semisolid brown residue, which was refiltered through a fresh pad of silica gel (55 g) using the same procedure (complete removal of polymeric material). The residue thus obtained was taken up in EtOAc (20 mL) and the solution was kept in a freezer at –20 °C for 5 h, whereupon the product crystallized as light orange prisms. The mother liquor was carefully decanted. The precipitate was dried by rotary evaporation under vacuum, followed by drying under high vacuum to constant mass to provide spectroscopically and microanalytically pure **2**.

14. The product is stable to storage at room temperature.

15. Physical properties and spectral data for **2** are as follows: mp 99–100 °C (uncorrected); ^{1}H NMR (400 MHz, acetone-d_6; the compound is poorly soluble in CDCl$_3$) δ: 1.88 (s, 3 H), 3.01 (s, 2 H), 3.62 (s, 3 H), 6.15 (d, J = 10.2 Hz, 2 H,), 7.18 (d, J = 10.2 Hz, 2 H,), 7.55 (s, 1 H); ^{13}C NMR (100 MHz, acetone-d_6) δ: 23.3, 42.5, 52.0, 54.2, 129.0, 149.6, 169.9, 170.3, 185.2; IR (NaCl, cm^{-1}): 1739, 1668, 1536; HRMS (ESI) m/z calculated for C$_{11}$H$_{13}$NO$_4$Na [M + Na]$^+$ 246.0742, found 246.0738.

Elemental analysis (average of duplicate runs): calculated for $C_{11}H_{13}NO_4$: C 59.19%, H 5.87%, N 6.27%, found C 59.13%, H 5.67%, N 6.11%.

Handling and Disposal of Hazardous Chemicals

The procedures in this article are intended for use only by persons with prior training in experimental organic chemistry. All hazardous materials should be handled using the standard procedures for work with chemicals described in references such as "Prudent Practices in the Laboratory" (The National Academies Press, Washington, D.C., 2011 www.nap.edu). All chemical waste should be disposed of in accordance with local regulations. For general guidelines for the management of chemical waste, see Chapter 8 of Prudent Practices.

These procedures must be conducted at one's own risk. *Organic Syntheses, Inc.*, its Editors, and its Board of Directors do not warrant or guarantee the safety of individuals using these procedures and hereby disclaim any liability for any injuries or damages claimed to have resulted from or related in any way to the procedures herein.

3. Discussion

The oxidative amidation of phenols[2] is a process that converts phenols **3** to dienones **5** through reaction with a hypervalent iodine reagent[3] such as (diacetoxyiodo)benzene [PhI(OAc)$_2$; DIB] in the presence of a suitable nitrogen nucleophile (N in **3-5**). The reaction may proceed through capture of an electrophile such as **4**. The dashed semicircles in **3-5** indicate that nucleophile and substrate may be tethered (intramolecular oxidative amidation) or may be independent (bimolecular oxidative amidation). While the process typically involves 4-substituted phenols such as **3** (*para*-oxidative amidation), reaction in the *ortho* mode is possible.[4] The nitrogen functionality in **5** emerges in the form of an amide (carboxamide, sulfonamide, or phosphoramide); hence the term "oxidative amidation." Intramolecular variants of the transformation include the oxidative cyclization of phenolic oxazolines,[5] sulfonamides,[4,6] and phosphoramides,[4,6] while the bimolecular reaction entails the Ritter-type interception of **4** with nitriles such as CH$_3$CN.[7,8] Dienones **5**, especially those arising from a bimolecular reaction, are valuable building blocks for the synthesis of various nitrogenous substances.[9,10]

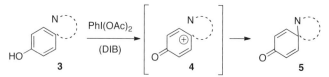

Figure 1. The Oxidative Amidation of Phenols

In its original form,[7] the bimolecular oxidative amidation of phenols required 1,1,1,3,3,3-hexafluoroisopropanol (HFIP) as a cosolvent, and it performed well (40–70 % yield of products) on scales up to ca. 300 mg of substrate. However, the cost of HFIP and the substantial amounts of polymeric matter generated when running the reaction on a larger scale rendered the original protocol impractical and uneconomical. The present procedure, substantially that of Liang and Ciufolini,[8] is simpler and more efficient, it suppresses the need for costly HFIP, and it can be more readily scaled up. The purification of the product is expediently achieved with no aqueous workup.

Optimization of the method revealed that *slow* addition of a CH_3CN solution of substrate to a *dilute* CH_3CN solution of DIB and TFA is crucial to minimize polymer formation. Best results were obtained when the final concentration of substrate was around 110 mmol/L. A slight molar excess of TFA relative to DIB is necessary for efficient conversion (1.3 equiv vs DIB for small scales, 1.5 equiv vs DIB for large scales). However, larger amounts of TFA lower the yield and complicate isolation procedures.

It should be noted that the use of $PhI(OCOCF_3)_2$ (PIFA, more costly) in lieu of DIB (less costly), with or without added TFA, produces generally inferior results. This raises questions about the precise role of TFA in the reaction. First, PIFA dissolves easily in CH_3CN, while DIB is poorly soluble in that solvent. Reactions run using suspensions of DIB in CH_3CN proceed poorly. Addition of slightly more than 1 equivalent of TFA to a CH_3CN suspension of DIB results in rapid and complete dissolution. It seems plausible that TFA promotes formation of a more soluble species such as $[PhI(OAc)(OCOCF_3)]$, *thereby releasing one equivalent of AcOH into the medium*. Second, the foregoing mixed iodonium carboxylate is likely to possess redox properties that are intermediate between those of DIB (weaker oxidant) and PIFA (stronger oxidant). Therefore, TFA probably increases the redox potential of the oxidant to an optimal level. Third, the oxidative amidation is definitely an acid-catalyzed process: in all likelihood, a small

194

excess of TFA provides more effective protonic catalysis than the AcOH released in the medium. Yet, as the reaction proceeds to completion, only ca. 1 equivalent of TFA is liberated, as opposed to the 2 equivalents released by PIFA. This minimizes undesirable acid-promoted side reactions. Fourth, the Ritter-type capture of an electrophile arising through oxidative activation of the phenol would lead to nitrilium ion **6**, which presumably must combine with a nucleophile. Undoubtedly, AcOH is more nucleophilic than TFA. If reactive intermediate **6** were form in a medium already containing AcOH, its conversion to **7**, and thence to **8**, would be more efficient.

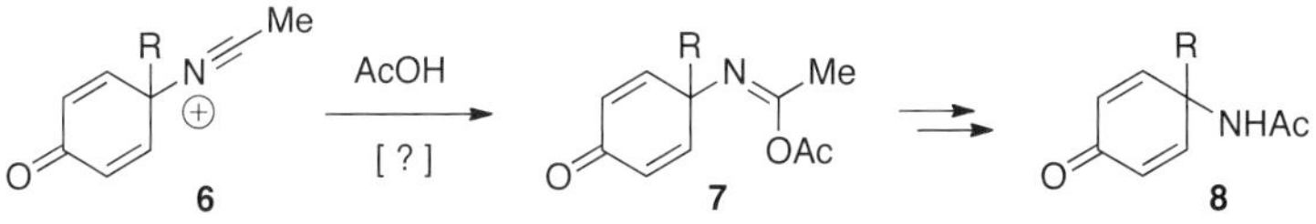

Figure 2. Possible Mechanism of the Bimolecular Oxidative Amidation of Phenols

The procedure described herein appears to have substantial scope with respect to the phenolic substrate. Representative examples are provided in Table 1.[8]

Table 1. Representative Examples of Bimolecular Oxidative Amidation of Phenols

Phenol	Product	Yield (%)[a]
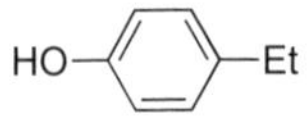		87
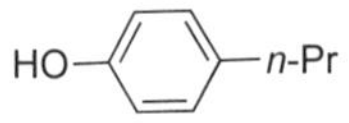		82
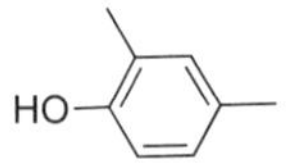	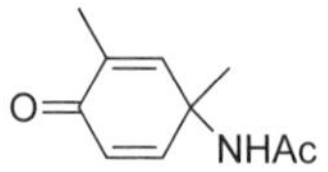	41
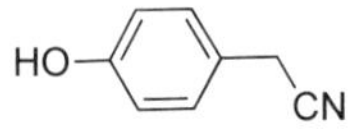		71
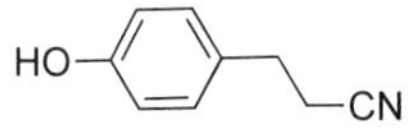	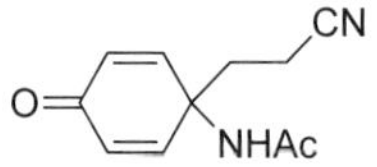	81
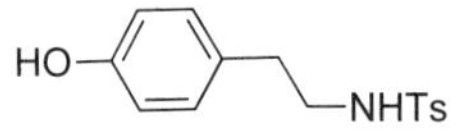	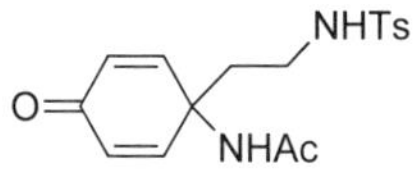	59
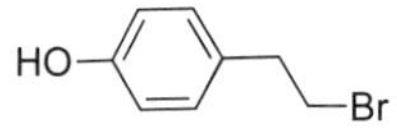	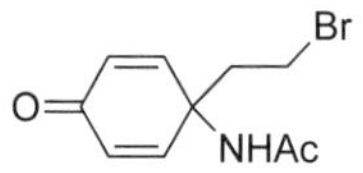	57
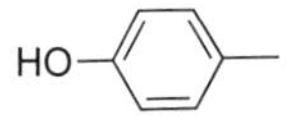		89
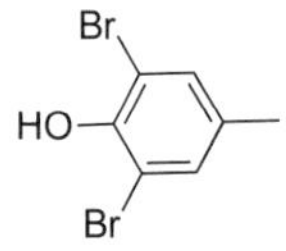	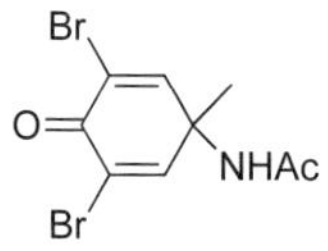	41
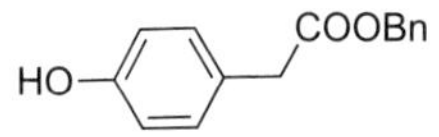		65

[a] Slow addition of 4 mmol of substrate to 1.2 equiv of DIB and 1.3 equiv of TFA versus DIB at rt, in a total of 500 mL of MeCN

1. Department of Chemistry, University of British Columbia, 2036 Main Mall, Vancouver, BC V6T 1Z1, Canada. E-mail: ciufi@chem.ubc.ca. We are grateful to the University of British Columbia, the Canada Research Chair Program, NSERC, CIHR, CFI, and BCKDF for support of our research program.

2. Reviews: (a) Ciufolini, M. A.; Canesi, S.; Ousmer, M.; Braun, N. A. *Tetrahedron* **2006**, *62*, 5318–5337. (b) Ciufolini, M. A.; Braun, N. A.; Canesi, S.; Ousmer, M; Chang, J; Chai, D. *Synthesis* **2007**, 3759–3772. (c) Liang, H.; Ciufolini, M. A. *Tetrahedron* **2010**, *66*, 5884–5892.

3. Leading reviews on the chemistry of hypervalent iodine reagents: (a) Varvoglis, A. *Hypervalent Iodine in Organic Synthesis*; Academic Press: San Diego, 1997. (b) Moriarty, R. M.; Prakash, O. *Org. React.* **1999**, *54*, 273–418. (c) Moriarty, R. M.; Prakash, O. *Org. React.* **2001**, *57*, 327–415. (d) Zhdankin, V. V.; Stang, P. J. *Chem. Rev.* **2002**, *102*, 2523–2584. (e) Stang, P. J. *J. Org. Chem.* **2003**, *68*, 2997–3008. (f) Wirth, T. *Angew. Chem. Int. Ed.* **2005**, *44*, 3656–3665. (g) Moriarty, R. M. *J. Org. Chem.* **2005**, *70*, 2893–2903. (h) Moriarty, R. M.; Prakash, O. *Hypervalent Iodine in Organic Chemistry: Chemical Transformations*; Wiley-Blackwell, UK, **2008**. (i) Zhdankin, V. V.; Stang, P. J. *Chem. Rev.* **2008**, *108*, 5299–5358. (j) Zhdankin, V. V. *Arkivoc* **2009**, 1–62. (k) Polysegu, L.; Deffieux, D.; Quideau, S. *Tetrahedron* **2010**, *66*, 2235–2261. (l) Silva, L. F., Jr.; Olofsson, B. *Nat. Prod. Rep.* **2011**, *28*, 1722–1754. The latter reference contains an extensive bibliography on the subject.

4. Liang, H.; Ciufolini, M. A. *Chem. Eur. J.* **2010**, *16*, 13262–13270, and references cited therein.

5. (a) Braun, N. A.; Ciufolini, M. A.; Peters, K.; Peters, E.-M. *Tetrahedron Lett.* **1998**, *39*, 4667–4670. (b) Braun, N. A.; Ousmer, M.; Bray, J. D.; Bouchu, D.; Peters, K.; Peters, E.; Ciufolini, M. A. *J. Org. Chem.* **2000**, *65*, 4397–4408.

6. Canesi, S.; Belmont, P.; Bouchu, D.; Rousset, L.; Ciufolini, M. A. *Tetrahedron Lett.* **2002**, *43*, 5193–5195.

7. Canesi, S.; Bouchu, D.; Ciufolini, M. A. *Org. Lett.* **2005**, *7*, 175–177.

8. Liang, H.; Ciufolini, M. A. *J. Org. Chem.* **2008**, *73*, 4299–4301.

9. (a) Ousmer, M.; Braun, N. A.; Ciufolini, M. A. *Org. Lett.* **2001**, *3*, 765–767. (b) Ousmer, M.; Braun, N. A.; Bavoux, C.; Perrin, M.; Ciufolini, M. A. *J. Am. Chem. Soc.* **2001**, *123*, 7534–7538. (c) Canesi, S.;

Bouchu, D.; Ciufolini, M. A. *Angew. Chem., Int. Ed.* **2004**, *43*, 4436–4438. (d) Mendelsohn, B. A.; Ciufolini, M. A. *Org. Lett.* **2009**, *11*, 4736. (e) Liang, H.; Ciufolini, M. A. *Org. Lett.* **2010**, *12*, 1760.

10. A variant of this chemistry enables the intramolecular oxidative *amination* of phenols with appropriate secondary amines as the nucleophiles; e.g.: (a) Scheffler, G.; Seike, H.; Sorensen, E. J. *Angew. Chem., Int. Ed.* **2000**, *39*, 4593–4596. (b) Mizutani, H.; Takayama, J.; Soeda, Y.; Honda, T. *Tetrahedron Lett.* **2002**, *43*, 2411–2414.

Appendix
Chemical Abstracts Nomenclature (Registry Number)

(Diacetoxy)iodobenzene: iodobenzene 1,1-diacetate, iodosobenzene 1,1-diacetate; (3240-34-4)

Trifluoroacetic acid; (76-05-1)

Methyl 4-hydroxyphenylacetate; (14199-15-6)

1,1,1,3,3,3-Hexafluoroisopropanol: hexafluoroisopropanol (920-66-1)

2,5-Cyclohexadiene-1-acetic acid, 1-(acetylamino)-4-oxo, methyl ester (837373-69-0)

Marco A. Ciufolini [B.S., Spring Hill College, Mobile, AL (1978), Ph.D., University of Michigan (1981, M. Koreeda), postdoc, Yale University (1982-84, S. Danishefsky)] has held academic positions at Rice University (Houston, TX, 1984-1997), the Ecole Supérieure de Chimie, Physique et Electronique de Lyon and the University of Lyon (Lyon, France, 1998-2004), and the University of British Columbia at Vancouver (2004-present), where he is currently the Canada Research Chair in Synthetic Organic Chemistry. His work focuses on the development of new synthetic methods and their application to the synthesis of nitrogenous substances.

 Jaclyn Chau [B.Sc., University of British Columbia, Vancouver, BC (2009)] joined the research group of Professor Marco A. Ciufolini in 2008 at the University of British Columbia. Her current research centers on the total synthesis of nitrogenous natural products.

 Aaron Bedermann was born in Madison, Wisconsin. He received his BS in Chemistry from the University of Wisconsin, where he performed undergraduate research under the supervision of Professor Richard Hsung. He is currently pursuing graduate research at Colorado State University under the guidance of Professor John L. Wood.

Nickel-Catalyzed Cross-Coupling of Aryl Halides with Alkyl Halides: Ethyl 4-(4-(4-methylphenylsulfonamido)-phenyl)butanoate

Submitted by Daniel A. Everson, David T. George, and Daniel J. Weix.[*1]
Checked by Jonas F. Buergler and John L. Wood.

1. Procedure

A.   *N-(4-Bromophenyl)-4-methylbenzenesulfonamide (1)* (Note 1). With no precautions to exclude air or moisture, a 500-mL round-bottomed flask is equipped with a PTFE-coated egg-shaped magnetic stir bar (38 x 16 mm). 4-Bromoaniline (Note 2) (15.1 g, 88.0 mmol, 1.00 equiv) and *p*-toluenesulfonyl chloride (Note 3) (17.1 g, 89.8 mmol, 1.02 equiv) are weighed into plastic weigh boats and added to the vessel through a powder funnel. Dichloromethane (Note 4) (230 mL) is then poured into the vessel from a graduated cylinder, followed by the addition of pyridine (Note 5) (7.8 mL, 97 mmol, 1.1 equiv) by syringe in two portions. The reaction vessel is stoppered with a rubber septum and an 18-gauge needle to vent the reaction to air before stirring (~500 rpm) at room temperature (20 °C) for 16 h. The reaction mixture turns from yellow to clear orange immediately after the addition of pyridine. After the reaction is judged complete by TLC analysis (Note 6), the reaction mixture is poured into a 2 L separatory funnel containing 800 mL of sat. NH$_4$Cl$_{(aq)}$ (Note 7). The organic layer is separated and the aqueous extracted with dichloromethane (2 x 50 mL). The combined organic layer is then washed with water (400 mL), brine (400 mL) and dried

200

over anhydrous $MgSO_4$ (10.0 g, ~1 min) (Note 8). The drying agent is removed by vacuum filtration through a course-fritted glass Büchner funnel. The filtrate is concentrated by rotary evaporation (30 °C, 74 mmHg) to yield 28.5–28.6 g (99% yield) of *N*-(4-bromophenyl)-4-methylbenzenesulfonamide as a white solid (Note 9).

B.   *Ethyl 4-(4-(4-methylphenylsulfonamido)phenyl)butanoate (2).* A 500-mL 3-necked Morton flask with 24/40 ground-glass joints is equipped with a nitrogen-gas inlet on the left neck, a glass stopper on the right neck, and the center neck is equipped for mechanical stirring (glass bearing, glass stirrer rod, 60 mm PTFE paddle, RW 20 Tekmar motor) (Note 10). $NiI_2 \cdot xH_2O$ (528 mg, 1.31 mmol, 0.051 equiv) (Note 11), 4,4'-di-methoxy-2,2'-bipyridine (284 mg, 1.31 mmol, 0.051 equiv) (Note 12), sodium iodide (1.25 g, 8.34 mmol, 0.33 equiv) (Note 13), and *N*-(4-bromophenyl)-4-methylbenzenesulfonamide (8.33 g, 25.5 mmol, 1.00 equiv) are transferred via powder funnel through the right neck of the Morton flask. To these solids, 1,3-dimethylpropyleneurea (DMPU, 105 mL) (Note 14) is added from a graduated cylinder. Then pyridine (105 µL, 1.31 mmol, 0.053 equiv) (Note 5) and ethyl 4-bromobutyrate (4.0 mL, 28 mmol, 1.1 equiv) (Note 15) are added by syringe. A gas outlet adapter with tubing attached to an oil-filled bubbler is placed in the right neck before submerging the reaction vessel up to the solvent line in an oil bath pre-equilibrated to 60 °C. The vessel headspace is purged with nitrogen for 15 min while the mixture is stirred (500-600 rpm) (Note 16). After the nitrogen purge is complete, the gas outlet on the right neck is replaced with the original glass stopper and the reaction mixture is stirred at 60 °C until the reaction mixture takes on a dark green color (15–30 min) (Note 17). Next, zinc powder (6-9 µm, 3.44 g, 52.6 mmol, 2.06 equiv) (Note 18) is added through a powder funnel attached to the right neck under a positive flow of nitrogen. The flask is kept under a slight positive pressure of nitrogen during the course of the reaction. Generally within 5-15 min after the addition of zinc, the reaction undergoes a characteristic dark green to orange-brown color change, indicating the reaction has begun. The reaction is judged complete when the color changes to black (Note 19).

Once the reaction mixture changes color to black, it is allowed to cool to room temperature before filtering though a pad of diatomaceous earth (15 g, wetted with 40 mL of diethyl ether and packed firmly in place) (Note 20) in a 350-mL course-fritted glass Büchner funnel. The reaction vessel is rinsed with ether (3 × 50 mL) (Note 21) and these rinses are filtered as well.

The filtrate is poured into a 1 L separatory funnel containing 150 mL of 1 M NH$_4$Cl$_{(aq)}$. The organic layer is separated and set aside. The aqueous layer is then extracted with additional ether (3 × 60 mL). The combined organic layers are washed with water (75 mL) and brine (75 mL), then dried over MgSO$_4$ (15 g, ~1 min). The solids are removed by vacuum filtration through a course-fritted glass Buchner funnel and washed with ether (20 mL). The filtrate is then concentrated by rotary evaporation (30 °C, 74 mmHg) to give a colorless oil. The aqueous layers are combined and set aside for later recovery of DMPU, if desired (Note 22).

The oil is purified by stepped-gradient flash chromatography on silica gel (column diameter 90 mm, 225 g SiO$_2$) (Note 23). The column is dry-packed with silica gel, equilibrated with hexanes (500 mL, passed through the column twice) (Note 24), and the compound is loaded directly onto the silica with dichloromethane (3 mL). The vessel containing the crude material is rinsed with 95:5 hexanes:EtOAc (5 mL), and additional dichloromethane (5 mL), which are then used to rinse the sides of the column (Note 25). After adding a layer of sand (~12 mm.) to the column, the compound is eluted with the following ratios (v:v) of hexanes:EtOAc: 95:5 (500 mL, discarded to forerun), 90:10 (500 mL, discarded to forerun), 85:15 (500 mL, discarded to forerun), 80:20 (500 mL), 75:25 (1000 mL), 70:30 (500 mL), 65:35 (500 mL), and 60:40 (500 mL) (Note 26). Fractions are collected in 25 mm x 150 mm test-tubes (~65-70 mL). The fractions containing only product as determined by TLC (Note 19) are rinsed with EtOAc into a tared 1 L recovery flask and concentrated by rotary evaporation (35 °C, 74 mmHg). A small, tared PTFE-coated magnetic stir bar (7 mm x 2 mm) is added to the viscous, colorless oil before drying on high vacuum (0.1 mmHg, 20 °C) for 1 h to remove any traces of solvent. A heat gun is used to gently warm the oil while on high vacuum and the heat is removed as soon as gas bubbles begin to evolve. After drying, the flask contained 7.16–7.70 g (78%–83% yield) (Note 27) of ethyl 4-(4-(4-methylphenylsulfonamido)phenyl)-butanoate as a colorless viscous oil.

2. Notes

1. This procedure is a modification of a literature method.[2]
2. 4-Bromoaniline (98+%, Alfa Aesar or 97%, Sigma-Aldrich) was used a received.

Org. Synth. **2013**, *90*, 200-214

3. *p*-Toluenesulfonyl chloride (98%, Alfa Aesar) was used as received. The submitters used 1.10 equivalents of *p*-toluenesulfonyl chloride, but in the hands of the checkers this lead to contaminated product; therefore, the checkers decreased the amount to 1.02 equivalents. This discrepancy may be due to the use of *p*-toluenesulfonyl chloride of different quality since the submitters used different vendors (98%, Lancaster or ≥98% Sigma-Aldrich).

4. Dichloromethane (certified ACS grade, Fischer Scientific) was used as received.

5. Pyridine (ultrapure, spectrophotometric grade 99.5+%, Alfa Aesar) was used as received.

6. The progress of the reaction can be followed by TLC analysis (eluent: 4:1 hexanes:EtOAc, visualization with 254 nm UV light). *p*-Toluenesulfonyl chloride has an R_f of 0.58, 4-bromoaniline has an R_f of 0.23, and *N*-(4-bromophenyl)-4-methylbenzenesulfonamide (**1**) has an R_f of 0.28. TLC analysis was performed on pre-coated glass plates (SiliaPlate, 60Å, 250 μm, F_{254}, SiliCycle). The submitters used EMD silica gel 60 F_{254} pre-coated glass plates.

7. Ammonium chloride (ACS certified, crystalline, Malinckrodt) was used as received.

8. Magnesium sulfate (anhydrous powder, Fisher Scientific) was used as received.

9. In the hands of the checkers the yield was greater than 99% and the product obtained as a white solid. The submitters report that in their case the yield of this procedure was consistently greater than 80%, but the color of the final product would vary from faint yellow to brown. The color of the material did not affect the outcome of the next reaction. The product exhibited the following physiochemical properties and was stable on the bench top for months: mp: 148–149 °C; ^{1}H NMR (400 MHz; CDCl$_3$) δ: 2.36 (s, 3 H), 6.96 (d, *J* = 8.6 Hz, 2 H), 7.22 (d, *J* = 7.8 Hz, 2 H), 7.31 (s, 1 H), 7.31 (d, *J* = 9.0 Hz, 2 H), 7.66 (d, *J* = 8.6 Hz, 2 H). ^{13}C NMR (101 MHz; CDCl$_3$) δ: 21.6, 118.4, 122.9, 127.2, 129.8, 132.3, 135.5, 135.7, 144.2. FTIR (film, cm^{-1}): 3254 (N-H), 1328 and 1162 (S=O). HRMS (*m/z*) (ESI) calc. for C$_{13}$H$_{13}$BrNO$_2$S (M+H$^+$) 327.9825, found 327.9823. Anal. Calcd. for C$_{13}$H$_{12}$BrNO$_2$S: C, 47.86; H, 3.71; N, 4.29; found: C, 47.96; H, 3.83; N, 4.40.

10. The stirrer paddle must not be in contact with the bottom or sides of the flask. The stirrer paddle was generally held ~10 mm above the bottom

of the flask to prevent mechanical activation of the reducing agent, which results in greater amounts of hydrodehalogenation by-products and lower product yields.

PTFE sleeves (ribbed, 24/40, Fischer Scientific) were used to seal all ground glass joints, and heavy mineral oil (mineral oil white, heavy, Malinckrodt) was used to lubricate the glass stir rod inside the 24/40 standard taper adapter (see photo). The checkers used thin PTFE sleeves.

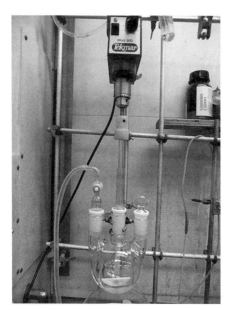

11. Nickel (II) iodide hydrate was purchased from Strem Chemicals and used as received. The submitters used elemental analysis to assess the water content to be $x = 3.5$, but, because a slight excess of nickel iodide does not change the outcome of these reactions, the molecular weight of the nickel (II) iodide was taken to be that of the pentahydrate (MW = 402.58). For further information, see discussion section.

12. 4,4'-Di-methoxy-2,2'-bipyridine (97%, Sigma-Aldrich) was used as received.

13. Sodium iodide (puriss. p.a., ≥99%, Sigma-Aldrich) was used as received. The submitters used Strem Chemicals (anhydrous, 99%).

14. The submitters state that 1,3-dimethylpropylene urea (DMPU) (99%, AK Scientific, Inc.) was used as received. The DMPU can be recovered and reused for subsequent cross-coupling reactions (see Note 22).

204

In the case of the checkers, the use of undistilled DMPU (99%, AK Scientific, Inc. or 98%, Alfa Aesar) resulted in significant amounts of a side product. According to NMR and mass analyses this contamination has been tentatively identified as 1,3,5- trimethyl-1,3,5-triazine-2,4,6-(1H,3H,5H)-trione (3) and is clearly arising from the DMPU, since it can be detected by simply extracting DMPU with water and diethyl ether. The fact that this contamination has the same R_f as 2 makes it difficult to remove by column chromatography. Distillation of the commercially available DMPU as described by the submitters (Note 22) decreased the amount of this side product significantly; however, trace amounts (0.5 mol % for the submitters) were always visible in the NMR spectrum (s, 3.35 ppm).

3

15. Ethyl 4-bromobutyrate (98%, Alfa Aesar) was used as received.

16. The submitters used argon instead of nitrogen and a rubber septum pierced with a needle attached to the oil bubbler to purge the vessel. Oxygen does not irreversibly decompose the catalyst because the stoichiometric reducing agent can reduce any oxidized catalyst. However, reactions with too much oxygen in the headspace often suffer from long induction periods that prevent the catalyst from reacting with starting materials until the oxygen in the headspace has finished reacting with the catalyst and reducing agent. Sometimes reactions do not proceed to completion in a reasonable amount of time (< 24 h) when there is a large headspace volume of air. See discussion section below.

17. In the hands of the submitters the color change to deep green was finished in about 15 minutes. The pre-stirring allows the nickel iodide to coordinate to the bipyridine complex and form the dark-green, tetrahedral (4,4'-dimethoxy-2,2'-bipyridine)NiI₂.

18. Zinc dust (<10 μm, ≥98%, Sigma-Aldrich) was used as received. The submitters used zinc powder (6-9 μm, 97.5% metal basis, Alfa Aesar).

19. The submitters have found the color change to be a reliable method for determining the end-point of the reaction. For this particular substrate the red-orange to black color change occurred in 4-6 h. No further product formation was observed if the reaction was allowed to stir longer (up to 48

h), and no decomposition of the product was observed either. In the hands of the checkers the color change occurred after 3-5 h. The progress of the reaction can be followed by TLC analysis (eluent: 4:1 hexanes:EtOAc, visualization with 254 nm UV light). N-(4-Bromophenyl)-4-methylbenzenesulfonamide (1) has an R_f of 0.23, ethyl 4-(4-(4-methylphenylsulfonamido)phenyl)butanoate (2) has an R_f of 0.15 and 4-methyl-N-phenylbenzenesulfonamide (hydrodehalogenated aryl bromide) has an R_f of 0.25. Since the starting material 1 and the hydrodehalogenated side product have very similar R_f judging the progress of the reaction by TLC is difficult; therefore, monitoring the color change is a good alternative method.

20. Filter agent, Celite® 545 was purchased from Sigma-Aldrich and used as received. The submitters used "Celite 545 filter aid" from Fischer Scientific.

21. Diethyl ether (certified ACS grade, 7 ppm BHT stabilizer, Fisher Scientific) was used as received.

22. To the combined aqueous layers from the workup, sodium chloride (Aldrich) was added until the solution was saturated (generally ~ 5 g, some undissolved salt does not pose a problem). DMPU was extracted from the salted aqueous layer with dichloromethane (3 × 75 mL). The combined organic layers were dried over magnesium sulfate (15 g, ~1 min), and the magnesium sulfate was removed by vacuum filtration through a course-fritted glass Büchner funnel. The filtrate was concentrated by rotary evaporation (35 °C, 73 mmHg) to give 90–95 mL (86%–90% recovery) of faintly yellow liquid. This material was then distilled from calcium hydride (Sigma-Aldrich, reagent grade, 95%) (60–61 °C, 0.05 mmHg)[3], dried to <1000 ppm water over 4 Å molecular sieves, and reused in subsequent cross coupling reactions. The submitters repeated the title cross coupling reaction on identical scale and obtained 8.63 g (94%) of ethyl 4-(4-(4-methylphenylsulfonamido)phenyl)butanoate (2) using DMPU recycled by this method.

23. Silica gel (SiliaFlash® P60, 230-400 mesh, SiliCycle) was used as received. The submitters used EMD silica gel 60 (mesh 230-400)

24. Hexanes (certified ACS grade, 4.2% various methylpentanes, Fisher Scientific) were used as received.

25. Ethyl acetate (EtOAc, certified ACS grade) was purchased from Fischer Scientific and used as received.

26. The separation between product and hydrodehalogenated aryl bromide is small ($\Delta R_f = 0.1$). The submitters point out, therefore, that the yield may vary 5–15% depending on the number of mixed fractions. The checkers obtained only few mixed fractions, not calculating the possible yield loss.

27. The submitters obtained the product as faintly yellow oil in 79–93% yield. The submitters also mention that the product oil will slowly solidify after scratching the sidewall of the vessel with a metal spatula. Both the oil and the resulting white solid had identical spectral properties. In the hands of the checkers the product solidified only in one case after a prolonged time in the freezer. The product exhibited the following physiochemical properties and was stable on the bench top for months: ^{1}H NMR (400 MHz; CDCl$_3$) δ: 1.22 (t, J = 7.2 Hz, 3 H), 1.85 (quintet, J = 7.5 Hz, 2 H), 2.24 (t, J = 7.4 Hz, 2 H), 2.34 (s, 3 H), 2.54 (t, J = 8.0 Hz, 2 H), 4.09 (q, J = 7.0 Hz, 2 H), 6.96-7.02 (m, 4 H), 7.05 (s, 1 H), 7.19 (d, J = 8.6 Hz, 2 H), 7.63 (d, J = 8.2 Hz, 2 H). ^{13}C NMR (101 MHz; CDCl$_3$) δ: 14.2, 21.5, 26.4, 33.5, 34.4, 60.3, 122.0, 127.2, 129.2, 129.6, 134.4, 136.1, 138.6, 143.7, 173.5. FTIR (film, cm^{-1}) 3255 (N-H); 1731 (C=O); 1337; 1162 (S=O). HRMS (m/z) (ESI) calc. for C$_{19}$H$_{27}$N$_2$O$_4$S (M+NH$_4^+$) 379.1687, found 379.1687. Anal. Calcd. for C$_{19}$H$_{23}$NO$_4$S: C, 63.13; H, 6.41; N, 3.88; found (oil): C, 62.97; H, 6.54; N, 3.98. mp (of crystallized material obtained by the submitters): 49-52 °C.

Handling and Disposal of Hazardous Chemicals

The procedures in this article are intended for use only by persons with prior training in experimental organic chemistry. All hazardous materials should be handled using the standard procedures for work with chemicals described in references such as "Prudent Practices in the Laboratory" (The National Academies Press, Washington, D.C., 2011 www.nap.edu). All chemical waste should be disposed of in accordance with local regulations. For general guidelines for the management of chemical waste, see Chapter 8 of Prudent Practices.

These procedures must be conducted at one's own risk. *Organic Syntheses, Inc.*, its Editors, and its Board of Directors do not warrant or guarantee the safety of individuals using these procedures and hereby disclaim any liability for any injuries or damages claimed to have resulted from or related in any way to the procedures herein.

3. Discussion

Very recently, our group reported a new nickel-catalyzed method to directly couple haloalkanes (iodides or bromides) with haloarenes (iodides, bromides, or electron-poor chlorides).[4-5] A distinguishing feature of this method is that the only organometallic intermediates are catalytic organonickel species – no organozinc or organomanganese reagents are involved. Previous transition metal-catalyzed methods to synthesize alkylated aromatic compounds have relied on the coupling of pre-formed aryl nucleophiles with alkyl electrophiles,[6-8] aryl electrophiles with pre-formed alkyl nucleophiles,[9] or *in situ* formation of a carbon nucleophile.[10-15] A major advantage of this new approach is that most carbon nucleophiles are synthesized from the corresponding organic halides,[16] and thus a synthetic step can be eliminated. Additionally, the absence of stoichiometric strong nucleophiles or bases to aid trans-metalation imparts excellent functional group compatibility. Concurrent with our work, Gosmini and Amatore developed a similar cobalt-catalyzed method, though the mechanism remains unclear.[17] Following our studies, Peng has studied the use of a simplified catalyst for inter- and intramolecular couplings under similar conditions[18] and Gong noted that the addition of $MgCl_2$ can improve yields with secondary alkyl halides.[19]

Acidic groups, such as N-arylsulfonamides (pKa similar to acetic acid in DMSO),[20] are well tolerated (title compound **2**). Substrates that are prone to β-elimination when metalated (**6**) couple cleanly with aryl bromides containing acidic protons (**5**) (Table 1, entry 1) in stark contrast to the difficulties conventional cross-coupling methods have with these types of substrates.[21] Carbon nucleophiles used for Suzuki,[22] Stille,[23] and Hiyama-Denmark[24] cross-coupling reactions are not reactive under these conditions, allowing for the straightforward synthesis of poly-substituted aromatic compounds (Table 1, entries 4-6). Electrophilic groups, such as alkyl-aryl ketones, trifluoromethanesulfonic acid esters, and acetylated phenols, are compatible with our method as well (Table 1, entries 7-9). Lastly, the coupling of **26** with **27** is the synthetic equivalent of the α-arylation of acetaldehyde (Table 1, entry 10).

Scaling our recent procedures from 1 mmol to 25 mmol required addressing two difficulties, the heterogeneous nature of the reduction and variable induction periods that resulted in unpredictable reaction times.

The best results are obtained with mechanical stirring, but care must be taken to avoid grinding the zinc dust with the stirrer. Mechanical activation of metal powders is well known[25-27] and in the present reaction results in hydrodehalogenation products. This is presumably due to direct insertion of the zinc into the starting materials. Keeping the stirrer above the bottom surface and using a Morton flask[28] to ensure turbulent mixing results in reliable, high yields on large scale. Zinc powder from Alfa Aesar has proven more reliable than from other suppliers. If very deactivated zinc must be used, then brief activation with hydrochloric acid can be effective.[5]

As we discovered when conducting kinetic studies on the present reaction,[5] oxygen in the headspace leads to long induction periods. Although the reactions proceed as expected after this induction period on small scale, in some large scale reactions we observed that the starting materials would decompose or the catalyst would become inactivated before reactions could complete. Sweeping most of the oxygen out with argon (or nitrogen) enables consistent reaction times and higher yields. These reactions are oxygen *tolerant* in that trace oxygen is not a concern, but large amounts of oxygen should be avoided.

Finally, although DMPU is routinely used on scale, its price is higher than other dipolar aprotic solvents routinely used in the lab. In order to conserve resources in our own lab, we worked out a simple procedure for recovering the DMPU that allows for efficient recycling of the solvent.

Table 1. Scope of Nucleophile Free Cross Coupling[a]

Ar–Br + Br–R' $\xrightarrow[\text{Zn}^0 \text{ (2 equiv), DMPU, 60 °C}]{\begin{array}{l}\text{5 mol \% NiI}_2\cdot x\text{H}_2\text{O}\\\text{5 mol \% Ligand (3) or (4)}\\\text{5 mol \% pyridine}\\\text{25 mol \% NaI}\end{array}}$ Ar–R'

Entry	Ligand	Ar–Br	Br–R'	Ar–R'	Yield(%)[b]
1	3	HO–⟨⟩–Br **5**	Br⌒OTBS **6**	HO–⟨⟩–⌒OTBS **7**	76
2	3	Me₂C=C(Me)Br **8**	Br⌒⌒CO₂Bn **9**	Me-alkene-CO₂Bn **10**	76[c] 91:11 (E):(Z)
3	3	**11**	**9**	**12**	58[c, d] 71:29 (Z):(E)
4	3	Me₂Si(OH)–⟨⟩–Br **13**	Br⌒⌒CO₂Et **14**	Me₂Si(OH)–⟨⟩–CO₂Et **15**	67
5	3	Me₃Sn–⟨⟩–Br **16**	**14**	Me₃Sn–⟨⟩–CO₂Et **17**	74
6	4	Bpin–⟨⟩–Br **18**	**14**	Bpin–⟨⟩–CO₂Et **19**	76
7	4	MeC(O)–⟨⟩–Br **20**	**14**	MeC(O)–⟨⟩–CO₂Et **21**	76
8	4	TfO–⟨⟩–Br **22**	**14**	TfO–⟨⟩–CO₂Et **23**	74
9	4	AcO–⟨⟩–Br **24**	**14**	AcO–⟨⟩–CO₂Et **25**	64
10	4	EtO₂C–⟨⟩–Br **26**	Br–dioxolane **27**	EtO₂C–⟨⟩–dioxolane **28**	60[e]

[a] Reaction conditions: organic halides (0.75 mmol each), NiI₂·xH₂O (0.054 mmol), ligand (0.05 mmol), pyridine (0.05 mmol), sodium iodide (0.19 mmol), zinc dust (>10 μm, 1.5 mmol), and DMPU (3 mL) were assembled on the bench in a 1 dram vial and heated for 5-41 h under air. Yields are of isolated and purified product. [b] Average of two runs. [c] Isolated as an inseparable mixture with benzyl butyrate, yields determined by NMR analysis of this mixture. [d] Starting material (2-bromo-2-butene) was an 88:12 ratio of (Z):(E) isomers. [e] Run at 80 °C and with 1 equiv sodium iodide.

Org. Synth. **2013**, *90*, 200-214

1. *daniel.weix@rochester.edu*. Department of Chemistry, University of Rochester, Rochester, New York U.S.A. 14627-0216. This work was supported by the University of Rochester, the NIH (R01 GM097243), and the NSF (Graduate Research Fellowship to DAE and Research Experience for Undergraduates Fellowship CHE-1156340 to DTG). Analytical data were obtained from the CENTC Elemental Analysis Facility at the University of Rochester, funded by the NSF (CHE-0650456).

2. McKeown, S. C.; Hall, A.; Blunt, R.; Brown, S. H.; Chessell, I. P.; Chowdhury, A.; Giblin, G. M. P.; Healy, M. P.; Johnson, M. R.; Lorthioir, O.; Michel, A. D.; Naylor, A.; Lewell, X.; Roman, S.; Watson, S. P.; Winchester, W. J.; Wilson, R. J. *Bioorg. Med. Chem. Lett.* **2007**, *17*, 1750–1754.

3. Li, C.-D.; Mella, S. L.; Sartorelli, A. C. *J. Med. Chem.* **1981**, *24*, 1089–1092.

4. Everson, D. A.; Shrestha, R.; Weix, D. J. *J. Am. Chem. Soc.* **2010**, *132*, 920–921.

5. Everson, D. A.; Jones, B. A.; Weix, D. J. *J. Am. Chem. Soc.* **2012**, *134*, 6146–6159.

6. Rudolph, A.; Lautens, M. *Angew. Chem., Int. Ed.* **2009**, *48*, 2656–2670.

7. Frisch, A. C.; Beller, M. *Angew. Chem., Int. Ed.* **2005**, *44*, 674–688.

8. Terao, J.; Kambe, N. *Bull. Chem. Soc. Jpn.* **2006**, *79*, 663–672.

9. Jana, R.; Pathak, T. P.; Sigman, M. S. *Chem. Rev.* **2011**, *111*, 1417–1492.

10. Amatore, M.; Gosmini, C. *Chem. Commun.* **2008**, 5019–5021.

11. Czaplik, W. M.; Mayer, M.; Jacobi von Wangelin, A. *Angew. Chem., Int. Ed.* **2009**, *48*, 607–610.

12. Czaplik, W. M.; Mayer, M.; Jacobi von Wangelin, A. *Synlett* **2009**, 2931–2934.

13. Krasovskiy, A.; Duplais, C.; Lipshutz, B. *J. Am. Chem. Soc.* **2009**, *131*, 15592–15593.

14. Duplais, C.; Krasovskiy, A.; Wattenberg, A.; Lipshutz, B. H. *Chem. Commun.* **2010**, *46*, 562–564.

15. Krasovskiy, A.; Duplais, C.; Lipshutz, B. H. *Org. Lett.* **2010**, *12*, 4742–4744.

16. Knochel, P., *Handbook of functionalized organometallics : applications in synthesis*. Wiley-VCH: Weinheim, 2005; p 653.

17. Amatore, M.; Gosmini, C. *Chem.–Euro. J.* **2010**, *16*, 5848–5852.

18. Yan, C.-S.; Peng, Y.; Xu, X.-B.; Wang, Y.-W. *Chem.–Eur. J.* **2012**, *18*, 6039–6048.
19. Wang, S.; Qian, Q.; Gong, H. *Org. Lett.* **2012**, *14*, 3352-3355.
20. The pKa of PhSO₂NHPh as been reported to be 11.9 in DMSO: Cheng, J.-P.; Zhao, Y. *Tetrahedron* **1993**, *49*, 5267−5276.
21. For an alternative method using potassium alkoxyethyl trifluoroborates, see *J. Org. Chem.* **2012**, *77*, 10399-10408.
22. Miyaura, N.; Suzuki, A. *Chem. Rev.* **1995**, *95*, 2457–2483.
23. Stille, J. K. *Angew. Chem., Int. Ed.* **1986**, *25*, 508–524.
24. Denmark, S. E.; Regens, C. S. *Acc. Chem. Res.* **2008**, *41*, 1486–1499.
25. Baker, K. V.; Brown, J. M.; Hughes, N.; Skarnulis, A. J.; Sexton, A. *J. Org. Chem.* **1991**, *56*, 698–703.
26. Tilstam, U.; Weinmann, H. *Org. Process Res. Dev.* **2002**, *6*, 906–910.
27. Girgis, M. J.; Liang, J. K.; Du, Z.; Slade, J.; Prasad, K. *Org. Process Res. Dev.* **2009**, *13*, 1094–1099.
28. Morton, A. A. *Ind. Eng. Chem. Anal. Ed.* **1939**, *11*, 170.

Appendix
Chemical Abstracts Nomenclature; (Registry Number)

4-Bromoaniline; (106-40-1)
p-Toluenesulfonyl chloride: 4-Methylbenzene-1-sulfonyl chloride; (98-59-9)
Pyridine; (110-86-1)
NiI₂·xH₂O: Nickel(II) iodide hydrate; (7790-34-3)
4,4'-Di-methoxy-2,2'-bipyridine: 4,4'-di-methoxy-2,2'-bipyridine; (17217-57-1)
Sodium iodide; (7681-82-5)
1,3-Dimethyl-propylene urea: 1,3-Dimethyltetrahydropyrimidin-2(*1H*)-one; (7226-23-5)
Ethyl 4-bromobutanoate; (2969-81-5)
Zinc powder: Zinc; (7440-66-6)

Daniel J. Weix was born and raised outside of Milwaukee, Wisconsin. He studied chemistry at Columbia University (BA, 2000), where he worked on helicene chemistry with Prof. Tom Katz. Graduate studies at the University of California, Berkeley (2000-2005) with Prof. Jon Ellman focused on the synthesis and application of *tert*-butanesulfinamide. After postdoctoral studies (2005-2008) on Ir-catalyzed allylation chemistry with Prof. John Hartwig at both Yale University and the University of Illinois, Daniel started his independent career at the University of Rochester in 2008. His research program focuses on the development of new concepts in catalysis, with a major current interest in cross-electrophile coupling reactions.

Daniel A. Everson was born in Minneapolis, Minnesota in 1985. He completed his B.S. in 2007 at the University of St. Thomas, St. Paul, Minnesota, working with Prof. J. Thomas Ippoliti. After working as a research associate for Prof. Ippoliti during 2007-2008 he joined the group of Daniel J. Weix at the University of Rochester, Rochester, New York in 2008, and was awarded an NSF graduate research fellowship in 2010. His research interests are in developing new transition metal mediated nucleophile free cross coupling methodologies.

David T. George was born in Highland Park, Illinois in 1990 and grew up in Middletown, New Jersey. He will complete his B.S. at the University of Rochester, Rochester, New York, in May 2013 and joined the group of Daniel J. Weix in 2011. In addition to research at the University of Rochester, David also worked for Burpee Materials Technology, Eatontown, New Jersey. His current research project is the application of reductive coupling reactions to natural product synthesis.

Jonas F. Buergler was born in Davos, Switzerland in 1982. He obtained his BSc and MSc in Chemistry (2006) from the Swiss Federal Institute of Technoloy (ETH) Zurich. In 2011 he was awarded his PhD from the ETH for his work on the synthesis and applications of P-stereogenic ferrocenyl phosphines under the supervision of Prof- Dr. A. Togni. Since September 2011 he is a postdoctoral fellow in the group of Prof. John L. Wood with fellowships from the Swiss National Science Foundation and the Novartis Jubilee Foundation, where he is currently working on the total synthesis of tetrapetalone A.

214

Discussion Addendum for:
Preparation of 4-Acetylamino-2, 2, 6, 6-tetramethylpiperidine-1-oxoammonium Tetrafluoroborate and the Oxidation of Geraniol to Geranial (2,6-Octadienal, 3,7-dimethyl-, (2e)-)

Prepared by James M. Bobbitt*,[1] Nicholas A. Eddy,[1] Jay J. Richardson,[1] Stephanie A. Murray,[2] and Leon J. Tilley.[2]
Original article: Bobbitt, J. M.; Merbouh, N. *Org. Synth.* **2005**, *82*, 80–83.

The useful nitroxide catalyst, 4-acetamido-TEMPO **2** and the oxoammonium salt, 4-acetamido-2,2,6,6-tetramethyl-1-oxopiperidinium tetrafluoroborate (4-acetylamino-2,2,6,6-tetramethylpiperidin-1-oxoammonium tetrafluoroborate, Bobbitt's salt) **3** have been prepared from 4-amino-2,2,6,6-tetramethylpiperidine **1** in molar amounts in an inexpensive and high yielding, revised procedure. Compound **1** is an industrial chemical available in quantity from TCI-America at a reasonable price, as well as from many other vendors. Using current prices, **2** and **3** can be prepared for less than $1.00 per gram. The procedure is shown in Scheme 1.

In addition to this revised preparation procedure (**A**), we will report in this addendum an overall view of stoichiometric oxoammonium chemistry (**B**), solubility properties of **2** and **3** and their relevance (**C**), the relative oxidation rates for hydroxyl groups in various chemical environments (**D**), examples of selective reactions (**E**), and some recent important developments from other laboratories (**F**).

A. The Revised Preparation of 2 and 3 (Scheme 1)[3]

Several changes have been made from the original procedure.[4-8] The acetylation of **1** to its acetamido derivative is now carried out in ice water and combined with base, catalysts, and peroxide to give in "one pot" the acetylated TEMPO derivative **2** in 90-93% yield. In the oxidation of **2** to **3** with HBF$_4$ followed by bleach (NaOCl), NaBF$_4$ is added to salt out the final product **3** in 90–93 % yield. The NaBF$_4$ addition results in common ion assisted precipitation, which provides an enhanced yield and product purity. The reduced solubilities are shown in Table 1. The procedure is entirely "green" since the only solvent is water.

Precise instructions are given for the recrystallization of **2** and **3**, if needed, and for the recovery of the leftover compounds after each step in the sequence.[3]

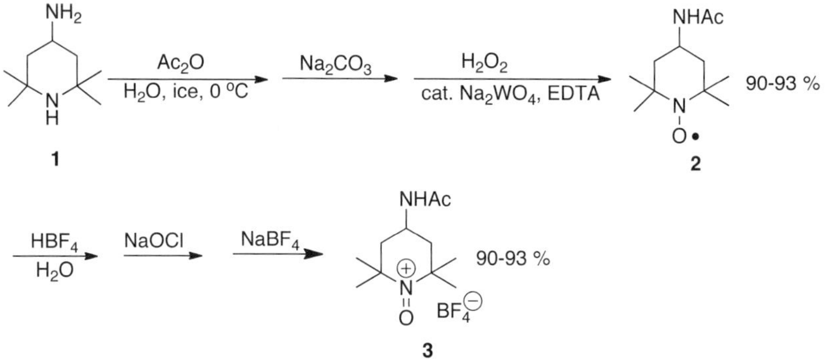

Scheme 1. Revised Procedure for the Preparation of **2** and **3**.

B. Overview of Oxidations with the Oxoammonium Ion

All of these reactions have been discussed in detail in an *Organic Reactions* chapter[9] and are shown in Scheme 2. The catalytic reaction (1) is certainly the most widely used since it is inexpensive and efficient. However, it is quite complex, has been discussed in detail[9] and will not be considered here. Variations on the stoichiometric reactions (2) are neutral reactions in DCM with silica gel (2a), reactions in the presence of pyridine bases (2b), and Golubev disproportionation oxidations (2c). The stoichiometric reactions are colorimetric and highly dependent on

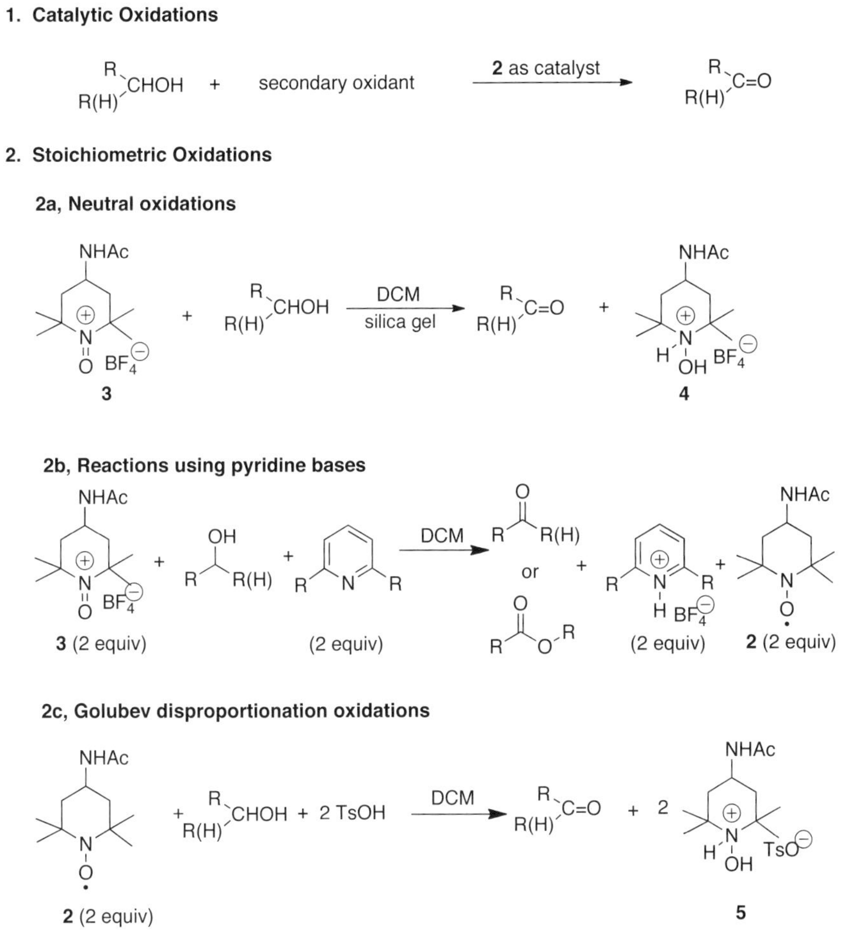

1. Catalytic Oxidations

2. Stoichiometric Oxidations

2a, Neutral oxidations

2b, Reactions using pyridine bases

2c, Golubev disproportionation oxidations

Scheme 2. Overall Stoichiometric Oxidations

solubilities, primarily in DCM. These solubilities have been measured and are also listed in Table 1.[3]

The stoichiometric reactions can be divided into three methods.

Method 2a. The simplest of the three methods is the neutral oxidation in DCM and silica gel.[5] The reaction is colorimetric in that a bright yellow slurry of **3** and silica gel is converted to a white slurry of the reduced oxidant **4**, the product and silica gel. The yields are near quantitative. Since both **3** and **4** are only slightly soluble in DCM (Table 1), the course of the reaction can be followed by filtering samples of the reaction mixture through a cotton wad and analyzing the samples by GC, TLC or even NMR.[10] A simple filtration through a 2-4 mm pad of silica gel yields a DCM solution of pure aldehyde or ketone that can be isolated or used in a tandem reaction. The method is ideal for the preparation of small amounts of volatile or hygroscopic aldehydes or ketones for tandem reactions. The major disadvantage of the method is that aliphatic alcohols react slowly, frequently requiring one or two days. Relative rates are presented in Table 2.

Method 2b. Oxidations in the presence of pyridine bases have been less explored, but this variation may become more important.[11-13] In general, oxidations in the presence of pyridine tend to give esters,[11] and oxidation in the presence of 2,6-lutidine tend to give aldehydes from primary alcohols. Again, the reactions are colorimetric, going from a yellow slurry to a bright orange red color due to the nitroxide **2**, which is the product of the reduction of **3**. The reaction is fast and is a more powerful oxidation system than method 2a. For example, trifluoromethylcarbinols are oxidized to ketones using method 2b,[12] but this reaction does not take place under neutral (2a) conditions. The disadvantage of this method is that both the pyridine base and the nitroxide must be removed for product isolation. This can be accomplished in several ways.[11-13]

Method 2c. Method 2c depends upon the *p*-toluenesulfonic acid-mediated disproportionation of two equivalents of **2** into a mixture of tosylate salts of **3** and **4**, which is the Golubev reaction.[14] The oxidation is carried out by the oxoammonium salt to give product and tosylate **5**. This method has been extensively used.[4,9] Once again, the reaction is colorimetric in that the yellow slurry is converted to the white reduced form, a tosylate salt **5**. This salt precipitates, but the precipitate is

218

somewhat more soluble than the tetrafluoroborate salt (Table 1). The reaction is fast, but sometimes the products require further purification.[4]

C. Solubility Studies and Colors of Oxoammonium Salts and Related Products

The solubility properties and the colors of the various compounds are important for the understanding of the reactions and are given in Table 1. The general discussion is given in the preceding section.

Table 1. Solubilities and Colors of the 4-Acetamido Compounds.

	Water	DCM (rt)	Diethyl Ether	Color
2	3 g/100 mL at 0 °C 3.5 g/100 mL at rt >50 g/100 mL at 100 °C 1 g/100 mL in 2 M NaCl[a]	soluble	0.28 g/100 mL at rt	orange
3	6 g/100 mL at 0 °C 8 g/100 mL at rt >50 g/100 mL at 100 °C 1.8 g/100 mL in 0.8M NaBF$_4$[a]	0.1 g/100 mL	insoluble	yellow
4	soluble	0.02 g/100 mL	insoluble	white

[a]These data are important for the revised preparation of **2** and **3**.[3]

D. Rate Studies of Type 2a Reactions

The rates of oxidations by method 2a are given in Table 2. The reference compound is benzyl alcohol. The rates were measured by oxidizing a mixture of two alcohols with half of the theoretical amount of oxidant followed by NMR analysis. These data are accurate to only one significant figure and are taken from two papers.[5, 15]

Table 2. Relative Rates of Oxidation of Various Alcohols with **3** Compared to the Oxidation of Benzyl Alcohol (rate = 1.0).

substrate	relative oxidation rate	substrate	relative oxidation rate
Allyl alcohols		**Benzyl alcohols (cont)**	

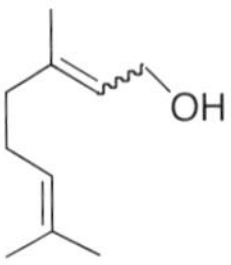

Relative oxidation rates	
Allyl alcohols (geranyl-type structure)	50-100
cinnamyl alcohol	6.4
allyl alcohol	0.8

Benzyl alcohols

methylenedioxybenzyl alcohol	7
4-MeO benzyl alcohol	6
benzyl alcohol	1.0
3-HO benzyl alcohol	1
4-F benzyl alcohol	0.8
4-Cl benzyl alcohol	0.5

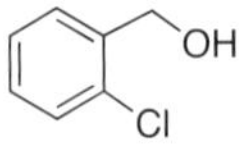

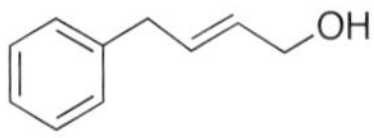

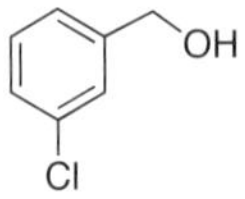

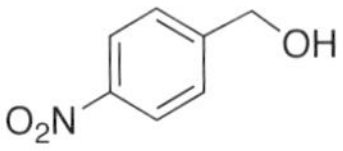

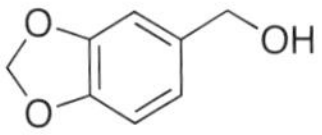

Benzyl alcohols (cont)

2-Cl benzyl alcohol	0.2
3-Cl benzyl alcohol	0.2
4-O₂N benzyl alcohol	0.09
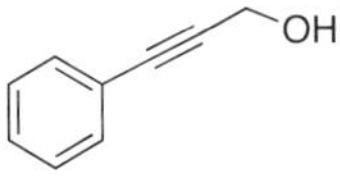 phenylpropargyl alcohol	0.1-0.2

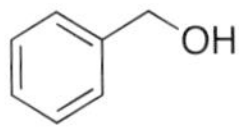

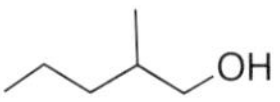

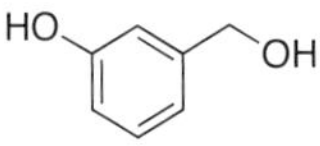

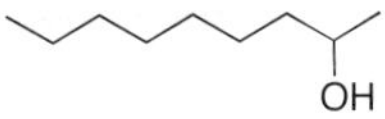

Aliphatic alcohols

octanol	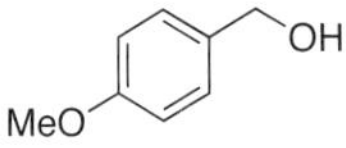 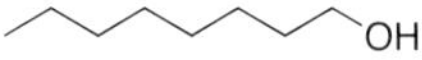 0.07-0.08
2-methyl primary alcohol	0.04
secondary octanol	0.1
neopentyl alcohol	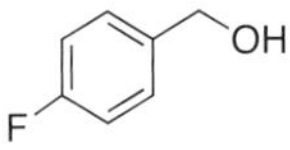 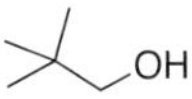 0.01-0.02
2-phenoxyethanol	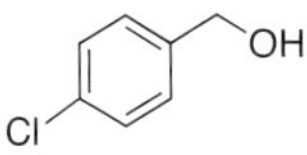 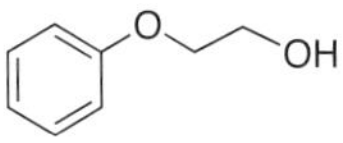 0

E. Examples of Selective Alcohol Oxidations of Polyalcohols

As shown above, changes in the substitution pattern of alcohols result in different rates of oxidation, and this selectivity can be quite useful (see Table 3). In general, allyl and benzyl alcohols can be oxidized selectively in

Org. Synth. **2013**, *90*, 215-228

the presence of aliphatic alcohol groups, and less sterically hindered aliphatic alcohols are easier to oxidize than more hindered alcohols. For each example the site of selective oxidation is indicated by an arrow, and the yield of the oxidation and primary reference are given below the structures.

Table 3. Selective Oxidations of Polyalcohols

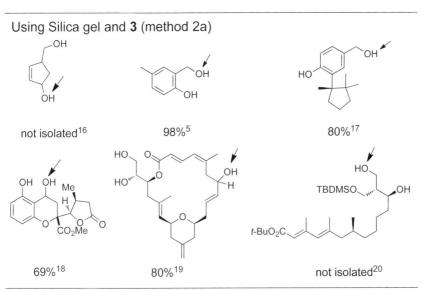

Using Silica gel and 3 (method 2a)

not isolated[16]   98%[5]   80%[17]

69%[18]   80%[19]   not isolated[20]

Using the Golubev disproportionation method (2c)

82%[21, 22]   94%[23]   75%[24]

88%[25]   98%[26]

F. Recent Literature and Some Unique New Reactions

Several recent reviews of oxoammonium salt and nitroxide-catalyzed oxidations have been published.[6,9,27,28] The following reactions are examples of some new chemistry of oxoammonium salts (other than the traditional oxidations of alcohols to aldehydes or ketones) which have been published since our *Organic Reactions* review (Miscellaneous Reactions on page 160).[9] Only one specific reaction is given in each case, although a number of similar substrates are described in each reference.

In entry 1, oxidations in water are described.[29]

In entries 2 and 3, methods for direct oxidation of alcohols to carboxylic acids are given.[15, 30] Entry 2 is of special interest, because an oxoammonium salt is used as a catalyst for oxidation rather than the usual nitroxide.

In the reactions shown in entry 4,[15] the remarkable selectivity of oxoammonium salts is illustrated: a benzyl alcohol can be oxidized in 98% yield in the presence of an aliphatic primary alcohol; in the second reaction, a primary alcohol can be oxidized to a dialdehyde in 97% yield; and in the last equation, an aliphatic aldehyde can be oxidized to a carboxylic acid in 95% yield in the *presence of an aromatic aldehyde*.

The synthesis of α,β-unsaturated ketones from silylenol ethers is described in entry 5.[31] Entry 6 involves a use of an oxoammonium salt as a cleavage reagent for benzyl ethers.[32] In entry. 7, a unique addition of an adjacent keto group to a cyclic ketone followed by the formation of a double bond is shown.[33] Entries 8, 9, and 10 show the imaginative use of metal ions as catalysts for interesting one-pot oxidations to form new carbon-carbon bonds.[34-36]

In entry 11, a "one-pot" reaction similar to entries 8-10 is shown with only an oxoammonium salt and no metals.[37]

In entries 12 and 13, the lutidine conditions (Scheme 2, eq 2b) for oxoammonium salt oxidation are presented. Neither reaction takes place under neutral conditions (Scheme 2, eq 2a). Vinyl, aryl, propargyl and aliphatic trifluoromethylcarbinols (entry 12) were oxidized to the trifluoromethyl ketones in fair to excellent yield.[12] Entry 13 shows the oxidation of a complex neopentyl alcohol, which was oxidized under these conditions.[38]

Org. Synth. **2013**, *90*, 215-228

Table 4. New Chemistry of Oxoammonium Salts

Entry	Reaction
1[29]	
2[30]	
3[15]	
4[15]	
5[31]	
6[32]	
7[33]	
8[34]	

Table 4 (cont)

Entry	Reaction

Entry 9[35]

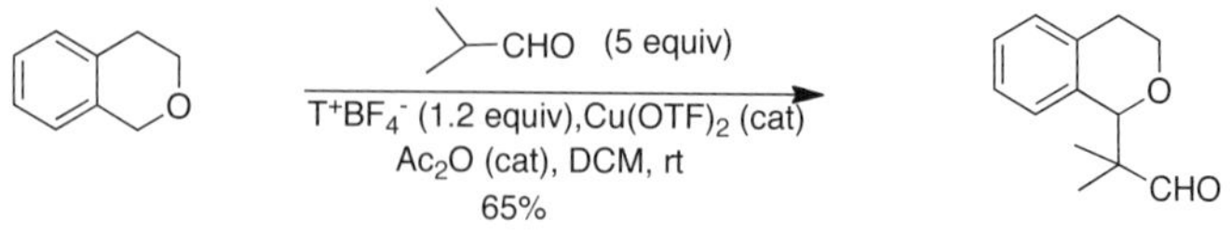

>⟍—CHO (5 equiv)
T⁺BF₄⁻ (1.2 equiv), Cu(OTF)₂ (cat)
Ac₂O (cat), DCM, rt
65%

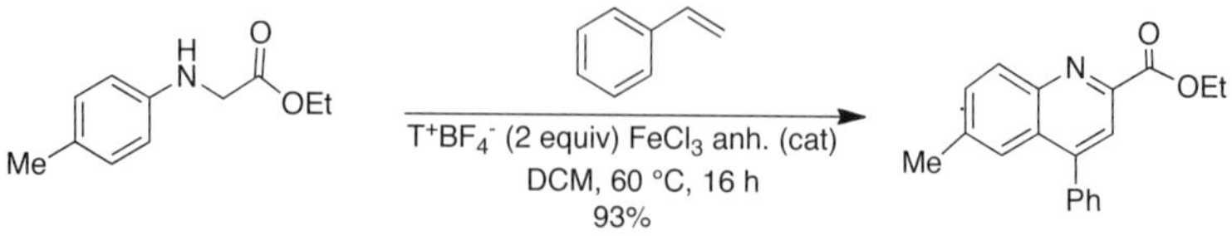

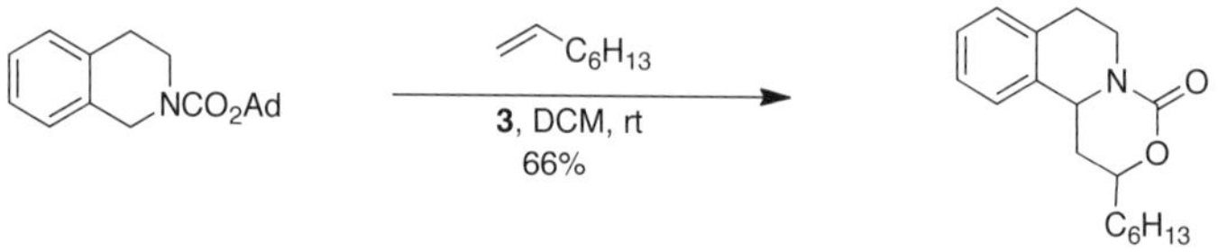

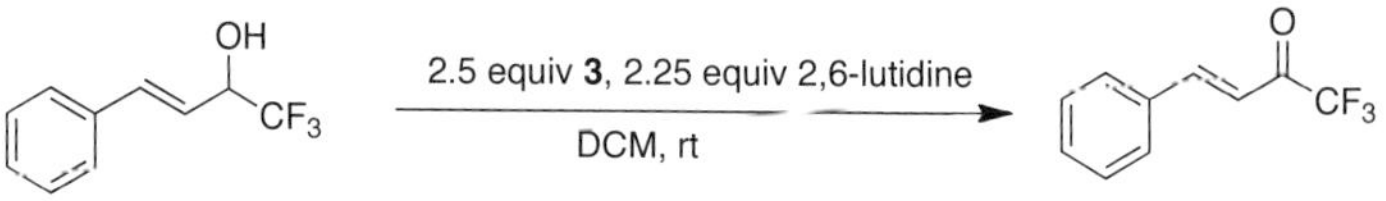

Entry 10[36]

T⁺BF₄⁻ (2 equiv) FeCl₃ anh. (cat)
DCM, 60 °C, 16 h
93%

Entry 11[37]

⟍⟍C₆H₁₃
3, DCM, rt
66%

Entry 12[12]

2.5 equiv 3, 2.25 equiv 2,6-lutidine
DCM, rt

Entry 13[38]

2.2 equiv 3, 2.2 equiv 2,6-lutidine
DCM, rt

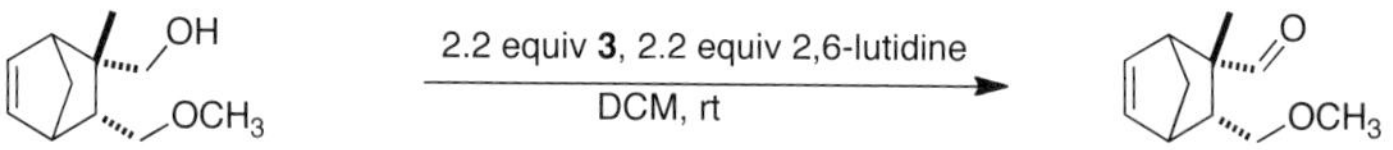

1. Department of Chemistry, University of Connecticut, Storrs, CT 06269-3060. E-mail, james.bobbitt@uconn.edu.
2. Shields Science Center, Stonehill College, Easton MA 02357
3. a) Tilley, L. J.; Bobbitt, J. M.; Murray, S. A.; Camire, C. E.; Eddy, N. A. *Synthesis* **2013**, *45*, 326-329. b) Mercadante, M.; Kelly, C. B.; Bobbitt, J. M.; Tilley, L. J.; Leadbeater, N. E. *Nature Protocols* **2013**, 8, 666–676.
4. Ma, Z.; Bobbitt, J. M. *J. Org. Chem.* **1991**, *56*, 6110–6114.
5. Bobbitt, J. M. *J. Org. Chem.* **1998**, *63*, 9367–9374.
6. Bobbitt, J. M. *TCIMAIL* **2011**, *No. 146,* 2–10.

Org. Synth. **2013**, *90*, 215-228

7. Merbouh, N.; Bobbitt, J. M.; Brückner, C. *Org. Prep. Proced. Int.* **2004**, *36*, 3–31.

8. Bobbitt, J. M.; Merbouh, N. *Org. Synth.* **2005**, *82*, 80–86.

9. Bobbitt, J. M.; Brückner, C.; Merbouh, N. *Org. React.* **2009**, *74*, 103–424.

10. This can be done as a No-D NMR using solvent presaturation. See Hoye, T. R.; Eklov, B M.; Ryba, T. D.; Voloshin, M.; Yac, L. *J. Org. Lett.*, **2004**, *6*, 953–956 and references therein. The procedure is found in footnote 8.

11. Merbouh, N.; Bobbitt, J. M.; Brückner, C. *J. Org. Chem.* **2004**, *69*, 5116–5119.

12. Kelly, C. B.; Mercadante, M. A.; Hamlin, T. A.; Fletcher, M. H.; Leadbeater, N. E. *J. Org. Chem.* **2012**, *77*, 8131–8144.

13. Merbouh, N.; Bobbitt, J. M.; Brückner, C. *Tetrahedron Lett.* **2001**, *42*, 8793–8796.

14. Golubev, V. A.; Zhdanov, R. I.; Gida, V. M.; Rozantsev, E. G. *Russ. Chem. Bull.* **1971**, 768.

15. Qiu, J. C.; Pradhan, P. P.; Blanck, N. B.; Bobbitt, J. M.; Bailey, W. F. *Org. Lett.* **2012**, *14*, 350–353.

16. Hudon, J.; Cernak, T. A.; Ashenhurst, J. A.; Gleason, J. L. *Angew. Chem. Int. Ed.* **2008**, *47*, 8885–8888.

17. Abad, A.; Agulló, C.; Cuñat, A. C.; Perni, R. H. *Tetrahedron-Asymmetry* **2000**, *11*, 1607–1615.

18. Qin, T.; Johnson, R. P.; Porco, Jr., J. A. *J. Am. Chem. Soc.* **2011**, *133*, 1714–1717.

19. Hoye, T. R.; Hu, M. *J. Am. Chem. Soc.* **2003**, *125*, 9576–9577.

20. Wovkulich, P. M.; Shankaran, K.; Kiegiel, J.; Uskokovié, M. R. *J. Org. Chem.* **1993**, *58*, 832–839.

21. Banwell, M. G.; Collis, M. P.; Mackay, M. F.; Richards, S. L. *J. Chem. Soc., Perkin Trans. 1* **1993**, 1913–20.

22. Banwell, M. G.; Edwards, A. J.; Harfoot, G. J.; Jolliffe, K. A. *J. Chem. Soc., Perkin Trans. 1* **2002**, 2439–2441.

23. Banwell, M. G.; Dupuche, J. R.; Gable, R. W. *Aust. J. Chem.* **1996**, *49*, 639–645.

24. Conrow, R. E. *Org. Lett.* **2006**, *8*, 2441–2443.

25. Findlay, A. D.; Gebert, A.; Cade, I. A.; Banwell, M. G. *Aust. J. Chem.* **2009**, *62*, 1173–1180.

26. Habel, L. W.; De Keersmaecker, S.; Wahlen, J.; Jacobs, P. A.; De Vos, D. E. *Tetrahedron Lett.* **2004**, *45*, 4057–4059.
27. Tebben, L.; Studer, A. *Angew. Chem. Int. Ed.* **2011**, *50*, 2–37.
28. Iwabuchi, Y. *J. Synth.. Org. Chem. Jpn.* **2008**, *66*, 1076–1084.
29. Mamros, A. N.; Sharrow, P. R.; Weller, W. E.; Luderer, M. R.; Fair, J. D.; Pazehoski, K. O.; Luderer, M. R. *ARKIVOC* **2011**, *(v)*, 23–33.
30. Shibuya, M.; Sato, T.; Tomizawa, M.; Iwabuchi, Y. *Chem. Commun.* **2009**, 1739–1741.
31. Hayashi, M.; Shibuya, M.; Iwabuchi, Y. *Org. Lett.* **2012**, *14*, 154–157.
32. Pradhan, P. P.; Bobbitt, J. M.; Bailey, W. F. *J. Org. Chem.* **2009**, *74*, 9524–9527.
33. Eddy, N. A.; Kelly, C. B.; Mercadante, M. A.; Leadbeater, N. E.; Fenteany, G. *Org. Lett.* **2012**, *14*, 498–501.
34. Richter, H.; García-Mancheño, O. *Eur. J. Org. Chem.* **2010**, 4460–4467.
35. Richter, H.; Rohlmann, R.; García-Mancheño, O. *Chem. Eur. J.* **2011**, *17*, 11622–11627.
36. Richter, H.; García-Mancheño, O. *Org. Lett.* **2011**, *13*, 6066–6069.
37. Richter, H.; Fröhlich, R.; Daniliuc, C.-G.; Garcia Manchẽno, O. *Angew. Chem. Int. Ed.* **2012**, *51*, 8656–8660.
38. Nicholas A. Eddy, Ph.D. dissertation, University of Connecticut, 2012, "Methodologies and Studies Directed Toward the Synthesis of Cucurbitacin I".

James M. Bobbitt was born in Charleston, WV in 1930. He received the B. Sc. in chemistry from West Virginia University in 1951 and the Ph .D. degree from The Ohio State University in 1955 with M. L. Wolfrom. He joined the University of Connecticut in 1956 after a year of postdoctoral work with Carl Djerassi. He started his career working on the structure elucidation and synthesis of natural products, then shifted to electroorganic oxidations with the intent to mimic natural synthetic reactions. In 1985, he discovered oxoammonium salt oxidations. In 1992, he formally retired from UCONN and since then has carried out bench chemistry in oxidation reactions.

Nicholas Eddy was born in Johnstown, PA in 1980. He obtained his B.S. degree in Chemistry from Indiana University of Pennsylvania in 2003, and M.S. degree in Chemistry in 2006 at Indiana University of Pennsylvania. He went on to pursue his Ph.D. in Chemistry at the University of Connecticut under the direction of Dr. Gabriel Fenteany, which was completed in November 2012. During his graduate studies, he worked on developing an oxidation with oxoammonium salts of cyclic 1,3-diketones leading to cyclohex-5-ene-1,2,4-triones.

Jay Richardson was born in Providence, RI in 1990. He studied chemical engineering at the University of Connecticut, where he pursued undergraduate research under the direction of Dr. Gabriel Fenteany and Nicholas Eddy. After completing his B.S. degree, he began work for Zeeco Corp. as a combustion research engineer.

Stephanie A. Murray was born in Holyoke, MA in 1991. She is currently working toward a B.S. degree in Chemistry at Stonehill College in Easton, MA which she expects to receive in May 2013. Under the direction of Dr. Leon Tilley, her undergraduate research focuses on trifluoromethyl directed gamma-silyl elimination in the cationic mediated synthesis of trifluoromethylcyclopropanes. She also spent time in the Hartwig group at the University of California, Berkeley where she worked on palladium-Josiphos catalyzed benzamide synthesis and rhodium-NHC catalyzed hydroamination of olefins.

Leon J. Tilley was born in New Haven, CT in 1968. He received a B.A. in chemistry and a B.A. in Russian language from Grinnell College, in Grinnell, Iowa in 1990 and a Ph.D. in chemistry from Indiana University, Bloomington in 1996. He joined the faculty at Stonehill College in 1996. His research interests include: silyl substituted cations; synthesis of strained hydrocarbons; fluoroalkyl compounds; oxidation reactions; and highly symmetric molecules. In 2011, he prepared 1-(trifluoromethyl)bicyclo[1.1.0]butane. He and his undergraduate students, in collaboration with colleagues and graduate students at UCONN are currently preparing a variety of trifluoromethyl- substituted strained hydrocarbons. Professor Tilley wishes to acknowledge the Office of Naval Research (ONR) (Award No.: N00014-11-1-0921) for partial funding of this work.

Preparation of 3-Oxocyclohex-1-ene-1-carbonitrile

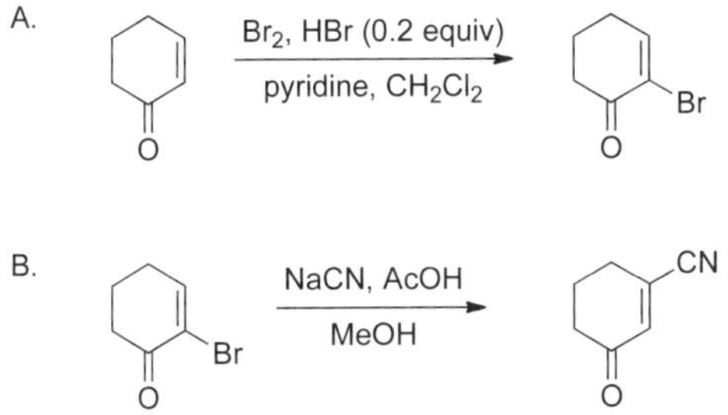

A.
Br_2, HBr (0.2 equiv)
pyridine, CH_2Cl_2
Br
O
B.
NaCN, AcOH
MeOH
CN

Submitted by Jesus Armando Lujan-Montelongo and Fraser F. Fleming.[1]
Checked by David Hughes.

1. Procedure

A. 2-Bromo-2-cyclohexen-1-one. A 3-necked, 500-mL, round-bottomed flask, equipped with a magnetic stir bar (PTFE-coated, oval, 4 cm), is charged with cyclohex-2-enone (15.0 g, 156 mmol, 1.0 equiv) and CH_2Cl_2 (150 mL) (Note 1). The center neck is fitted with a 25-mL addition funnel equipped with a gas inlet adapter connected to a nitrogen line and gas bubbler. One outer neck is sealed with a glass stopper; the other neck is capped with a rubber septum through which a thermocouple probe is inserted (Note 2). The flask is immersed in a dry ice/acetonitrile bath and stirring is begun. After the internal temperature reaches -45 °C, HBr (48% aq. solution, 3.6 mL, 32 mmol, 0.2 equiv) is added dropwise via syringe through the septum in 1 min. The addition funnel is charged with neat Br_2 (9.0 mL, 175 mmol, 1.1 equiv), which is then added dropwise until the orange-red color indicative of excess bromine persists (Note 3). After 10 min, the addition funnel is replaced by a clean 125-mL addition funnel charged with pyridine (25.5 mL, 316 mmol, 2.0 equiv), which is then added dropwise over 15 min (Note 4). After the addition, the cold bath is removed and the dropping funnel is replaced by a reflux condenser equipped with a gas inlet adapter, which is connected to a nitrogen line and gas bubbler. The mixture is allowed to warm to room temperature over 30 min, then heated to reflux (43 °C) for 1 h using a heating mantle. After cooling to room temperature the flask contents are transferred to a 1-L separatory funnel, the

Published on the Web 4/12/2013

reaction flask is washed with CH$_2$Cl$_2$ (50 mL), and the additional solution is added to the separatory funnel. The mixture is washed with an aqueous solution of sodium thiosulfate (0.7 M, 150 mL). The organic phase is separated and the aqueous phase extracted with CH$_2$Cl$_2$ (50 mL). The combined organic layers are washed with 1M HCl (100 mL). The organic layer is removed and the acidic, aqueous phase is extracted with CH$_2$Cl$_2$ (50 mL). The combined organic fraction is washed with water (50 mL), brine (50 mL), and then filtered through a bed of Na$_2$SO$_4$ (50 g) into a tared 1-L round-bottomed flask. The solution is concentrated by rotary evaporation (40 °C, 200 mmHg) to about 60 mL (84 g), then hexanes (100 mL) are added and the concentration continued to dryness (Note 5). Vacuum drying (10 mmHg) to constant weight (1.5 h) affords crude 2-bromo-2-cyclohexenen-1-one (26.7 g, 98% yield) as a slightly yellow powder. The crude 2-bromo-2-cyclohexen-1-one is suitable for the cyanation-elimination reaction without purification (Notes 6 - 8).

> *Caution! Sodium cyanide and HCN are extremely toxic. The experimentalist should use sufficient personal protection and only handle sodium cyanide solid and solutions in a well-ventilated fume hood. The aqueous waste solutions containing cyanide should be treated with excess bleach before disposal.[2] Good housekeeping is essential - all spills around balances or work areas should be immediately cleaned up and the area washed down with bleach.*

B. *3-Oxocyclohex-1-enecarbonitrile.* A 500-mL round-bottomed flask is charged with MeOH (120 mL) (Note 9), crude 2-bromo-2-cyclohexen-1-one (18.6 g, 106 mmol, 1.0 equiv), and a magnetic stir bar (PTFE-coated, oval, 3 cm). The flask is capped with a rubber septum and inserting an 18-gauge needle through the septum provides a nitrogen inlet. A thermocouple probe is also inserted through the septum (Note 2). The reaction flask is immersed in a tap-water bath and then neat AcOH (6.5 mL, 114 mmol, 1.1 equiv) is added dropwise to the stirred solution via syringe over 1 min. After 5 min, the septum is briefly removed, solid NaCN (5.68 g, 116 mmol, 1.1 equiv) is added in one portion, and the septum is replaced (Note 10). After 10 min a second portion of NaCN (2.73 g, 56 mol, 0.5 equiv) is added, followed, after another 10 min, by a third portion of NaCN (1.25 g, 26 mmol, 0.25 equiv), replacing the septum after each addition. After 20 min (Note 11), the water bath is replaced with an ice-water bath. Once the flask contents equilibrate to 5 °C, the septum is removed, and an aqueous solution of Na_2CO_3 (0.5 M, 100 mL) is added. After 1 min, solid NaCl (10 g) is added and the mixture is stirred for 5 min. The flask contents are transferred to a 1-L separatory funnel, the reaction flask is washed with EtOAc (100 mL), and the additional solution is added to the separatory funnel. The organic phase is separated and the aqueous phase is extracted with EtOAc (3 x 100 mL). The combined organic phase is washed with 5 °C aqueous NaOH (1M, 100 mL) (Note 12) and then washed with brine (2 x 50 mL). The organic phase is dried by filtering through a bed of Na_2SO_4 (50 g) into a tared 1-L round-bottomed flask. The solvent is removed by rotary evaporation (40 °C, 10 mmHg) to afford an orange oil (12.6 g), which is purified by column chromatography on Florisil (Notes 13 and 14) to afford 3-oxocyclohex-1-enecarbonitrile (8.07 g, 63% yield) as a pale yellow oil spectrally identical to material previously reported (Note 15).[3]

2. Notes

1.   The following reagents and solvents were used as received for Step A: cyclohex-2-enone (Sigma-Aldrich, 95+%), HBr (48% aq., Sigma-Aldrich), CH_2Cl_2 (Fischer Certified ACS, stabilized), Br_2 (Sigma-Aldrich, reagent grade), and pyridine (Sigma-Aldrich, ACS reagent, >99.0%).

2.   The internal temperature was monitored using a J-Kem Gemini digital thermometer with a Teflon-coated T-Type thermocouple probe (12-inch length, 1/8 inch outer diameter, temp range –200 to +250 °C).

3.   Bromine is extremely corrosive and an irritant and must be transferred within a well-ventilated fume hood. Dropwise addition of bromine at intervals of one drop every 2-3 sec, slowing to one drop every 5 sec near the end point, gave the best results. The entire addition time was 40 min. The internal temperature was maintained at –45 to –48 °C. An excess of three drops of bromine is typically added after the end point to ensure the orange-red color indicative of excess bromine persists. Unreacted starting material carries through to the next step and cannot be removed in the chromatographic purification of the nitrile product of step B.

4.   Pyridine was added at a rate of about 1 drop/sec with the temperature rising to –39 °C. The reaction was monitored by ^{1}H NMR by removing a 0.1 mL reaction aliquot and quenching into 0.5 mL EtOAc/0.5 mL 0.7M sodium thiosulfate solution, separating the

organic layer, filtering through a cotton plug, and concentrating to dryness. Diagnostic ^{1}H NMR resonances: starting material, δ: 6.01 (dt, J = 2.0, 10.3 Hz, 1 H) and 7.03 (dt, J = 4.3, 10.3, 1 H); product, δ: 7.42 (t, J = 4.5 Hz, 1 H).

5. The flush with hexanes results in crystallization during the concentration, which provides a product with greater stability. The solid obtained by concentration of the dichloromethane solution was found to be less stable.

6. The crude material was 95% pure based on GC analysis (t_R = 9.7 min; conditions: Agilent DB35MS column; 30 m x 0.25 mm; initial temp 60 °C, ramp at 20 °C/min to 280 °C, hold 15 min). The submitters carried out a recrystallization of 18.0 g of 2-bromo-2-cyclohexen-1-one by dissolution in 25 mL of boiling EtOAc, filtration through a pre-wetted funnel with filter paper, followed by the addition of hot hexane (~4 mL) to induce turbidity, cooling, filtration on a Büchner funnel, and drying for 20 min at 0.01 mmHg, providing 13.3 g of white crystals (73% yield, mp 74–75 °C).[4] The checker purified the product as follows. To a 100-mL round-bottomed flask equipped with a 2-cm oval PTFE-coated magnetic stir bar was added crude 2-bromo-2-cyclohexen-1-one (7.80 g) and t-BuOMe (10 mL). The slurry was stirred for 5 min. The flask was fitted with a 25-mL addition funnel through which was added n-heptane (10 mL) over a 10 min period. The slurry was stirred for 20 min at room temperature, then filtered and washed with n-heptane (10 mL) to afford 2-bromo-2-cyclohexen-1-one as an off-white solid (5.45 g, 70% recovery, mp 72–74 °C, >99% pure by GC).

7. 2-Bromo-2-cyclohexen-1-one has the following spectroscopic properties: ^{1}H NMR (400 MHz, CDCl$_3$) δ: 2.04–2.10 (m, 2 H), 2.43–2.47 (m, 2 H), 2.60–2.64 (m, 2 H), 7.42 (t, J = 4.5 Hz, 1 H); ^{13}C NMR (100 MHz, CDCl$_3$) δ: 22.8, 28.5, 38.5, 124.0, 151.3, 191.4; IR (neat): 3043, 1680, 1598 cm^{-1}; HRMS (ESI) calculated for C$_6$H$_7$BrO 196.9572; found 196.9570 (M+Na)$^+$; GC-MS m/z (rel intensity): 176 (M$^+$, 78), 174 (M$^+$, 77), 148 (87), 146 (88), 135 (24), 133 (24), 120 (38), 118 (38), 67 (100), 55 (42).

8. The checker found both crude and recrystallized 2-bromo-2-cyclohexen-1-one to be unstable at room temperature, decomposing within 3 days to a purple gum with release of gas. Material stored in the freezer darkened but remained unchanged by NMR for a period of 2 weeks.

9. The following reagents and solvents were used as received for Step B: methanol (Fisher Optima, 99.9%), HOAc (Fisher Certified ACS,

232

100.0%), NaCN (Sigma-Aldrich, ACS certified, 95+%), EtOAc (Fisher, ACS reagent), Florisil (Sigma-Aldrich, 100-200 mesh), hexanes (Fisher, ACS reagent, >98.5%), methyl t-butyl ether (Sigma-Aldrich, >98.5%).

10. The temperature rose as follows after the NaCN additions: #1, 18 to 28 °C; #2, 25 to 28 °C; #3, 24 to 25 °C.

11. Progress of the reaction was monitored by TLC (1:1 MTBE/hexanes) and visualized with UV light; $R_f = 0.4$ for product and 0.5 for starting material. The checker found the reaction achieved >95% completion by the current protocol. The submitters suggest that if complete conversion is not achieved, an additional portion of NaCN (0.25 equiv) and AcOH (3.2 mL) can be added and the reaction checked again after 20 min. If necessary, the water bath temperature can be raised to 50–55 °C in order to facilitate complete conversion.

12. Washing the organic phase with NaOH is designed to purge the solution of residual HCN. These precautions minimize the potential for contact with HCN during removal of solvent on a rotary evaporator that must be vented to, or located in, a fume hood. Residual NaCN is removed in the aq. washes, which should be treated with bleach prior to disposal.

13. Column chromatography on silica gel causes significant adsorption of the oxonitrile resulting in a low yield. However, the submitters found that 2 g of cyclohex-2-enone provided crude keto nitrile that was readily purified by radial chromatography (4 mm SiO_2 plate, EtOAc/hexanes as eluent, 1:19) without significant adsorption. Concentration on a rotary evaporator provided pure 3-oxocyclohex-1-ene-carbonitrile (1.77 g, 70% yield over two steps).

14. Florisil (325 g) was wet-packed in a 7-cm diameter column using MTBE:hexanes (1:3) with a 0.5 cm bed of sand topping the column. The crude oil was loaded neat and rinsed onto the column with 2 x 5 mL MTBE. Elution was carried out with 1 L MTBE:hexanes (1:3), 1 L MTBE:hexanes (1:2), and 1 L of MTBE:hexanes (2:3). Column flow was 40 mL/min and 50 mL fractions were collected. Product elution was followed by TLC as indicated in Note 11. Fractions 33-50 were combined and concentrated by rotary evaporation (40 °C bath, 100 mmHg to 10 mmHg), then dried under vacuum (10 mmHg for 2 h) to constant weight to afford 7.20 g of a pale yellow oil. Purity by GC was 95 % (same conditions as Note 6, t_R = 8.9 min). Fractions 30-32 and 51-60 were combined and likewise concentrated to 1.50 g. Purity by GC was 78%. These fractions were re-chromatographed using 100 g of Florisil, eluting with 300 mL

MTBE:hexanes (1:3) and 600 mL MTBE:hexanes (1:2), collecting 25 mL fractions. Fractions 22-33 were combined and concentrated by rotary evaporation (40 °C bath, 100 mmHg to 10 mmHg), then vacuum dried to constant weight to afford 0.87 g of product. Purity by GC was 92%. The rich cuts from the two chromatographies were combined to afford product (8.07 g, 63% yield, GC purity 95%) as a pale yellow oil.

15. 3-Oxocyclohex-1-enecarbonitrile has the following spectroscopic properties: ^{1}H NMR (400 MHz, CDCl$_3$) δ: 2.10–2.17 (m, 2 H), 2.49–2.53 (m, 2 H), 2.57 (td, J = 6.0, 2.0 Hz, 2 H), 6.52 (t, J = 2.0 Hz, 1 H); ^{13}C NMR (100 MHz, CDCl$_3$) δ: 22.3, 27.8, 37.4, 117.2, 131.2, 138.9, 196.5. IR (neat) 3059, 2232, 1682, 1606 cm^{-1}; HRMS calculated for C$_7$H$_7$NO 144.0420; found 144.0414 (M+Na)$^+$. GC-MS m/z (rel intensity): 121 (M$^+$, 68), 93 (100), 66 (27), 65 (38), 64 (30). After 4 months of storage at 4 °C the submitters found the sample underwent modest discoloration from light yellow to orange, although no loss of sample integrity was observed by ^{1}H NMR and there was no change in reactivity.

Safety and Waste Disposal Information

The procedures in this article are intended for use only by persons with prior training in experimental organic chemistry. All hazardous materials should be handled using the standard procedures for work with chemicals described in references such as "Prudent Practices in the Laboratory" (The National Academies Press, Washington, D.C., 2011 www.nap.edu). All chemical waste should be disposed of in accordance with local regulations. For general guidelines for the management of chemical waste, see Chapter 8 of Prudent Practices.

These procedures must be conducted at one's own risk. *Organic Syntheses, Inc.*, its Editors, and its Board of Directors do not warrant or guarantee the safety of individuals using these procedures and hereby disclaim any liability for any injuries or damages claimed to have resulted from or related in any way to the procedures herein.

3. Discussion

Cyclic oxoalkenenitriles juxtapose three orthogonal functionalities capable of selective functionalization; an olefin, a ketone, and a nitrile functionality.[5] 3-Oxocycloalkenecarbonitriles, in particular, are excellent scaffolds for diversity oriented synthesis,[6] feature strategically in syntheses,[7] and as precursors for mechanistic studies.[8] The electron deficient olefin in 3-oxocyclohex-1-ene-1-carbonitrile (Scheme 1, **1**) is an excellent participant in [2+2][9] (**1** → **2** and **1** → **3**) and [4+2][10] cycloadditions (**1** → **4**), and in conjugate additions (**1** → **5**).[11] Sequential addition of two Grignard reagents to 3-oxocyclohex-1-ene-1-carbonitrile (**1**) allows stepwise 1,2-1,4-additions to afford the extremely versatile *C*-magnesiated nitrile **5**.[11] Stereoselective annulations[12] provide access to bicyclic and tricyclic nitriles **6**[11] and **7**,[6] whereas complementary alkylation strategies access nitriles **8** and **9** bearing diastereomeric, quaternary centers,[13] and allow *N*-alkylation to enamides **10**.[14]

Scheme 1. Synthetic Applications of 3-Oxocyclohex-1-enecarbonitrile (**1**)

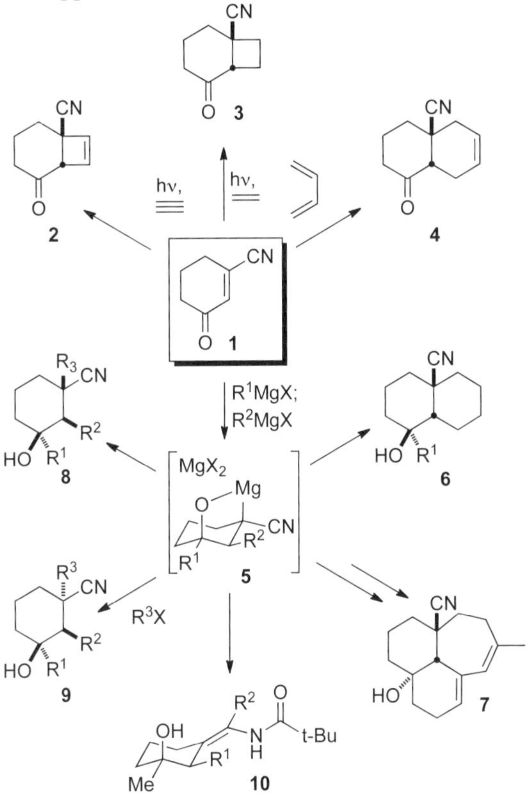

The value of 3-oxocyclohex-1-ene-1-carbonitrile (**1**) has stimulated several syntheses: cyanide addition to 1,3-cyclohexandione monoethylene ketal,[15] Et$_2$AlCN addition to 3-methoxycyclohexenone,[16] or to cyclohexenone followed by periodinane oxidation,[17] oxidative transposition of an allylic cyanohydrin,[18] and allylic oxidation of cyclohexenecarbonitrile with PhI(OAc)$_2$ and t-BuOOH,[19] or with CrO$_3$ and 3,5-dimethylpyrazole.[3] Among strategies for synthesizing **1**,[3,14-19] and related sequences employed in total synthesis campaigns,[7] the bromination-cyanation of cyclic enones[20] is conspicuous for efficiency, cost, and operational simplicity (Scheme 2). Formation of 2-bromocyclohex-2-enone (**12**) from cyclohexenone (**11**) is fast and virtually quantitative. The subsequent conjugate addition of cyanide provides an intermediate ketone **13** from which dehydrohalogenation is readily achieved simply through modest heating. The synthesis of 3-oxocyclohex-1-ene-1-carbonitrile (**1**) is very expedient and inexpensive. Operationally, the reaction uses standard glassware and affords gram quantities of 3-oxocyclohex-1-ene-1-carbonitrile (**1**) from cyclohexenone in 2-steps that can be completed in one day.

Scheme 2. Synthesis of 3-Oxocyclohex-1-enecarbonitrile (**1**)

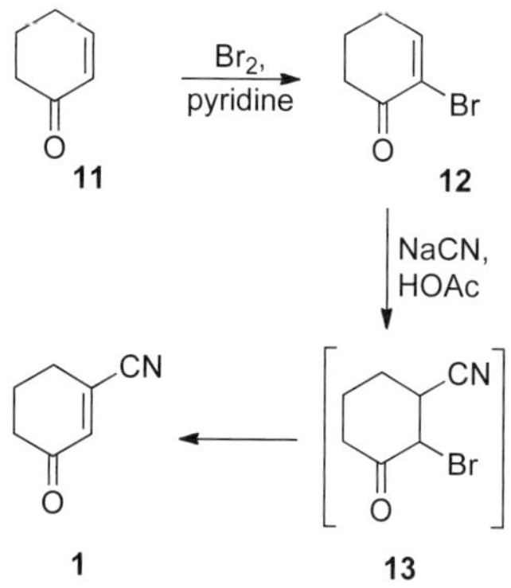

Employing the bromination-cyanation sequence with the homolog cycloheptenone (**14**) affords the seven-membered 3-oxocyclohept-1-ene-1-carbonitrile (Scheme 3, **16**).[8b,18,21] The bromination of cycloheptenone (**14**) parallels the bromination of cyclohexenone (**11**), although the elimination of the intermediate dibromide requires heating the reaction mixture for 30 min at 70 °C to form 2-bromocyclohept-2-enone (**15**). Control over the temperature required for the elimination of HBr during the formation of 2-bromocyclohept-2-enone (**15**) is critical. The temperature was conveniently controlled using microwave irradiation,[22] which reproducibly affords very

236

pure 2-bromocyclohept-2-enone. Subsequent cyanation-elimination affords 3-oxocyclohept-1-enecarbonitrile (**16**) in 71% yield over two steps. The resulting 3-oxocyclohept-1-enecarbonitrile functionality has been strategically employed in total synthesis[20] and is a valuable partner for sequential 1,2-1,4-addition-alkylations.[21]

Scheme 3. Synthesis of 3-Oxocyclohept-1-ene-1-carbonitrile (**16**)

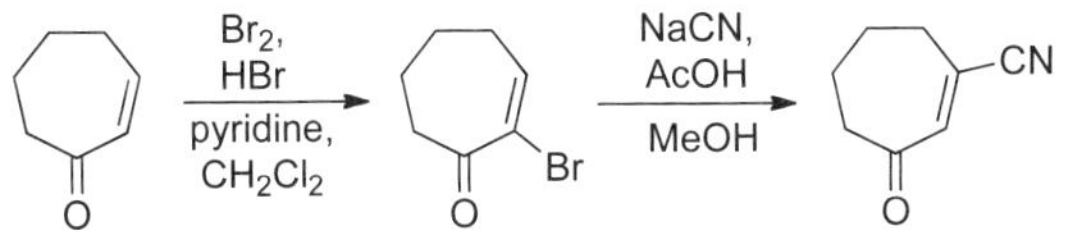

Collectively, the bromination-cyanation of cyclic enones provides efficient, cost-effective syntheses of 3-oxocycloalkene-1-carbonitriles. The syntheses are rapid and provide access to functionalized building blocks ideally suited for synthetic applications.

1. Department of Chemistry, Duquesne University, Pittsburgh, PA 15282-1530, email: flemingf@duq.edu. This work was supported by the National Science Foundation (CHE 1111406) and in part by CONACYT (J. A. L.-M.). Drew Davic is thanked for assistance in performing GCMS.
2. For the reaction of hypochlorite solutions with cyanide, see the following reference and references therein: Gerritsen, C. M.; Margerum, D. W. *Inorg. Chem.* **1990**, *29*, 2757–2762.
3. Fleming, F. F.; Zhang, Z.; Wei, G. *Synthesis* **2005**, 3179–3180.
4. For a previously reported mp of 72.5–75 °C see: Shih, C. *J. Org. Chem.* **1980**, *45*, 4462–4471; mp 75–76 °C reported by Kowalski, C. J.; Weber, A. E.; Fields, K. W. *J. Org. Chem.* **1982**, *47*, 5088–5093.
5. Fleming, F. F.; Iyer, P. S. *Synthesis* **2006**, 893–913.
6. (a) Oguri, H.; Hiruma, T.; Yamagishi, Y.; Oikawa, H.; Ishiyama, A.; Otoguro, K.; Yamada, H.; Omura, S. *J. Am. Chem. Soc.* **2011**, *133*, 7096–7105. (b) Oguri, H.; Yamagishi, Y.; Hiruma, T.; Oikawa, H. *Org. Lett.* **2009**, *11*, 601–604.
7. (a) Fleming, F. F.; Wei, G.; Steward, O. W. *J. Org. Chem.* **2008**, *73*, 3674–3679. (b) Morita, M.; Sone, T.; Ymatsugu, K.; Sohtome, Y.;

Matsunaga, S.; Kanai, M.; Watanabe, Y.; Shibasaki, M. *Bioorg. Medicinal Chem. Lett.* **2008**, *18*, 600–602. (c) Yamatsugu, K.; Kamijo, S.; Suto, Y.; Kanai, M.; Shibasaki, M. *Tetrahedron Lett.* **2007**, *48*, 1403–1406. (d) Fukuta, Y.; Mita, T.; Fukuda, N.; Kanai, M.; Shibasaki, M. *J. Am. Chem. Soc.* **2006**, *128*, 6312–6313. (e) Jansen, J. H. M.; Lugtenburg, J. *Eur. J. Org. Chem.* **2000**, 829–836. (f) Kutney, J. P.; Gunning, P. J.; Clewly, R. G.; Somerville, J.; Rettig, S. J. *Can. J. Chem.* **1992**, *70*, 2094–2114.

8. (a) Klärner, F.-G.; Wurche, F.; von E. Doering, W.; Yang, J. *J. Am. Chem. Soc.* **2005**, *127*, 18107–18113. (b) Mease, R. C.; Hirsch, J. A. *J. Org. Chem.* **1984**, *49*, 2925–2937.

9. (a) Cantrell, T. S. *Tetrahedron* **1971**, *27*, 1227–1237. (b) Agosta, W. C.; Lowrance, W. W. Jr. *Tetrahedron Lett.* **1969**, 3053–3054.

10. Yang, W. Q.; Chen, S. Z.; Huang, L. *Chin. Chem. Lett.* **1998**, *9*, 233–234.

11. (a) Fleming, F. F.; Wei, Y.; Liu, W.; Zhang, Z. *Tetrahedron* **2008**, *64*, 7477–7488. (b) Fleming, F. F.; Wei, Y.; Liu, W.; Zhang, Z. *Org. Lett.* **2007**, *9*, 2733–2736. (c) Fleming, F. F.; Zhang, Z.; Wang, Q.; Steward, O. W. *Angew. Chem., Int. Ed.* **2004**, *43*, 1126–1129.

12. Fleming, F. F.; Gudipati, S. *Eur. J. Org. Chem.* **2008**, 5365–5374.

13. (a) Fleming, F. F.; Zhang, Z.; Wei, G.; Steward, O. W. *J. Org. Chem.* **2006**, *71*, 1430–1435. (b) Fleming, F. F.; Zhang, Z.; Wei, G.; Steward, O. W. *Org. Lett.* **2005**, *7*, 447–449.

14. Fleming, F. F.; Wei, G.; Zhang, Z.; Steward, O. W. *Org. Lett.* **2006**, *8*, 4903–4906.

15. (a) Wang, Y.; Doering, W. v. E.; Staples, R. J. *J. Chem. Crystallogr.* **1999**, *29*, 977–982. (b) Cronyn, M. W.; Goodrich, J. E. *J. Am. Chem. Soc.* **1952**, *74*, 3331–3333.

16. Agosta, W. C.; Lowrance, W. W. *J. Org. Chem.* **1970**, *35*, 3851–3856.

17. Nicolaou, K. C.; Gray, D. L. F.; Montagnon, T.; Harrison, S. T. *Angew. Chem., Int. Ed.* **2002**, *41*, 996–1000.

18. Hudlicky, J. R.; Werner, L.; Semak, V.; Simionescu, R.; Hudlicky, T. *Can. J. Chem.* **2011**, *89*, 535–543.

19. Zhao, Y.; Yeung, Y.-Y. *Org. Lett.* **2010**, *12*, 2128–2131.

20. (a) Campos, K. R.; Klapars, A.; Kohmura, Y.; Pollard, D.; Ishibashi, H.; Kato, S.; Takezawa, A.; Waldman, J. H.; Wallace, D. J.; Chen, C.-y.; Yasuda, N. *Org. Lett.* **2011**, *13*, 1004–1007. (b) Ohmori, N. *J. Chem. Soc., Perkin Trans. 1*, **2002**, 755–767. (c) Magnus, P.; Waring,

M. J.; Ollivier, C.; Lynch, V. *Tetrahedron Lett.* **2001**, *42*, 4947–4950. (d) Ohmori, N. *Chem. Commun.* **2001** 1552–1553. (e) Isobe, M.; Nishikawa, T.; Pikul, S.; Goto, T. *Tetrahedron Lett.* **1987**, *28*, 6485–6488. (f) Audenaert, F.; De Keukeleire, D.; Vandewalle, M. *Tetrahedron* **1987**, *43*, 5593–5604.

21. Fleming, F. F.; Wei, G.; Zhang, Z.; Steward, O. W. *J. Org. Chem.* **2007**, *72*, 5270–5275.

22. A Biotage® microwave reactor (Model: Initiator) and a 20 mL reaction tube was used in a reaction employing 1.8 g of cyclohept-2-enone. The radiation absorption parameter was set to NORMAL, with the apparatus initially recording a slight increase of pressure (2 - 3 Psi).

Appendix
Chemical Abstracts Nomenclature; (Registry Number)

2-Bromo-2-cyclohexen-1-one; (50870-61-6)
Cyclohex-2-enone; (930-68-7)
3-Oxocyclohex-1-enecarbonitrile; (CAS# 25017-78-1)

Fraser Fleming earned his B. Sc. (Hons.) at Massey University, New Zealand, in 1986 and a Ph. D. under the direction of Edward Piers at the University of British Columbia, Canada, in 1990. After postdoctoral research with James D. White at Oregon State University he joined the faculty at Duquesne University, Pittsburgh, in 1992. His research interests lie in stereochemistry and organometallics, particularly as applied to alkenenitrile conjugate additions and metalated nitrile alkylations.

Jesus Armando Lujan-Montelongo, a native of Queretaro, Mexico, completed his BS at the National University Autonomous of Mexico (UNAM) in 2003. He continued his studies at UNAM, earning his Ph. D. in Organic Chemistry in 2005 under the direction of Jose G. Avila-Zarraga. After postdoctoral studies with Luis Miranda at the Institute of Chemistry (UNAM) developing free-radical based synthetic methods, he joined Fraser Fleming at Duquesne University in Pittsburgh, where he is working on nitrile-based methodology. His research interests lie in the total synthesis of natural products and in organometallic chemistry.

Air Oxidation of Primary Alcohols Catalyzed by Copper(I)/TEMPO. Preparation of 2-Amino-5-bromo-benzaldehyde

Submitted by Jessica M. Hoover and Shannon S. Stahl.[1]
Checked by Guido P. Möller and Erick M. Carreira.

1. Procedure

A. *2-Amino-5-bromobenzyl alcohol* (**2**). To a one-necked 1-L round-bottomed flask equipped with a Teflon-coated magnetic stir bar (5 cm x 7 mm) is added 2-amino-5-bromobenzoic acid (**1**) (9.87 g, 45.7 mmol, 1.0 equiv) and dry THF (400 mL) (Note 1). The flask is fitted with a septum and nitrogen inlet needle. The solution is cooled in an ice bath under an atmosphere of nitrogen gas. Lithium aluminum hydride (5.00 g, 132 mmol, 2.9 equiv) (Note 2) is added portion-wise (0.5 g portions) over the course of 1 h by temporarily removing the septum. The reaction mixture is allowed to warm slowly to room temperature overnight with stirring (20 h). When the reaction is complete as determined by TLC (Note 3) the crude reaction mixture is poured slowly into ethyl acetate (400 mL) in a 2 L Erlenmeyer flask equipped with a magnetic stir bar cooled in an ice bath (Note 4). The excess LiAlH₄ is quenched by the slow addition of water (50 mL) to the stirred mixture over 30 min. Additional water is added (450 mL) and the mixture is stirred until two distinct layers form (~30 min). The mixture is transferred to a 2 L separatory funnel and the layers separated. The aqueous layer is extracted twice with ethyl acetate (2 x 500 mL) (Note 5). The combined organic layers are transferred to a 4 L separatory funnel, washed

Org. Synth. **2013**, *90*, 240-250
Published on the Web 4/23/2013
© 2013 Organic Syntheses, Inc.

with brine (600 mL) and dried for 30 min over Na$_2$SO$_4$ (100 g). After filtration, the solvent is removed by rotary evaporation (25 °C, 30 mmHg) to give a light yellow solid. Analytically pure material is obtained after recrystallization. Into a one-necked, 250-mL round-bottomed flask equipped with a Teflon-coated magnetic stir bar (3 cm x 5 mm) the crude material is dissolved in a minimum amount of refluxing ethyl acetate (15 mL. Hexanes (~100 mL) is added over 10 min to the stirred and refluxing mixture until the product precipitates (Note 6). The mixture is allowed to cool to room temperature and then stored in the freezer at –15 °C for 3 h. The product is isolated by suction filtration on a Büchner funnel, washed with hexanes (50 mL), and dried by vacuum (0.01 mmHg, 23 °C) to provide a first crop (6.53–7.01 g). A second crop is obtained from the combined filtrates after removal of the solvent by rotary evaporation (25 °C, 30 mmHg). The resulting solids are dissolved in a one–necked, 100-mL round-bottomed flask equipped with a Teflon-coated magnetic stir bar (3 cm x 5 mm) in a minimum amount of refluxing ethyl acetate (5 mL). To the refluxing, stirred solution is added hexanes (50 mL) over 10 min until precipitation begins. After cooling to room temperature the suspension is stored in a freezer at –15 °C overnight for collection of a second crop (0.91–1.09 g) by suction filtration on a Büchner funnel. The alcohol **2** is obtained in 80–88% yield (7.44–8.10 g,) as a light tan powder (Note 7).

B. *2-Amino-5-bromobenzaldehyde* (**3**). To a one-necked 500-mL round-bottomed flask equipped with a Teflon-coated magnetic stir bar (5 cm x 7 mm) is added 2-amino-5-bromobenzyl alcohol (**2**) (6.10 g, 30.0 mmol, 1 equiv) and MeCN (60 mL) (Note 8). [Cu^I(MeCN)$_4$](OTf) (569 mg, 1.51 mmol, 0.05 equiv), 2,2'-bipyridine (bpy) (236 mg, 1.51 mmol, 0.05 equiv), and 2,2,6,6-tetramethylpiperidine-*N*-oxyl (TEMPO) (236 mg, 1.51 mmol, 0.05 equiv) (Note 9) are each added as a solid, and the weighing vessel (often a small test tube) is rinsed with MeCN (30 mL each) to ensure complete delivery of the reagents and a total reaction volume of 150 mL. After the addition of *N*-methyl imidazole (NMI) (240 µL, 248 mg, 3.02 mmol, 0.1 equiv) (Note 10), the dark red/brown reaction mixture is stirred open to air at 500 rpm (Note 11) at room temperature until the starting material is consumed as determined by TLC (2–3 h) (Note 3) (Note 12). Upon completion, the reaction mixture is diluted with ethyl acetate (300 mL), filtered through a plug of silica (Note 13), and washed with ethyl acetate (400 mL). The solvent is concentrated by rotary evaporation (20 °C, 0.04 mmHg) to afford the crude aldehyde as a yellow brown solid (Note 14).

Analytically pure material is obtained after purification by silica column chromatography. The solid is dissolved in a minimum amount of CHCl₃ (20 mL) and loaded onto a column prepared from silica gel (150 g, Silicycle SiliaFlash® P60, 230-400 mesh) slurried in 10% EtOAc-hexanes (column dimensions: 5.5 cm diameter x 30 cm height, 18 cm packed height). Elution with 10% EtOAc-hexanes (200 mL initial collection followed by 30 mL fractions, 1.4 L total solvent volume eluted) affords the product in fractions 19-35. The product containing fractions are combined and the solvent is removed by rotary evaporation (0.1 mmHg, 20 °C). Subsequent drying at 0.01 mmHg (22 °C) provides aldehyde **3** in 89–91% yield (5.37–5.50 g) as a bright yellow powder (Notes 15 and 16).

2. Notes

1.   Checkers used 2-amino-5-bromobenzoic acid (98%) from ABCR-Chemicals and used it as received. Submitters purchased 2-amino-5-bromobenzoic acid from Aldrich and used it as received. Inhibitor free THF was purchased from Sigma-Aldrich and passed through a column of alumina before use.

2.   Checkers used lithium aluminum hydride powder purchased from Acros Organics and used it as received. Submitters purchased lithium aluminum hydride from Sigma-Aldrich as 0.5 g pellets and used it as received. Lithium aluminum hydride is pyrophoric and reacts violently with water.

3.   TLC conditions: hexanes:ethyl acetate = 2:1, plates were visualized by UV and KMnO₄ stain, silica gel stationary phase, R$_f$(acid) = 0.1–0.2 (streak), R$_f$(alcohol) = 0.2, R$_f$(aldehyde) = 0.7.

4.   Ethyl acetate was purchased from Sigma-Aldrich and used as received.

5.   The addition of a small amount of brine, ~50-100 mL, improves the separation of the layers.

6.   Hexanes was purchased from Sigma-Aldrich Chemical Company and used as received.

7.   2-Amino-5-bromobenzyl alcohol (**2**) has the following physical and spectroscopic properties: mp 112 – 113 °C; ¹H NMR (acetone-d_6, 400 MHz) δ: 4.19 (t, J = 5.5, 1 H), 4.55 (d, J = 5.5, 2 H), 4.82 (br s, 2 H), 6.65 (d, J = 8.5, 1 H), 7.12 (dd, J = 8.5, 2.4, 1 H), 7.22 (d, J = 2.4, 1 H). ¹³C NMR (acetone-d_6, 100 MHz) δ: 62.6, 108.3, 117.6, 128.7, 131.1,

242

131.3, 146.8. HRMS (ESI-TOF) m/z: Calcd. for $C_7H_8BrNNaO$ [M+Na]: 223.9681, found: 223.9687. IR (ATR, cm^{-1}): 3381, 3201 (br), 1473, 1408, 1340, 1268, 1192, 1079. Anal. calcd for C_7H_8BrNO: C, 41.61; H, 3.99 N, 6.93. Found: C, 41.78; H, 3.94; N, 6.91.

8.   Acetonitrile was purchased from Sigma-Aldrich and passed through a column of activated alumina using a solvent purification system. Although the procedure reported here involves the use of dry acetonitrile (MeCN) solvent, untreated MeCN shows similar yields and reaction times.

9.   Tetrakisacetonitrile copper(I) trifluoromethanesulfonate and 2,2'-biyridine were purchased from Sigma-Aldrich and used as received. Checkers purchased 2,2,6,6-tetramethylpiperidine-N-oxyl from ABCR-Chemicals and used it as received. Submitters purchased 2,2,6,6-tetramethylpiperidine-N-oxyl from Sigma-Aldrich and used it as received.

10. N-Methylimidazole was purchased from Sigma-Aldrich and used as received.

11.   For the synthesis of volatile aldehydes, the neck should be fitted with a water condenser and a septum with a balloon of house air fitted with a needle. For all alcohols, the reactions proceed more rapidly if an O_2 balloon is employed instead of ambient air. *Users should be aware that the use of pure O_2 with organic solvents is potentially explosive.*

12.   Most reactions will change from the initial dark red/brown color to a dark green color upon completion. In the case of aldehyde **3**, the yellow/brown color of the crude product masks the typical color change of the reaction.

13.   The silica plug is composed of 100 g of silica (Silicycle SilicaFlash® P60, 230-400 mesh) in a 250 mL M porosity fritted funnel.

14.   The crude product is >95% pure by ^{1}H NMR spectroscopy, but contains small amounts of TEMPO.

15.   2-Amino-5-bromobenzaldehyde (**3**) has the following physical and spectroscopic properties: mp 74–76 °C; ^{1}H NMR (CDCl$_3$, 400 MHz) δ: 6.14 (br s, 2 H), 6.56 (d, $J = 8.8$, 1 H), 7.37 (dd, $J = 8.8$, 2.3, 1 H), 7.58 (d, $J = 2.4$, 1 H), 9.79 (s, 1 H). ^{13}C NMR (CDCl$_3$, 100 MHz) δ: 107.4, 118.1, 120.1, 137.5, 138.0, 148.8. 192.9. HRMS (EI) m/z: Calcd. for C_7H_6BrNO [M]$^+$: 198.9633, found: 198.9628. IR (ATR, cm^{-1}): 3424, 3322, 1649, 1614, 1545, 1468, 1390, 1311, 1185. Anal. calcd for C_7H_6BrNO: C, 42.03; H, 3.02; N, 7.00. Found: C, 42.30; H, 3.02; N, 7.03. This aldehyde appears to be bench-stable and no decomposition was observed after storing the aldehyde on the bench under ambient conditions for weeks.

16. Submitters reported two runs. Starting from **2** (6.10 g, 30.2 mmol) **3** was obtained in 97% (5.81 g). Starting from **2** (5.8 g, 29 mmol) **3** was obtained in 97% (5.57 g).

Handling and Disposal of Hazardous Chemicals

The procedures in this article are intended for use only by persons with prior training in experimental organic chemistry. All hazardous materials should be handled using the standard procedures for work with chemicals described in references such as "Prudent Practices in the Laboratory" (The National Academies Press, Washington, D.C., 2011 www.nap.edu). All chemical waste should be disposed of in accordance with local regulations. For general guidelines for the management of chemical waste, see Chapter 8 of Prudent Practices.

These procedures must be conducted at one's own risk. *Organic Syntheses, Inc.*, its Editors, and its Board of Directors do not warrant or guarantee the safety of individuals using these procedures and hereby disclaim any liability for any injuries or damages claimed to have resulted from or related in any way to the procedures herein.

3. Discussion

Aldehydes are useful synthetic intermediates commonly employed to access a variety of complex molecules. Unfortunately traditional methods for the selective oxidation of primary alcohols to aldehydes often involve the use, separation, and disposal of expensive or toxic stoichiometric reagents, particularly hypervalent iodine reagents[2] and metal oxides.[3] Other common methods require the careful maintenance of low temperature conditions (as in a Swern oxidation[4]), or the cautious handling of sensitive materials (such as 2-iodoxybenzoic acid [IBX]).

The use of molecular oxygen is an attractive alternative and significant progress has been made in the development of catalytic methods for the aerobic oxidation of alcohols.[5] For aerobic methods to compete with traditional routes, however, they must afford a broad scope of aldehydes (or ketones) in high yields, be operationally simple, and use inexpensive, readily available reagents and solvents. Unfortunately, few existing aerobic methods satisfy these criteria. For example, catalyst systems derived from Pd[6] and Ru[7] are often inhibited by heterocycles and other nitrogen-, oxygen-, and

244

sulfur-containing functional groups, or they promote oxidation of other functional groups (such as the Pd-catalyzed oxidation of alkenes[8]). Cu-based[9,10] catalysts often demonstrate broader functional group tolerance; however, several features of these systems limit their widespread adoption. For example, some catalysts exhibit low activity with aliphatic alcohols, thereby restricting their utility to the oxidation of 1° benzylic or allylic alcohols,[9] or they require pure O_2 as the oxidant, in some cases using non-traditional halogenated solvents (e.g., fluorobenzene).[10] The Cu[I]/TEMPO catalyst system described here[11] enables efficient aerobic oxidation of aliphatic alcohols while maintaining a broad substrate scope and functional-group compatibility, employs O_2 from ambient air as the oxidant, and utilizes common, readily available reagents and solvent. The practical features of this method, including its operational simplicity, predictability, reliability and chemoselectivity, make it a compelling alternative to traditional methods for the oxidation of primary alcohols.

Using this aerobic Cu[I]/TEMPO system, benzylic and allylic alcohols typically undergo complete oxidation within several hours. The oxidation of aliphatic alcohols often requires longer reaction times (20 – 24 h). Representative examples of the substrate scope are shown in Table 1. The method tolerates diverse functional groups including heterocycles such as pyridines, furans, and thiophenes (entries 2 and 6), in addition to alkenes (entry 3) and alkynes. Alcohols containing free anilines and aryl halides (entry 1) also undergo facile oxidation to the corresponding aldehyde, as do ethers, esters, thioethers, and acetals, although not included here.[11] The reaction conditions are sufficiently mild that Z-allylic alcohols (entry 3) and alcohols with adjacent stereocenters (entry 5), proceed efficiently without isomerization of alkene or stereocenter. A small number of functional groups remain challenging for this copper-based catalyst system. Alcohols bearing phenols or terminal alkynes do not yield the corresponding aldehydes, and alcohols containing a vicinal coordinating group, such as an ether or free amine, can be problematic.

Table 1. Aldehydes obtained by Cu^{I}/TEMPO catalyzed aerobic oxidation of primary alcohols

entry	aldehyde	time (h)	% Yield^a
1	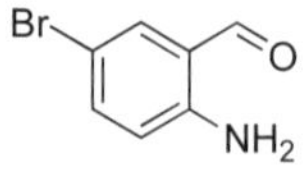	2	89-91%^{b}
2	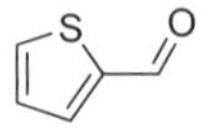	3	83%
3		2	>98%^{c} 20:1 Z:E
4	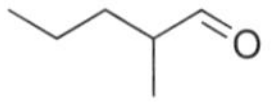	24 11	83% 98%^{d}
5	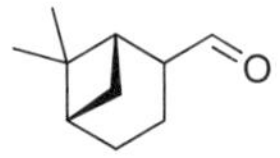	24	>98%
6	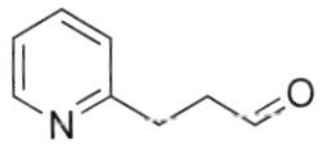	3-4?	95%
7	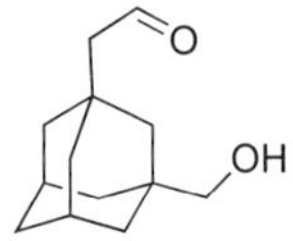	2.5	>98%
8	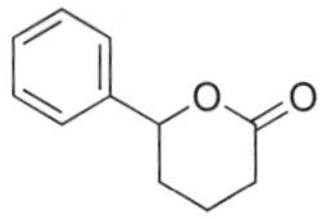	24	88%^{e}

^{a} for 1 mmol scale, ref 11. Standard conditions: [Cu(MeCN)$_4$](OTf) (0.01 M), bpy (0.01 M), TEMPO (0.01 M), NMI (0.02 M), 0.2 M alcohol in MeCN, rt, ambient air ^{b} this work ^{c} with air balloon ^{d} with O$_2$ balloon ^{e} reaction at 50 °C

In addition to the broad substrate scope, this new Cu^{I}/TEMPO catalyst system has many appealing practical characteristics. The reaction setup is straightforward, employs standard glassware and commercially available reagents, and most reactions can be carried out in open reaction vessels employing ambient air as the source of oxidant. During the course of our

246

studies, we found that larger scale (>1 g) oxidations of aliphatic alcohols are sensitive to the reaction vessel and an oversized flask enables reproducible reaction times (1 L flask for 50 mmol reaction). Aliphatic alcohols with substituents in the alpha position (Table 1, entries 4 and 5) react more slowly and may need increased reaction temperatures (50 °C) or the use of an O_2 balloon in order to reach completion within 24 h. Because these reactions operate under ambient air conditions, low-boiling aldehydes can be lost to evaporation over the course of the reaction, in which case a sealed reaction vessel equipped with a balloon of house air (or O_2) enables the aldehydes to be obtained in high yields (Table 1, entries 3 and 4). Detailed reaction conditions accounting for these substrate variations have been presented elsewhere.[12]

The separation and isolation of the aldehyde product is also straightforward, in most cases requiring only filtration of the reaction mixture through a silica plug or an aqueous extraction to remove the Cu salts to provide aldehyde product that is pure, based on ^{1}H NMR spectroscopic analysis. Given the volatility of many aldehydes and their propensity to undergo side reactions, the ability to achieve high purity with minimal handling is advantageous. This aspect of the Cu[I]/TEMPO system has recently been used to access amines from alcohols via a multi-step procedure.[13]

Finally, this Cu[I]/TEMPO catalyst system shows high selectivity for the oxidation of 1° alcohols in the presence of 2° alcohols, allowing for the selective oxidation of unprotected diols, to yield hydroxyaldehydes (Table 1, entry 7) or lactones (Table 1, entry 8). In most cases, the optimized Cu[I]/TEMPO catalyst system described here is suitable for achieving highly selective oxidations, enabling oxidation of 1° over 2° alcohols and showing high preference for sterically accessible alcohols. In some cases, the use of a milder catalyst system employing either a Cu[I]Br or Cu[II]Br$_2$ salt in place of [Cu[I](MeCN)$_4$](OTf) is required for achieving selectivity and these methods have been outlined elsewhere.[11,12]

1. Department of Chemistry, University of Wisconsin-Madison, 1101 University Avenue, Madison, WI 53706. E-Mail: stahl@chem.wisc.edu, Phone: (608) 265-6288, Fax: (608) 262-6143.
We are grateful to the NIH (RC1-GM091161), the ACS Green Chemistry Institute Pharmaceutical Roundtable. NMR spectroscopy

facilities were partially supported by the NSF (CHE-9208463) and NIH (S10 RR08389).

2. (a) Dess, D. B.; Martin, J. C. *J. Org. Chem.* **1983**, *48*, 4155–4156. (b) Frigerio, M.; Santagostino, M. *Tetrahedron Lett.* **1994**, *35*, 8019–8022. (c) Uyanik, M.; Ishihara, K. *Chem. Commun.* **2009**, 2086–2099.
3. (a) Corey, E. J.; Suggs, J. W. *Tetrahedron Lett.* **1975**, *16*, 2647–2650. (b) Ladbury, J. W.; Cullis, C. F. *Chem Rev.* **1958**, *58*, 403–438. (c) Fatiadi, A. J. *Synthesis* **1976**, 65–104.
4. (a) Mancuso, A. J.; Huang, S.-L.; Swern, D. *J. Org. Chem.* **1978**, *43*, 2480–2482. (b) Mancuso, A. J.; Brownfain, D. S.; Swern, D. *J. Org. Chem.* **1979**, *44*, 4148–4150. (c) Tidwell, T. T. *Synthesis* **1990**, 857–870.
5. (a) Arends, I. W. C. E.; Sheldon, R. A. In *Modern Oxidation Methods* (ed. Bäckvall, J.-E.) 83–118 (Wiley-VCH, Weinheim, 2004). (b) Sheldon, R. A.; Arends, I. W. C. E.; ten Brink, G.-J.; Dijksman, A. *Acc. Chem. Res.* **2002**, *35*, 774–781. (c) Zhan, B.-Z.; Thompson, A. *Tetrahedron* **2004**, *60*, 2917–2935. (d) Mallat, T.; Baiker, A. *Chem. Rev.* **2004**, *104*, 3037–3058. (e) Markó, I. E.; Giles, P. R.; Tsukazaki, M.; Chellé-Regnaut, I.; Gautier, A.; Dumeunier, R.; Philippart, F.; Doda, K.; Mutonkole, J.-L.; Brown, S. M.; Urch, C. J. *Adv. Inorg. Chem.* **2004**, *56*, 211–240. (f) Schultz, M. J.; Sigman, M. S. *Tetrahedron* **2006**, *62*, 8227–8241.
6. (a) Gligorich, K. M.; Sigman, M. S. *Chem. Commun.* **2009**, 3854–3867. (b) Stahl, S. S. *Angew. Chem. Int. Ed.* **2004**, *43*, 3400–3420. (c) Nishimura, T.; Onoue, T.; Ohe, K.; Uemura, S. *J. Org. Chem.* **1999**, *64*, 6750–6755. (d) Schultz, M. J.; Park, C. C.; Sigman, M. S. *Chem. Commun.* **2002**, 3034-3035. (e) Schultz, M. J.; Hamilton, S. S.; Jensen, D. R.; Sigman, M. S. *J. Org. Chem.* **2005**, *70*, 3343–3352.
7. (a) Lenz R.; Ley, S. V. *J. Chem. Soc., Perkin Trans. 1* **1997**, 3291–3292. (b) Dijksman, A.; Marino-González, A.; Payeras, A. M.; Arends, I. W. C. E.; Sheldon, R. A. *J. Am. Chem. Soc.* **2001**, *123*, 6826–6833. (c) Hasan, M.; Musawir, M.; Davey, P. N.; Kozhevnikov, I. V. *J. Mol. Catal. A* **2002**, *180*, 77–84. (d) Mizuno, N.; Yamaguchi, K. *Catal. Today* **2008**, *132*, 18–26.
8. (a) Nishimura, T.; Kakiuchi, N.; Onoue, T.; Ohe, K.; Uemura, S. *J. Chem. Soc., Perkin Trans. 1* **2000**, 1915–1918. (b) Mifsud, M.; Parkhomenko, K. V.; Arends, I. W. C. E.; Sheldon, R. A. *Tetrahedron* **2010**, *66*, 1040–1044.

9. (a) Semmelhack, M. F.; Schmid, C. R.; Cortés, D. A.; Chou, C. S. *J. Am. Chem. Soc.* **1984**, *106*, 3374–3376. (b) Gamez, P.; Arends, I. W. C. E.; Reedijk, J.; Sheldon, R. A. *Chem. Commun.* **2003**, 2414–2415.

10. (a) Markó, I. E.; Giles, P. R.; Tsukazaki, M.; Brown, S. M.; Urch, C. J. *Science* **1996**, *274*, 2044–2046. (b) Markó, I. E.; Gautier, A.; Mutonkole, J.-L.; Dumeunier, R.; Ates, A.; Urch, C. J.; Brown, S. M. *J. Organomet. Chem.* **2001**, *624*, 344–347. (c) Markó, I. E., Gautier, A.; Dumeunier, R.; Doda, K.; Philippart, F.; Brown, S.M.; Urch, C. J. *Angew. Chem. Int. Ed.* **2004**, *43*, 1588–1591. (d) Ragagnin, G.; Betzemeier, B.; Quici, S.; Knochel, P. *Tetrahedron* **2002**, *58*, 3985–3991. (e) Kumpulainen, E. T. T.; Koskinen, A. M. P. *Chem. Eur. J.* **2009**, *41*, 10901–10911.

11. Hoover, J. M.; Stahl, S. S. *J. Am. Chem. Soc.* **2011**, *133*, 16901–16910.

12. Hoover, J. M.; Steves, J. E.; Stahl, S. S. *Nat. Protoc.* **2012**, *7*, 1161–1166.

13. Redford, J. E.; McDonald, R. I.; Rigsby, M. L.; Wiensch, J. D.; Stahl, S. S. *Org. Lett.* **2012**, *14*, 1242–1245.

Appendix
Chemical Abstracts Nomenclature; (Registry Number)

2-Amino-5-bromobenzyl alcohol; (20712-12-3)
2-Amino-5-bromobenzaldehyde; (21924-57-0)
2-Amino-5-bromobenzoic acid; (5974-88-7)
Lithium aluminum hydride (LiAlH$_4$); (16853-85-3)
2,2'-Bipyridine (bpy); (366-18-7)
2,2,6,6-Tetramethylpiperidine-*N*-oxyl (TEMPO); (2564-83-2)
N-Methylimidazole (NMI); (616-47-7)

Shannon S. Stahl is a Professor of Chemistry at the University of Wisconsin-Madison, where he began his independent career in 1999. His research group specializes in homogeneous catalysis, with an emphasis on aerobic oxidation reactions and their applications in organic chemistry. He was an undergraduate at the University of Illinois at Urbana-Champaign, and subsequently attended Caltech (Ph.D., 1997), where he was an NSF predoctoral fellow with Prof. John Bercaw. From 1997–1999, he was an NSF postdoctoral fellow with Prof. Stephen Lippard at MIT.

Jessica Hoover received her B.S. degree in Chemistry from Harvey Mudd College in 2004 where she conducted research with Professor Adam Johnson. She then moved to Seattle where she received her Ph.D. degree from the University of Washington in 2009 under the mentorship of Professors James Mayer and Forrest Michael. She was a postdoctoral fellow with Professor Shannon Stahl at the University of Wisconsin Madison (2009-2012) before beginning her independent career as Assistant Professor of Chemistry at West Virginia University in Fall 2012.

Guido Möller studied chemistry at the Westfälische Wilhelms-Universität Münster (Dipl.-Chem. 2011). In the same year, he joined the research group of Prof. Erick M. Carreira at the ETH Zurich to pursue his Ph.D.

250

One-Pot Preparation of Cyclic Amines from Amino Alcohols

Submitted by Feng Xu[1] and Bryon Simmons.
Checked by Tomoaki Maehara and Tohru Fukuyama.

1. Procedure

Indoline oxalic acid salt. A 500-mL, 3-necked, round-bottomed flask equipped with a septum through which is inserted a thermocouple probe, an overhead stirrer with a paddle size of 5 cm, and a 50-mL pressure-equalizing addition funnel fitted with a nitrogen inlet, is charged with anhydrous DME (80 mL) (Note 1) and $SOCl_2$ (6.2 mL, 0.087 mol, 1.2 equiv) (Note 2) at ambient temperature. A solution of 2-aminophenethyl alcohol (10.0 g, 0.070 mol, 1.0 equiv) (Note 3) in DME (20 mL) is added dropwise to the stirred solution via the additional funnel over 1–1.5 h, maintaining the internal temperature at 20–30 °C with an external cooling bath (Notes 3 and 4). After addition, the batch is further stirred for 6–7 h at ambient temperature (Note 5). Sodium hydroxide (2.5 N, 128 mL, 0.32 mol, 4.4 equiv), followed by water (16 mL), is added to the reaction mixture via the addition funnel over 30 min, maintaining the internal temperature at <35 °C with an ice/water cooling bath (Note 6). The reaction mixture is then warmed to 60 °C and stirred for 10 h (Note 7). The reaction mixture is cooled to ambient temperature and transferred to a 1-L separatory funnel. *tert*-Butyl methyl ether (MTBE, 100 mL) and water (56 mL) (Note 8) are added. The organic phase is retained and the separated aqueous phase is back extracted with MTBE (56 mL) (Note 9). The combined organic phase is washed with brine (43 mL) (Note 10), dried over sodium sulfate (Notes 11 and 12), and concentrated by rotary evaporation to dryness under reduced pressure (35 °C bath, 60 mmHg). The resulting crude product is dissolved with ethyl acetate (ca. 90 mL) to a volume of 100 mL (Note 13).

A 500-mL, 3-necked, round-bottomed flask equipped with a septum through which is inserted a thermocouple probe, an overhead stirrer with a paddle size of 5 cm, and a 100-mL pressure-equalizing addition funnel fitted

Published on the Web 5/1/2013
© 2013 Organic Syntheses, Inc.

with a nitrogen inlet, is charged with oxalic acid dihydrate (9.3 g, 0.073 mol, 1.04 equiv) (Note 14) and methanol (14 mL). The resulting stirred solution is warmed to ambient temperature (Note 15). About 30 mL of the above crude product solution in ethyl acetate is added dropwise via the additional funnel at ambient temperature over 15 min. The batch is seeded with crystalline product (3 mg) (Note 16) and stirred for 30 min to form a seed bed slurry. Then, the rest of the product solution in ethyl acetate is added dropwise over 2 h. The slurry is stirred at ambient temperature for 15 h, then filtered through a 100-mL sintered glass funnel (Note 17). The wet cake is washed with 10% methanol in ethyl acetate (2 x 15 mL). Air suction drying affords the oxalic acid salt of indoline (12.0–12.1 g, 79%) as a white crystalline solid (Note 18).

2. Notes

1.   Anhydrous 1,2-dimethoxyethane (DME) was obtained from Sigma–Aldrich and used as received. All solvents (*tert*-butyl methyl ether, ethyl acetate, methanol) were obtained from Fisher Scientific and used as received.

2.   Thionyl chloride was obtained from Sigma–Aldrich and used as received.

3.   2-Aminophenethyl alcohol (97%) was obtained from Sigma–Aldrich and used as received.

4.   The addition of 2-aminophenethyl alcohol was mildly exothermic. A cold water bath was used to maintain the internal temperature between 20–30 °C.

5.   Typically, a slurry forms within 1–2 h after addition of thionyl chloride. The submitters report that the reaction progress can be monitored by HPLC. After stirring the reaction mixture for 6–7 h, >99% conversion was achieved as determined by HPLC analysis: YMC Pro Pack C18 column, 4.6 x 250 mm, 5 μm particle size, 40 °C, mobile phase: MeCN/10 mM, pH 6.5 phosphate buffer; MeCN increased from 30% to 70% over 18 min. Flow rate: 1.0 mL/min; UV detector at 210 nm. Retention times: 2-aminophenethyl alcohol, 4.5 min; 2-(2-chloroethyl)aniline, 12.7 min.

6.   The addition of NaOH was mildly exothermic. The pH of the quenched solution containing 2-(2-chloroethyl)aniline was ~13–14.

7.   By HPLC analysis, >99% of 2-(2-chloroethyl)aniline was converted to the desired cyclized product, indoline. Retention time of

Org. Synth. **2013**, *90*, 251-260

indoline: 9.4 min. The aqueous phase pH decreased to ~9 as HCl formed during the cyclization neutralized a portion of the excess NaOH.

8. The addition of 56 mL water dissolved the precipitated inorganic salts.

9. By HPLC analysis, the product loss to the first aq. phase (~280 mL) was ~5%; the loss to the back-extracted aq. phase was <0.5%.

10. The HPLC assay yield of the final organic phase (~230 mL) was 95%. No product was lost to the brine wash.

11. Anhydrous sodium sulfate was obtained from Sigma–Aldrich and used as received.

12. Sodium sulfate was filtered through a medium porosity sintered glass funnel.

13. Alternatively, the wet organic phase after aqueous workup could be azeotropically dried and solvent-switched to ethyl acetate under reduced pressure.

14. Oxalic acid dihydrate was obtained from Fisher Scientific and used as received.

15. Dissolution of oxalic acid dihydrate in methanol was endothermic.

16. It is recommended to seed the batch to relieve super saturation for a robust crystallization. The seed could be prepared by subdividing 1.6 mL of the crude indoline solution in EtOAc used for salt formation, which is then mixed with a solution of oxalic acid dihydrate (0.15 g) in MeOH (1.6 mL). The mixture is concentrated by rotary evaporation to dryness under reduced pressure (35 °C bath, 60 mmHg). The resulting crude product is triturated with 3.2 mL of 20% MeOH in EtOAc at ambient temperature to give the crystalline indoline oxalic acid salt, which can be used 'as is' to seed the batch or can be filtered and air-suction dried.

17. Typical supernatant concentration of indoline free base: 7–8 mg/mL by HPLC analysis.

18. Indoline oxalic acid salt has the following physical and spectroscopic properties: 99.5% purity (HPLC conditions in note 5, retention times: oxalic acid, 2.2 min; indoline, 9.4 min); mp = 128–129 °C (Lit.[2] mp = 128 °C); ^{1}H NMR (500 MHz, d_6-DMSO) δ: 2.90 (t, J = 8.3 Hz, 2 H), 3.40 (t, J = 8.3 Hz, 2 H), 6.55 (m, 2 H), 6.91 (m, 1 H), 7.03 (m, 1 H), 10.91 (s, br, 1.5 H); ^{13}C NMR (125 MHz, d_6-DMSO) δ: 29.1, 46.0, 110.4, 119.1, 124.3, 126.9, 129.9, 149.3, 161.6; Anal. Calcd for $C_{10}H_{11}ClNO_4$: C, 57.41; H, 5.30; N, 6.70. Found: C, 57.37; H, 5.28; N, 6.67.

Handling and Disposal of Hazardous Chemicals

The procedures in this article are intended for use only by persons with prior training in experimental organic chemistry. All hazardous materials should be handled using the standard procedures for work with chemicals described in references such as "Prudent Practices in the Laboratory" (The National Academies Press, Washington, D.C., 2011 www.nap.edu). All chemical waste should be disposed of in accordance with local regulations. For general guidelines for the management of chemical waste, see Chapter 8 of Prudent Practices.

These procedures must be conducted at one's own risk. *Organic Syntheses, Inc.*, its Editors, and its Board of Directors do not warrant or guarantee the safety of individuals using these procedures and hereby disclaim any liability for any injuries or damages claimed to have resulted from or related in any way to the procedures herein.

3. Discussion

Many methods[3,4] have been developed to prepare cyclic amines through cyclodehydration of amino alcohols. Classical indirect cyclodehydration of amino alcohols typically involves a tedious sequence of protection/activation/cyclization/deprotection. Although commonly implemented,[3,4] these indirect approaches require multiple chemical steps that reduce the overall efficiency of the transformation.

Direct cyclodehydration of amino alcohols is one of the most straight-forward approaches to prepare cyclic amines.[5-9] However, direct chlorination of an amino alcohol free base with $SOCl_2$, which was discovered several decades ago, has not been well studied.[10] Its application to prepare cyclic amines is underutilized due to the expected competition[10] between N- and O-sulfinylation and subsequent 'inevitable' side reactions. Low yields are typically an issue for this reaction[12-14] when $SOCl_2$, as reported,[13,14] is typically added to a solution of the amino alcohol in the presence or absence of a base.

The development of the one-pot process described here is based on a rational mechanistic understanding of the chlorination pathway (Scheme 1), which is further confirmed by NMR studies.[15] Unlike the prevailing literature procedure, a clean cyclodehydration transformation is achieved by 'inverse' addition of a solution of the free amino alcohol in an appropriate solvent (such as DME, *i*-PrOAc, and CH_2Cl_2) to a solution of $SOCl_2$. As such, the amino alcohol becomes instantly protonated upon contact with HCl as it is generated in the $SOCl_2$ solution. Because the amino alcohol is added slowly to keep a low concentration of its protonated salt in the reaction mixture, the protonated amino alcohol would be expected immediately to react with excess $SOCl_2$ and to retain a kinetically favorable, homogenous reaction solution before it could crystallize.[12,15] Thus, complete conversion could be achieved. In addition, the minor *N*-sulfinylated intermediates (such as sulfamic chloride **1**) that could be formed by reacting with $SOCl_2$ are also preserved and further converted to the corresponding chlorides (such as **2** and **3**, X = Cl) in the acidic inverse-addition reaction media, because the nucleophilic amine species that could react with these *N*-sulfinylated intermediates are either quenched by HCl or converted to sulfamic acids with $ClSO_2H$.

Scheme 1. Reaction Pathways for Chlorination with $SOCl_2$

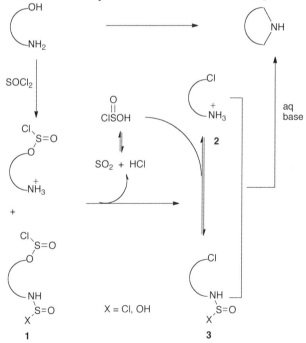

The examples shown in Table 1 illustrate the scope of this simple, practical one-pot process. We were gratified to observe that all of the amino alcohols examined were cleanly and efficiently converted to the

Table 1. Direct Chlorination/Cyclization of Amino Alcohols

Entry	Starting Material	Conditions[a]		Product	Yield (%)[b]
		Chlorination	Cyclization		
1	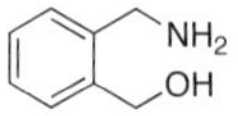	$(MeOCH_2)_2$ rt	rt	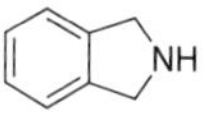	83[c]
2	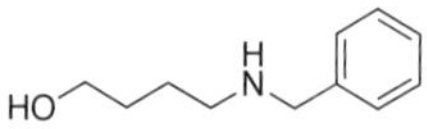	$(MeOCH_2)_2$ rt	rt	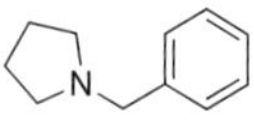	94
3	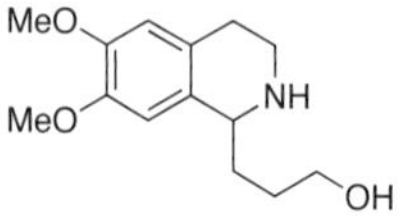	CH_2Cl_2 40 °C	$(MeOCH_2)_2$ 40 °C, 3 h	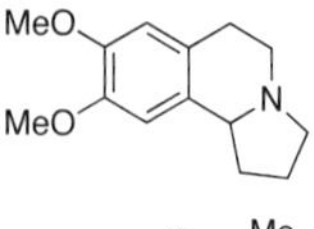	95
4	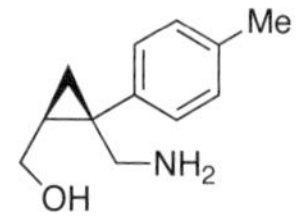	i-PrOAc rt	rt	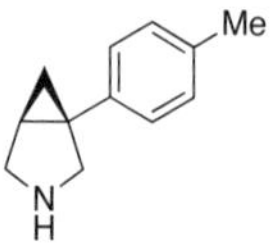	92[c]
5	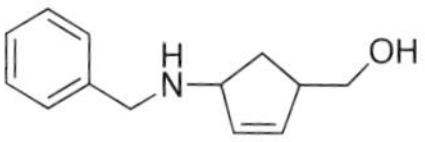	CH_2Cl_2 rt	$(MeOCH_2)_2$ 40 °C, 3 h	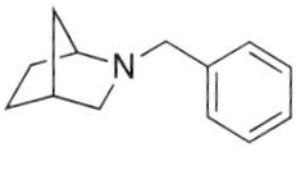	90
6	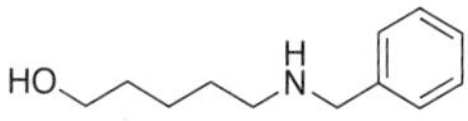	$(MeOCH_2)_2$ rt	rt	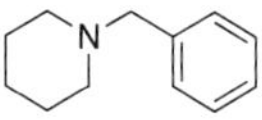	92[c]
7	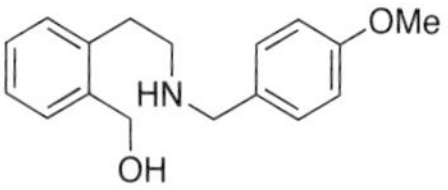	$(MeOCH_2)_2$ rt	40 °C	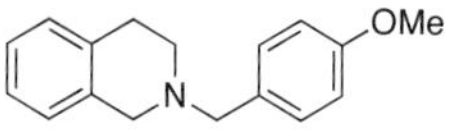	96[c]
8	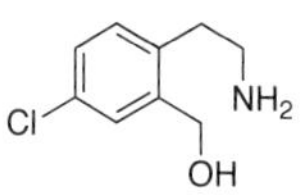	$(MeOCH_2)_2$ rt	rt	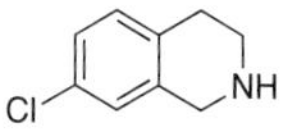	79
9	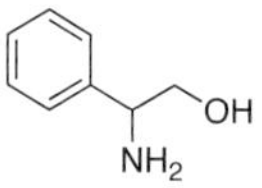	$(MeOCH_2)_2$ rt		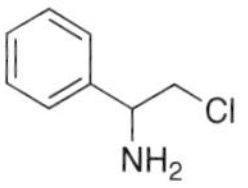	99
10	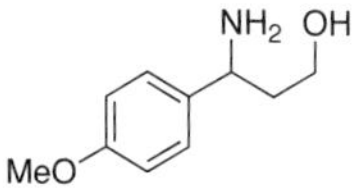	$(MeOCH_2)_2$ 40 °C		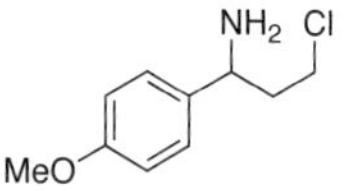	99

[a] Unless otherwise noted, the cyclization was carried out in the same solvent as the chlorination. [b] Unless otherwise noted, isolated yield through SiO_2 column chromatography purification. [c] Isolated as its HCl salt.

corresponding chloroamine in nearly quantitative yield. Full conversion to the desired chloride was observed within 5 h for most examples; however, some substrates (for example, Table 1, entries 3 and 10) required heating at 40 °C for several hours to achieve complete conversion. For a direct cyclodehydration transformation, the crude chloroamine intermediates were then treated with base. The intramolecular cyclization rate is dependent on the substrates as expected. Attempts to cyclize the readily formed 1,2- and 1,3-chloroamines (Table 1, entries 9 and 10) even at elevated temperatures resulted in complicated mixtures containing only small amounts of desired cyclized product.[16]

1. Department of Process Chemistry, Merck Research Laboratory, Rahway, NJ 07065, USA. E-mail: feng_xu@merck.com

2. Huisgen, R.; König, H. *Chem. Ber.* **1959**, *92*, 203.

3. For recent reviews, see: (a) Buffat, M. G. P. *Tetrahedron* **2004**, *60*, 1701. (b) Larock, R. C. *Comprehensive Organic Transformations* 2nd Ed. Wiley–VCH: New York, 1999. pp 689–702; pp 779–784.

4. For recent examples and applications, see: (a) Smith, A. B., III; Kim, D.-S. *J. Org. Chem.* **2006**, *71*, 2547. (b) Trost, B. M.; Tang, W.; Toste, F. D. *J. Am. Chem. Soc.* **2005**, *127*, 14785. (c) Pyne, S. G.; Davis, A. S.; Gates, N. J.; Hartley, J. P.; Lindsay, K. B.; Machan, T.; Tang, M. *Synlett* **2004**, 2670. (d) Kan, T.; Kobayshi, H.; Fukuyama, T. *Synlett* **2002**, 697. (b) Ina, H.; Kibayashi, C. *J. Org. Chem.* **1993**, *58*, 52. (c) Burgess, K.; Chaplin, D. A.; Henderson, I.; Pan, Y. T.; Elbein, A. D. *J. Org. Chem.* **1992**, *57*, 1103. (d) Schink, H. E.; Pettersson, H.; Bäckvall, J.–E. *J. Org. Chem.* **1991**, *56*, 2769.

5. For recent applications of direct cyclodehydration methods, see: (a) de Figueiredo, R. M.; Fröhlich, R.; Christmann, M. *J. Org. Chem.* **2006**, *71*, 4147. (b) Wu, T. R.; Chong, J. M. *J. Am. Chem. Soc.* **2006**, *128*, 9646. (c) Pulz, R.; Al–Harrasi, A.; Reissig, H.–U. *Org. Lett.* **2002**, *4*, 2353 and references cited therein.

6. For examples of cyclodehydration via organometallic redox method: (a) Nota, T.; Murahashi, S.–I. *Synlett.* **1991**, 693; (b) Murahashi, S.–I; Kondo, K.; Hakata, T. *Tetrahedron Lett.* **1982**, *23*, 229; (c) Grigg, R.; Mitchell, T. R. B.; Sutthivaiyakit S.; Tongpenyai, N. *Chem. Commun.* **1981**, 611.

7. For examples of acidic cyclodehydration, see: (a) Tanaka, H.; Murakami, Y.; Aizawa, T.; Torii, S. *Bull. Chem. Soc. Jpn.* **1989**, *62*, 3742; (b) Papageorgiou, C.; Borer, X. *Helv. Chim. Acta* **1988**, *71*, 1079

8. For recent reviews on application of phosphorus-assisted Mitsunobu reactions, see: (a) Hughes, D. L. *Org. Prep. Proc. Int.* **1996**, *28*, 127; (b) Hughes, D. L. *Org. Reactions* **1992**, *42*, 335; (c) Mitsunobu, O. *Synthesis* **1981**, 1.

9. For reviews on application of the Appel reaction and its variants, see: (a) Castro, B. R. *Org. Reactions* **1983**, *29*, 1. (b) Appel, R. *Angew. Chem., Int. Ed. Engl.* **1975**, *14*, 801.

10. Chlorination of alcohols, excluding amino alcohols, with SOCl₂ is thoroughly studied in terms of both mechanism and practical applications. For recent reviews regarding SOCl₂, see: (a) El–Sakka, I. A.; Hassan, N. A. *J. Sulfur Chem.* **2005**, *26*, 33; (b) Wirth, D.D. In *Encyclopedia Reagents Org. Synth.*; Paquette, L. A. Ed.; John Wiley & Sons: New York, **1995**, pp 4873; (c) Oka, K. *Synthesis* **1981**, 661.

11. For example, see: Pilkington, M.; Wallis, J. D. *Chem. Commun.* **1993**, 1857.

12. Low yields are typically an issue for the chlorination of amino alcohol salts. For examples, see: (a) Back, T. G.; Nakajima, K. *J. Org. Chem.* **2000**, *65*, 4543. (b) Bubnov, Y. N.; Zykov, A. Y.; Ignatenko, A. V.; Mikhailovsky, A. G.; Shklyaev, Y. V.; Shklyaev, V. S. *Izv. Akad. Nauk., Ser. Khim.* **1996**, 935. (c) Dobler, M.; Beerli, R.; Weissmahr, W. K.; Borschberg, H–J. *Tetrahedron: Asymmetry* **1992**, *11*, 1411. (d) Kóbor, J.; Fülöp, F.; Bernáth, G.; Sohár, P. *Tetrahedron* **1987**, *43*, 1887. (e) Bäckvall, J. E.; Renko, Z. D.; Byström, S. E. *Tetrahedron Lett.* **1987**, *28*, 4199. (f) Piper, J. R.; Johnston, T. P. *J. Org. Chem.* **1963**, *28*, 981. (g) Norton, T. R.; Seibert, R. A.; Benson, A. A.; Bergstrom, F. W. *J. Am. Chem. Soc.* **1946**, *68*, 1572.

13. For examples of chlorination of amino alcohols with SOCl₂ mediated by base, see: (a) Muratake, H.; Natsume, M. *Tetrahedron Lett.* **2002**, *43*, 2913. (b) Howarth, N. M.; Malpass, J. R.; Smith, C. R. *Tetrahedron* **1998**, *54*, 10899. (c) Bosch, J.; Mestre, E.; Bonjoch, J.; López, F.; Granados, R. *Heterocycles* **1984**, *22*, 767. (d) Mousseron, M.; Winternitz, F.; Mousseron–Canet, M. *Bull. Soc. Chim. Fr.* **1953**, 737.

14. For examples of chlorination of amino alcohols with SOCl₂ in the absence of base by adding SOCl₂ to amino alcohol solution, see: (a) Gursky, M. E.; Ponomarev, V. A.; Pershin, D. G.; Bubnov, Y. N.;

Antipin, M. Y.; Lyssenko, K. A. *Russ. Chem. Bull.* **2002**, *51*, 1562. (b) Foubelo, F.; Gómez, C.; Gutiérrez, A.; Yus, M. *J. Heterocycl. Chem.* **2000**, *37*, 1061. (c) Howarth, N. M.; Malpass, J. R.; Smith, C. R. *Tetrahedron* **1998**, *54*, 10899. (d) Gursky, M.E.; Potapova, T. V.; Bubnov, Y. N. *Izv. Akad. Nauk., Ser Khim.* **1998**, 1450. (e) Bubnov, Y. N.; Zykov, A. Y.; Ignatenko, A. V.; Mikhailovsky, A. G.; Shklyaev, Y. V.; Shklyaev, V. S. *Izv. Akad. Nauk., Ser Khim.* **1996**, 935. (f) Granier, C.; Guilard, R. *Tetrahedron* **1995**, *51*, 1197. (g) Grigg, R.; Santhakumar, V.; Sridharan, V.; Thornton–Pett, M.; Brideg, A. W. *Tetrahedron* **1993**, *49*, 5177. (h) Takano, S.; Inomata, K.; Sato, T.; Ogasawara, K. *Chem. Commun.* **1989**, 1591. (i) Barbry, D.; Couturier, D.; Ricart, G. *Synthesis* **1980**, 387. (j) Sturm, P. A.; Cory, M.; Henry, D. W. *J. Med. Chem.* **1977**, *20*, 1333. (k) Giacet, C.; Bécue, G. *C. R. Acda. Sc. Paris, C* **1967**, *264*, 103.

15. Xu, F.; Simmons, B.; Reamer, R. A.; Corley, E.; Murry, J.; Tschaen, D. *J. Org. Chem.* **2008**, *73*, 312.

16. For examples of relative rates of cyclization as a function of ring size: see (a) Mandolini, L. *Adv. Phys. Org. Chem.* **1986**, *22*, 1. (b) Winnik, M.A. *Chem. Rev.* **1981**, *81*, 491. (c) Galli, C.; Illuminati, G.; Mandolini, L.; Tamborra, P. *J. Am. Chem. Soc.* **1977**, *99*, 2591.

Appendix
Chemical Abstracts Nomenclature; (Registry Number)

Thionyl chloride; (7719-09-7)
2-Aminophenethyl alcohol; (5339-85-5)
2-(2-Chloroethyl)aniline; (762177-99-1)
Oxalic acid dihydrate; (6153-56-6)
Indoline; (486-15-1)

Dr. Feng Xu obtained his Ph.D. at Shanghai Institute of Organic Chemistry (SIOC), Chinese Academy of Sciences in 1989 where he worked on the total synthesis of complex natural products. He joined SIOC before moving to the USA. After he undertook a postdoctoral fellow with Professors Martin Kuehne and James Dittami, and completed the total syntheses of several complex indole alkaloids, he joined Merck Process Research Department in 1996.

Dr. Bryon Simmons received a B.S. degree in chemistry in 2002 from Brigham Young University and a Ph.D. degree in 2010 with Professor David W. C. MacMillan at Princeton University. He currently works at Merck in the Process Research Department.

Tomoaki Maehara was born in Shizuoka, Japan in 1989. He received his B.S.in 2012 from the University of Tokyo. In the same year, he began his graduate studies at the Graduate School of Pharmaceutical Sciences, the University of Tokyo, under the guidance of Professor Tohru Fukuyama. His research interests are in the area of the total synthesis of natural products.

260

Preparation of Tetrabutylammonium (4-fluorophenyl)trifluoroborate

Submitted by Fabrizio Pertusati, Parag V. Jog, and G. K. Surya Prakash.*[1]
Checked by Changming Qin and Huw M. L. Davies.

1. Procedure

A.   *Tetrabutylammonium   (4-fluorophenyl)trifluoroborate   3.*
4-Fluorophenyl boronic acid **1** (4.00 g, 28.6 mmol, 1.00 equiv) (Note 1) is placed in an open single-necked 1000-mL round-bottomed flask equipped with a magnetic stir bar and suspended in a mixture of chloroform (400 mL) (Note 2) and water (80 mL) (Note 3). Tetrabutylammonium bifluoride (TBABF) **2** (24.2 g, 85.8 mmol, 3.00 equiv) (Note 4) is directly added to a 250-mL pressure-equalizing addition funnel and dissolved with H_2O (120 mL). Open to air, the addition funnel is fitted to the reaction flask and the clear solution of tetrabutylammonium bifluoride **2** is added dropwise to the stirred boronic acid suspension over a period of 1 h (Note 5). Upon addition of the bifluoride solution, the boronic acid dissolves and the biphasic mixture becomes transparent. After 2.5 h (Note 6) the reaction mixture is transferred to a 1000-mL separatory funnel and the two layers are separated. The water layer is extracted with chloroform (4 x 100 mL) and the combined organic phases are washed with water (3 x 200 mL), brine (2 x 150 mL), and dried over anhydrous magnesium sulfate (30 g, 1 h) (Note 7). Magnesium sulfate is removed by vacuum filtration on a Büchner funnel (Note 8) and the filtrate is evaporated by rotary evaporation (bath temp 40 °C), with the final concentration in a 250-mL round bottom flask. The resulting clear transparent oil is placed under high vacuum (0.5 mmHg) for 3 h to afford a white solid (14.5 g). The crude product is purified by recrystallization from ethanol/water (Note 9) as follows. The solid trifluoroborate in the 250-mL

Org Synth. **2013**, *90*, 261-270
Published on the Web 5/7/2013
© 2013 Organic Syntheses, Inc.

round-bottomed flask is dissolved in ethanol (9.0 mL) upon heating at 50 °C in a water bath. After cooling to 23 °C, water (4.0 mL) is added dropwise to the flask with swirling until the solution becomes cloudy and remains persistently cloudy. This solution is placed in a refrigerator at 0 °C for 3 h. A white solid precipitates from the mixture and is collected by gravity filtration (Note 10). Water (3 x 20 mL) is used to rinse residual solids from the 250 mL flask, and all the rinses are used to wash the solid, which is then dried under high vacuum (0.2 mmHg) over P_2O_5 (30 g) (15 h) to remove traces of ethanol and water, affording a white crystalline solid (11.2 g, 97%) (Note 11).

2. Notes

1. 4-Fluorophenylboronic acid (Sigma-Aldrich) was used as received.

2. Stir bar (VWR) 2x5/16 inch-octagonal was used. Chloroform (≥99.8%, HPLC grade, Sigma-Aldrich) was used as received.

3. ASTM type II water was produced by RO/DI (HARLECO).

4. Tetrabutylammonium bifluoride (> 95.0%, TCI America, Inc.) was used as received. Use of excess reagent was necessary for complete conversion.

5. Rate of addition was approximately 70 drops/min.

6. The reaction was monitored by ^{19}F NMR according to the following procedure: 0.5 mL of the chloroform layer was withdrawn from the mixture, the solvent was evaporated on a rotary evaporator, and the crude mixture was examined by ^{19}F NMR using $CDCl_3$ as solvent. ^{19}F NMR showed complete absence of fluorine signals corresponding to 4-fluorophenyl boronic acid **1** after 2.5 h reaction time.

7. Magnesium sulfate (anhydrous, EMD Chemicals) was used as received.

8. 150 mL Büchner funnel, medium Frit.

9. Ethanol (200 proof, DECON Laboratories, Inc) was used as received.

10. Fisherbrand filter paper, Qualitative P8, Porosity-coarse, 24 cm diameter.

11. Tetrabutylammonium (4-fluorophenyl)trifluoroborate **3** has the following physical and spectroscopic properties: mp 83–84 °C; IR (neat):

2964, 2876, 1588, 1489, 1186, 1003, 973, 952, 825, 739 cm^{-1}; ^{1}H NMR (400 MHz, CDCl$_3$) δ: 0.94 (t, J = 7.2 Hz, 12 H), 1.26–1.35 (m, 8 H), 1.39–1.45 (m, 8 H), 2.94–2.99 (m, 8 H), 6.86 (t, J = 8.8 Hz, 2 H), 7.55 (t, J = 7.4 Hz, 2 H); ^{13}C NMR (100 MHz, CDCl$_3$) δ: 13.3, 19.2, 23.4, 57.8, 112.9 (d, J_{C-F} = 18.6 Hz), 132.9 (d, J_{C-F} = 6.0 Hz), 161.6 (d, J_{C-F} = 238.9 Hz); ^{19}F NMR (376 MHz, CDCl$_3$) δ: –119.0 (brs, 1F), –141.6 (br, s, 3F); ^{11}B NMR (192 MHz, CDCl$_3$) δ: 3.34; MS (ESI): m/z (%) = 163.0349 (100) [M-NBu$_4$]$^-$; HRMS (ESI) m/z [M-NBu$_4$]$^-$ calculated for C$_6$H$_4$BF$_4$: 163.0347 found: 163.0349; Anal. Calcd C$_{22}$H$_{40}$BF$_4$N: C, 65.18; H, 9.95; N, 3.46. Found: C, 65.13; H, 9.80; N, 3.47 (after first recrystallization).

Handling and Disposal of Hazardous Chemicals

The procedures in this article are intended for use only by persons with prior training in experimental organic chemistry. All hazardous materials should be handled using the standard procedures for work with chemicals described in references such as "Prudent Practices in the Laboratory" (The National Academies Press, Washington, D.C., 2011 www.nap.edu). All chemical waste should be disposed of in accordance with local regulations. For general guidelines for the management of chemical waste, see Chapter 8 of Prudent Practices.

These procedures must be conducted at one's own risk. *Organic Syntheses, Inc.*, its Editors, and its Board of Directors do not warrant or guarantee the safety of individuals using these procedures and hereby disclaim any liability for any injuries or damages claimed to have resulted from or related in any way to the procedures herein.

3. Discussion

Since the development of Suzuki–Miyaura cross coupling, boronic acids have gained enormous importance. Despite their widespread use, boronic acids have several distinct drawbacks and limitations. Some boronic acids (cyclopropyl-, heteroaryl-, and vinylboronic acids) show instability upon storage. Boronic acids easily lose water and hence are not monomeric species but, rather, exist as dimeric and cyclic trimeric anhydrides. While this does not affect the cross coupling reactions, uncertainties about the exact stoichiometry of the reaction being performed can be an issue.

Furthermore, under certain reaction conditions, boronic acids are prone to protiodeboronation, resulting in reduced yields. To circumvent these limitations, Vedejs[2] introduced potassium trifluoroborate salts as stable and easy to handle boronic acid substitutes. Although the potassium salts have been synthetically useful, they present the disadvantage of having poor solubility in organic solvents other than methanol, acetonitrile, or water. This characteristic may constitute a problem when apolar substrates, for example, hydrophobic polymeric boronic acids, have to be transformed into the corresponding trifluoroborates. Tetrabutylammonium trifluoroborates[3] are soluble in common organic solvents such as chloroform and dichloromethane, are air-stable, have long shelf-life, and are very easy to handle. Prior to our successful report[4] of one-pot syntheses of a variety of tetrabutylammonium trifluoroborates directly from boronic acids, they were prepared by one of two general methods: by ion exchange from the corresponding potassium salt or from a concentrated methanol solution of a boronic acid treated with three equiv of 48% aqueous hydrofluoric acid (HF), followed by neutralization of the corresponding hydronium trifluoroborate with tetrabutylammonium hydroxide (TBAH).[5] Both methods present some disadvantages. The first requires the preparation of the potassium salts and subsequent cation exchange, while the second requires use of corrosive and hazardous aqueous HF.

The present methodology is a straightforward, one-pot procedure that avoids the use of noxious and highly corrosive HF. By contrast, TBABF is not corrosive and is commercially available in 25 g lots. The reaction can be performed in an open flask without requiring solvent purification and the product is usually isolated in good to excellent yields after purificiation by simple extraction/washing and simple recrystallization procedures. Most importantly, the present protocol is diversely applicable to a variety of substrates as illustrated in Table 1.

Table 1. Synthesis of aromatic and alkyl tetrabutylammonium trifluoroborates

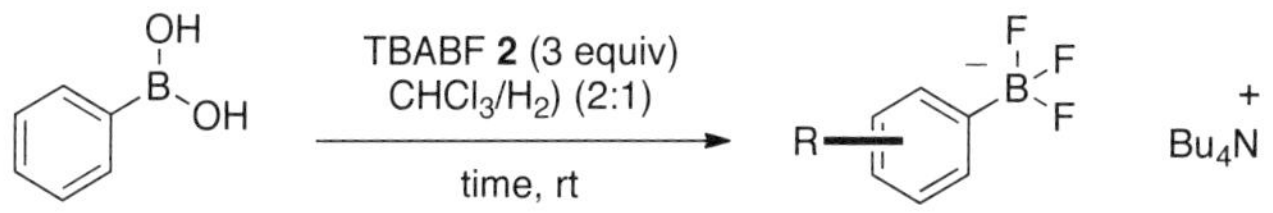

Entry	Substrate	Time	Product	% Yield
1	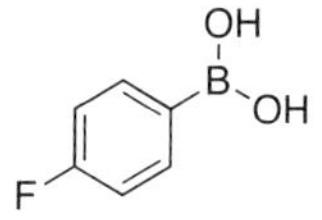	1	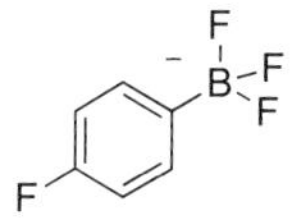	84
2	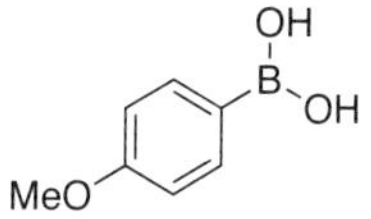	1	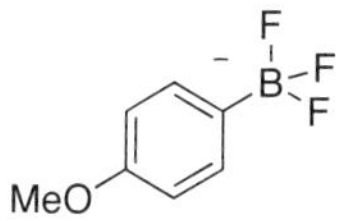	75
3	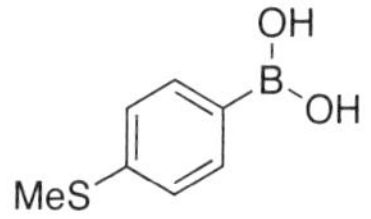	1.5	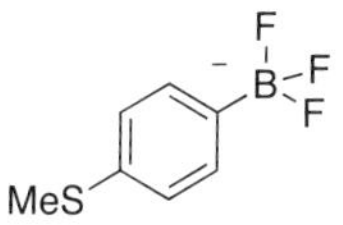	91
4	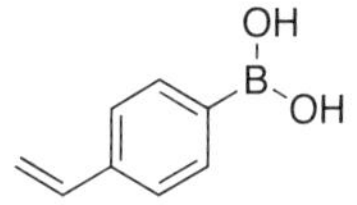	1.5	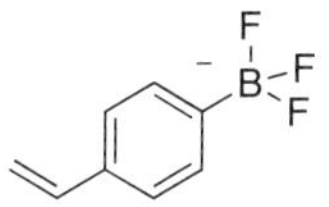	93
5	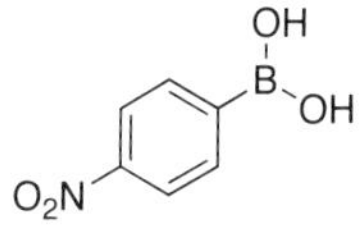	1	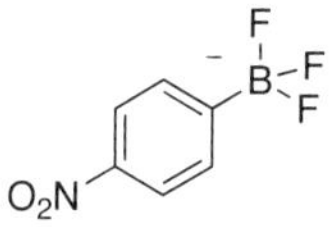	96
6	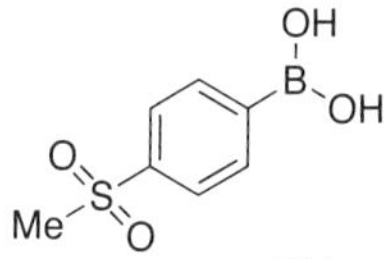	1	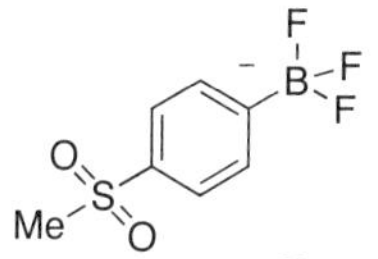	97
7	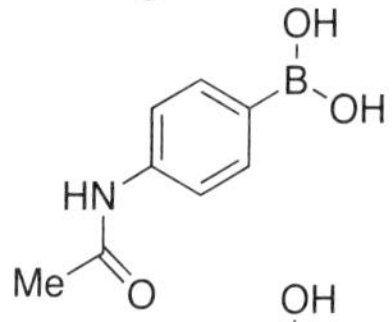	1	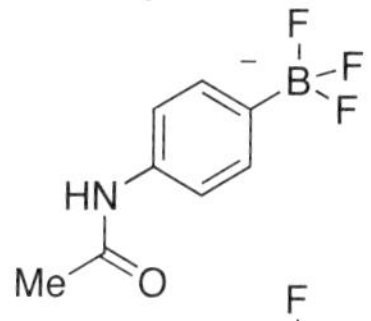	95
8	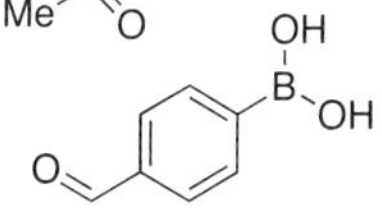	3	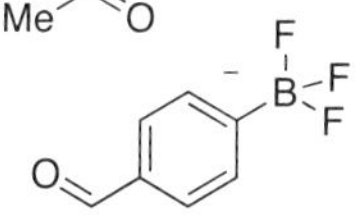	96

Table 1. (continued)

Entry	Substrate	Time	Product	% Yield
9		45 min		97
10		1		96
11		1		96
12		1		96
13		2		70
14		1.5		97
15		4		74
16		4		80
17		5		50[a,b]
18		5		50[a,b]
19		30 min		86[b]
20		30 min		98

[a] Reaction performed under a nitrogen atmosphere
[b] Crude yield determined by ^{19}F NMR spectroscopic data

The present methodology is quite tolerant of functional groups. Both electron-withdrawing and electron-donating substituents give excellent yields of corresponding trifluoroborates. Even aliphatic boronic acids afford good yields and the difficult to access cyclopropyl, alkynyl, styrenyl and cyclohexyl trifluoroborates are now available in moderate to excellent yields. Boronic esters also appear to react, albeit in moderate conversions as determined by ^{19}F NMR analysis using an internal standard. This protocol can be extended to heteroaromatic systems as exemplified in Table 2. Although the corresponding yields of the trifluoroborates are moderate, the fact that such molecules can be synthesized by this straightforward methodology is important.

Table 2. Synthesis of Heteroaromatic Tetrabutylammonium Trifluoroborates[a]

Entry	Substrate	Time	Product	% Yield
1	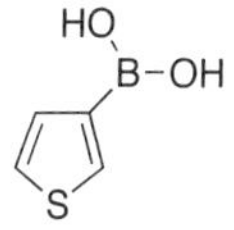	6	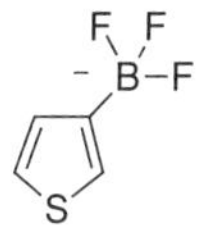	64
2	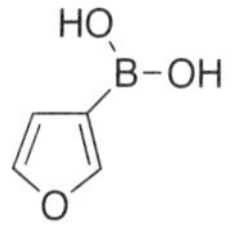	24	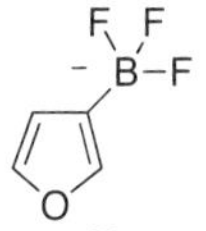	66[b]
3	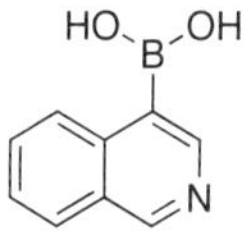	7	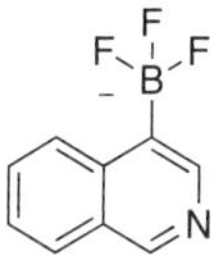	81[b]
4	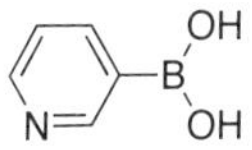	6	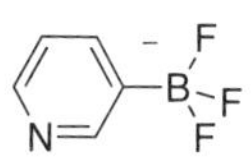	50
5	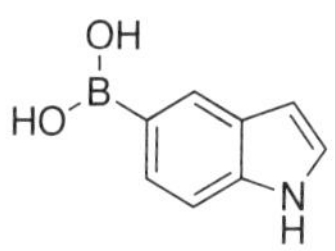	6	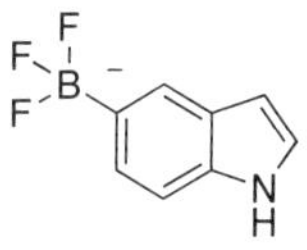	50
6	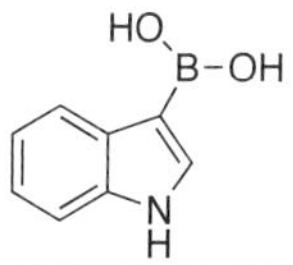	2.5	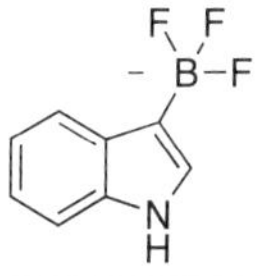	94

[a] Reaction conditions: HetAr (1 equiv), 2 (3 equiv), CHCl$_3$-H$_2$O (2:1, v:v), rt
[b] Crude yield determined from 19F NMR spectroscopic data

All of the heterocyclic trifluoroborates synthesized by this method can be used as coupling partners in cross-coupling reactions to synthesize a variety of heterocyclic structural motifs that are useful for biological screening.

Potassium trifluoroborates have been used in many transition-metal catalyzed cross-coupling reactions, as illustrated in the literature.[6] Importantly, palladium-catalyzed cross-coupling reactions of tetrabutylammonium trifluoroborates with a variety of aryl halides have also been reported.[5] Lipophilic benzyltrimethylammonium trifluoroborates have also been reported to provide better stereocontrol in alkylation reactions as compared to their potassium counterparts.[7] Applications such as these suggest that tetrabutylammonium trifluoroborates have good, but largely unexplored, potential as coupling partners, especially where the corresponding potassium salts show less promise.

Appendix
Chemical Abstracts Nomenclature; (Registry Number)

4-Fluorophenyl boronic acid: Boronic acid, B-(4-fluorophenyl)-, (CAS 1765-93-1),

Tetrabutylammonium bifluoride: 1-Butanaminium, N,N,N-tributyl-, (hydrogen difluoride) (1:1) (CAS 23868-34-0),

Tetrabutylammonium (4-Fluorophenyl)trifluoroborate): 1-Butanaminium, N,N,N-tributyl-, (T-4)-trifluoro(4-fluorophenyl)borate(1-) (1:1) , (CAS 1291068-40-0).

1. Loker Hydrocarbon Research Institute, University of Southern California, Los Angeles, CA 90089-1919, USA.
2. Vedejs, E.; Chapman, R. W.; Fields, S. C.; Lin, S.; Schrimpf, M. R. *J. Org. Chem.* **1995**, *60*, 3020–3027.
3. Tetrabutylammonium bifluoride is commercially available from TCI America. Otherwise, it can be easily prepared from readily available chemicals (KHF$_2$, Bu$_4$NHSO$_4$, KHCO$_3$) according to the literature procedure: Landini, D.; Molinari, H.; Penso, M.; Rampoldi, A., *Synthesis* **1988**, 953–955.
4. Prakash, G. K. S.; Pertusati, F.; Olah, G. A. *Synthesis*, **2011**, 292–302.

5. Batey, R. A.; Quach, T. D. *Tetrahedron Lett.* **2001,** 42, 9099–9103.
6. (a) Darses, S.; Michaund, G.; Genet, J.-P. *Eur. J. Org. Chem.* **1999,** 1875–1883. (b) Molander, G. A.; Ito, T. I. *Org. Lett.* **2001,** *3,* 393–396. (c) Molander, G. A.; Yun, C. S.; Ribagorda, M.; Biolatto, B. *J. Org. Chem.* **2003,** *68,* 5534–5539. (d) Saevmarker, J.; Rydfjord, J.; Gising, J.; Odell, L. R.; Larhed, M. *Org. Lett.* **2012,** *14,* 2394–2397. (e) Yao, B.; Liu, Y.; Wang, M.-K.; Li, J.-H.; Tang, R.-Y.; Zhang, X.-G.; Deng, C.-L. *Adv. Synth. Catal.* **2012,** *354,* 1069–1076. (f) Colombel, V.; Presset, M.; Oehlrich, D.; Rombouts, F.; Molander, G. A. *Org. Lett.* **2012,** *14,* 1680–1683. (g) Masuyama, Y.; Sugioka, Y.; Chonan, S.; Suzuki, N.; Fujita, M.; Hara, K.; Fukuoka, A. *J. Mol. Catal. A: Chem.* **2012,** *352,* 81–85. (h) Engle, K. M.; Thuy-Boun, P. S.; Dang, M.; Yu, J. -Q. *J. Am. Chem. Soc.* **2011,** *133,* 18183–18193. (i) Shintani, R.; Takeda, M.; Soh, Y. -T.; Ito, T.; Hayashi, T. *Org. Lett.* **2011,** *13,* 2977–2979. (j) Wu, X.-F.; Neumann, H.; Beller, M. *Adv. Synth. Catal.* **2011,** *353,* 788–792. (k) Li, M.; Wang, C.; Ge, H. *Org. Lett.* **2011,** *13,* 2062–2064.
7. Vedejs, E.; Fields, S. C.; Hayashi, R.; Hitchcock, S. R.; Powell, D. R.; Schrimpf, M. R. *J. Am. Chem. Soc.* **1999,** 121, 2460–2470.

G. K. Surya Prakash is a Professor of Chemistry and the Director of the Loker Hydrocarbon Research Institute at the University of Southern California holding the George A. and Judith A. Olah Nobel Laureate Chair in Hydrocarbon Chemistry with research contributions and interests in selective fluorination methods, new synthetic methods, mechanistic studies of organic reactions, electrochemistry, superacid chemistry and hydrocarbon chemistry. He has received many honors and accolades including 2004 ACS award for Creative Work in Fluorine Chemistry, the 2006 George A. Olah Award in Hydrocarbon or Petroleum Chemistry and the 2006 Richard C. Tolman Award.

Fabrizio Pertusati was born in Turin (Italy) in 1972. He did his undergraduate work at the Turin University on surfactant chemistry and after three years in PCB industry obtained his doctorate in organic chemistry at Cardiff University. After a postdoctoral work at Emory University with Professor Fred Menger, he then joined the Prakash group at the Loker Hydrocarbon Institute, University of Southern California, in 2008 working on organotrifluoroborate chemistry. He is currently at the School of Pharmacy at Cardiff University working in the laboratory of Professor Chris McGuigan on the diastereoselective synthesis of phosphoroamidate prodrugs as anti-HCV agents.

Parag V. Jog was born in Pune, India in 1976. He received his Bachelors (Chemistry, 1996) and Masters (Analytical Chemistry, 1998) at Bombay University. He earned his Ph.D. from Michigan Technological University in organo-sulfur chemistry (Prof. Dallas K. Bates, 2005). After doing his postdoctoral research at University of Urbana-Champaign (Synthetic Ion Channels, Prof. Mary S. Gin) and California Institute of Technology (Conformational Analysis of small organic molecules using NMR, Prof. John D. Roberts), he is currently working at University of Southern California under Prof. G. K. Surya Prakash in the field of organo-fluorine chemistry, specifically, developing direct trifluoromethylation methods.

Changming Qin was born in Shandong, China in 1982. He did his undergraduate work at Ludong University on the preparation of polymer nanocomposites under the guidance of Prof. Yucai Hu. He obtained his Masters degree in organic chemistry in 2008 at Wenzhou University under the supervision of Prof. Huayue Wu, working on palladium-catalyzed transformations of aryl boronic acids. After graduation, he worked at the University of Hong Kong with Prof. Chi-Ming Che in 2008-2009, and then joined Prof. Huw Davies' group at Emory University in 2010. His current research is focused on design and synthesis of chiral dirhodium catalysts and their application in novel asymmetric carbeniod transformations.

270

Allyl Cyanate-To-Isocyanate Rearrangement: Preparation of *tert*-Butyl 3,7-Dimethylocta-1,6-dien-3-ylcarbamate

Submitted by Yoshiyasu Ichikawa,[1] Noriko Kariya, and Tomoyuki Hasegawa.

Checked by Eiji Yoshida and Tohru Fukuyama.

1. Procedure

A. (E)-3,7-Dimethylocta-2,6-dien-1-ylcarbamate (**3**): An oven-dried 500-mL, three-necked, round-bottomed flask (Note 1) equipped with a magnetic stir bar (8 × 30 mm), a thermometer with a Teflon holder, a rubber septum and an inlet adapter with a three-way stopcock fitted with an argon inlet is charged with geraniol (**1**) (11.0 g, 71.3 mmol, 1.0 equiv) (Note 2) and dry dichloromethane (180 mL) (Note 3). The flask is immersed in a sodium chloride/ice bath (internal temperature –15 °C). Trichloroacetyl isocyanate (9.30 mL, 14.7 g, 78.0 mmol, 1.10 equiv) (Note 4) is added slowly via a syringe through the rubber septum over a 10 min period at a rate to maintain the internal temperature below –10 °C. After stirring at –15 °C (internal temperature) for 20 min (Note 5), methanol (0.90 mL, 0.71 g, 22 mmol, 0.30 equiv) (Note 6) is added via a syringe inserted through the rubber

Published on the Web 5/21/2013

© 2013 Organic Syntheses, Inc.

septum to quench excess trichloroacetyl isocyanate (Note 7). The solution is transferred to a 1-L recovery flask with the aid of dichloromethane (15 mL), and then concentrated by rotary evaporation (200 mmHg, water bath temperature 40 °C) and dried under vacuum (1.0 mmHg) (Note 8) to yield trichloroacetyl carbamate **2** as a pale yellow oil (25.9 g). The crude **2** in the 1-L recovery flask is dissolved in methanol (240 mL) (Note 6) and a magnetic stir bar (8 × 30 mm) is placed in the flask. Aqueous K_2CO_3 (2 M, 140 mL) is added to the stirred mixture in portions at room temperature over a 10 min period, resulting in formation of a cloudy solution (Note 9). After stirring at room temperature for 3 h while checking the consumption of trichloroacetyl carbamate **2** by TLC analysis (Note 5), the reaction mixture is concentrated by rotary evaporation (70 mmHg, water bath temperature 40 °C) to about half the original volume. The resulting cloudy solution is transferred to a 1-L separatory funnel and extracted with diethyl ether (3 × 50 mL) (Note 10). The combined organic extracts are washed with water (50 mL) and brine (50 mL), dried over anhydrous $MgSO_4$ and filtered through cotton. After removal of the solvent by rotary evaporation (70 mmHg, water bath temperature 40 °C), the oily residue is dried under vacuum (1.0 mmHg) (Note 8) to constant weight, affording crude carbamate **3** (13.93 g, 99%) (Note 11) as a white solid, which is used in Step B without further purification.

   B. *t-Butyl 3,7-dimethylocta-1,6-dien-3-ylcarbamate* (**6**). The crude carbamate **3** (13.93 g, 70.6 mmol, 1.0 equiv) in a 300-mL recovery flask is dissolved in toluene (10 mL) (Note 12) and the solvent is removed azeotropically by rotary evaporation (70 mmHg, 50 °C). After repeating this procedure once more followed by drying under vacuum (1.0 mmHg) (Note 8) for 30 min, crude **3** is dissolved in dry dichloromethane (2 × 10 mL) and transferred by a pipette to a 500-mL three-necked, round-bottomed flask equipped with a stir bar (8 × 30 mm), a thermometer with a Teflon holder, a 50-mL pressure-equalizing addition funnel with a rubber septum, and an inlet adapter with a three-way stopcock fitted with an argon inlet. The flask is charged with dry dichloromethane (180 mL), triphenylphosphine (28.0 g, 107 mmol, 1.5 equiv) (Note 13) and triethylamine (20.0 mL, 14.5 g, 143 mmol, 2.0 equiv) (Note 14). The flask is cooled with a sodium chloride/ice bath (internal temperature –15 °C). Carbon tetrabromide (37.8 g, 114 mmol, 1.6 equiv) (Note 15) in dry dichloromethane (20 mL) is added via the dropping funnel over a 30 min period, maintaining the internal temperature below –10 °C. The dropping funnel is rinsed with dry dichloromethane (5 mL). After stirring at –15 to –10 °C (internal temperature) for 20 min (Note 16), the cooling bath is removed and the yellow reaction mixture is diluted with *n*-hexane (150 mL) (Note 17). After stirring at room temperature for 30 min, the light-brown solution is poured into aqueous KHSO$_4$ (1 M, 300 mL) in a 1-L Erlenmeyer flask equipped with a stir bar (8 × 30 mm). The three-necked flask is rinsed with *n*-hexane (2 × 10 mL). After stirring vigorously for 30 min, the solution is transferred into a 1-L separatory funnel. The Erlenmeyer flask is rinsed with *n*-hexane (2 × 10 mL). The organic layer is separated, washed with water (200 mL), saturated aqueous NaHCO$_3$ (100 mL) and brine (200 mL), dried over anhydrous MgSO$_4$ and filtered through cotton. Concentration by rotary evaporation (200 mmHg, 40 °C) gives a crude mixture containing allyl isocyanate **5** and triphenylphosphine oxide as a yellow solid (72.8 g), to which is added *n*-hexane (200 mL). The resulting mixture is swirled in a sonicator for 5 min to form a suspension. After filtration through a pad of Celite (Note 18) and washing the filter cake with *n*-hexane (3 × 50 mL), the combined golden yellow filtrate is concentrated by rotary evaporation (90 mmHg, water bath temperature 40 °C), and then dried under vacuum (1.5 mmHg) for 30 min to provide crude isocyanate **5** (25.4 g) as a pale yellow oil (Notes 19 and 20).

An oven-dried 1-L, three-necked, round-bottomed flask equipped with a stir bar (8 × 30 mm), a thermometer with a Teflon holder, a 100-mL dropping funnel with a rubber septum, and an inlet adapter with a three-way stopcock fitted with an argon inlet is charged with dry tetrahydrofuran (280 mL) (Note 21) and *tert*-butyl alcohol (68.0 mL, 52.3 g, 711 mmol, 10.0 equiv) (Note 22). After cooling to 0 °C in an ice bath, lithium hexamethyldisilazide (1 M solution in tetrahydrofuran, 71.3 mL, 71.3 mmol, 1.0 equiv) (Notes 23 and 24) is added via the dropping funnel over a 10 min period, maintaining the internal temperature below 5 °C. The cooling bath is removed and the solution is stirred at room temperature for 30 min. The dropping funnel is quickly replaced with a 50-mL dropping funnel and the solution of lithium *t*-butoxide is cooled with a sodium chloride/ice bath (internal temperature –15 °C). A solution of crude **5** (25.4 g) in dry tetrahydrofuran (15 mL) is added through the dropping funnel over a 10 min period, keeping the internal temperature below –10 °C. After rinsing the dropping funnel with dry tetrahydrofuran (5 mL), the reaction mixture is stirred below –10 °C for 20 min (Note 25). Saturated aqueous NH$_4$Cl (100 mL) and water (100 mL) are added to quench the reaction and the mixture is transferred into a 1-L separatory funnel. The separated aqueous layer is extracted with diethyl ether (2 × 100 mL). The combined organic layers are washed with water (2 × 100 mL), brine (200 mL), and dried over anhydrous MgSO$_4$, filtered through cotton, and then concentrated by rotary evaporation (70 mmHg, water bath temperature 40 °C) to afford crude carbamate **6**, which is dried under vacuum (0.3 mmHg) to constant weight (21.15 g) (Note 26). The residual orange liquid is subjected to silica gel chromatography (Note 27). The combined fractions are concentrated by rotary evaporation (70 mmHg, water bath temperature 40 °C) and dried under vacuum (1.0 mmHg) (Note 8) to furnish Boc-carbamate **6** as a light-yellow oil (12.44 g, 69% overall yield from **1**) (Notes 28 and 29).

2. Notes

1. All glass apparatus was dried in an oven at 120 °C for 2 h and cooled in a drying desiccator.

2. Geraniol (**1**) (97%) was purchased from Wako Pure Chemical Industries, Ltd. and used directly without purification.

3. The submitters purchased dichloromethane from Wako Pure Chemical Industries, Ltd. and dried over 4Å molecular sieves. The water content determined by Karl-Fischer titration was <10 ppm. The molecular sieves were activated in a round-bottomed flask by heating in a household microwave oven (500 W) at full power for 5 min, and then cooled to room temperature under vacuum (1.8 mmHg) for 10 min. This procedure was repeated 3 times. The checkers purchased dichloromethane from Kanto Chemicals Co., Inc. and purified by a Glass Contour Solvent Dispensing System. The water content by Karl-Fischer titration was <10 ppm.

4. Trichloroacetyl isocyanate (97%) was purchased from Wako Pure Chemical Industries, Ltd. and used as received.

5. The progress of the reaction was monitored by thin-layer chromatography (TLC) using E. Merck pre-coated 0.25 mm thick silica gel 60 F254 plates. The plates were eluted with v/v 3:1 *n*-hexane/ethyl acetate and visualized with 254 nm UV light followed by dipping in a phosphomolybdic acid solution (w/v 3:20 in ethanol) and heating on a hot plate. The product trichloroacetyl carbamate **2** had an R$_f$ = 0.58 and the starting material geraniol (**1**) an R$_f$ = 0.33. Trichloroacetyl carbamate **2** undergoes gradual hydrolysis to carbamate **3** on TLC resulting in formation of a new spot with an R$_f$ = 0.29.

6.   Methanol (99.8%) was purchased from Wako Pure Chemical Industries, Ltd. and used as received.

7.   A slight temperature increase (ca 1–2 °C) was observed.

8.   The submitters dried the material at 0.3 mmHg on a vacuum line.

9.   After stirring at room temperature for 15 min, the solution became clear.

10.   When using ethyl acetate instead of diethyl ether, the crude product contained some byproducts, which were detrimental in step B.

11.   An analytically pure sample was obtained by dissolving the crude carbamate **3** (8.35 g) in *n*-hexane (50 mL) and cooling in an ice bath for one hour. The precipitated white solid was quickly filtered and washed with chilled *n*-hexane to provide pure carbamate **3** (4.63 g 55%). (*E*)-3,7-Dimethylocta-2,6-dien-1-ylcarbamate (**3**) exhibits the following properties: mp 30–31 °C; IR (NaCl): 3464, 3346, 2921, 1712, 1602 cm^{-1}; ^{1}H NMR (400 MHz, CDCl$_3$) δ: 1.59 (s, 3 H), 1.69 (s, 3 H), 1.71 (s, 3 H), 2.05–2.12 (m, 4 H), 4.56 (br s, 2 H), 4.59 (d, *J* = 6.9 Hz, 2 H), 5.09 (br t, *J* = 6.4 Hz, 1 H), 5.36 (br t, *J* = 6.9 Hz, 1 H); ^{13}C NMR (100 MHz, CDCl$_3$) δ: 16.3, 17.6, 25.6, 26.2, 39.4, 61.8, 118.4, 123.6, 131.7, 142.0, 157.4; Anal. calcd for C$_{11}$H$_{19}$NO$_2$: C, 66.97; H, 9.71; N, 7.10. Found C, 66.90; H, 9.48; N, 7.10; HRMS (ESI): *m/z* calcd for C$_{11}$H$_{20}$NO$_2$ (M+H$^+$ 198.1494, found 198.1495.

12.   Toluene (99.5%) was purchased from Wako Pure Chemical Industries, Ltd. and used as received.

13.   Triphenylphosphine (Extra Pure grade), purchased from Kanto Chemical Co., Inc., was dried in a desiccator over phosphorus pentoxide under vacuum (ca. 20 mmHg) prior to use.

14.   Triethylamine, purchased from Wako Pure Chemical Industries, Ltd., was distilled from CaH$_2$ and stored over potassium hydroxide pellets.

15.   Carbon tetrabromide (99%) was purchased from Tokyo Chemical Industry Co., Ltd. and used as received.

16.   The reaction was monitored by TLC. Carbamate **3** has an R$_f$ = 0.29 eluting with v/v 3:1 *n*-hexane/ethyl acetate and isocyanate **5** has an R$_f$ = 0.40 (*n*-hexane) and 0.80 (*n*-hexane/diethyl ether 20:1).

17.   *n*-Hexane (95%) was purchased from Wako Pure Chemical Industries, Ltd. and used as received.

18.   A porcelain Büchner funnel (4 cm diameter × 5 cm height) filled with 20 g of Celite was employed.

19.   The submitters found that crude **5** can be purified by dissolution in

n-hexane followed by passing the solution through a short column of silica gel. After elution with *n*-hexane followed by collecting and concentrating the fractions, isocyanate **5** was obtained as an oil, which was distilled using a Kugelrohr (0.83 mmHg, 50–55 °C). Although the resulting clear oil was homogeneous according to proton and carbon NMR spectroscopy, a correct elemental analysis could not be achieved. Spectral data for 3-isocyanato-3,7-dimethylocta-1,6-diene (**5**) are as follows: IR (NaCl): 2977, 2931, 2257 cm^{-1}; ^{1}H NMR (400 MHz, CDCl$_3$) δ: 1.41 (s, 3 H), 1.56–1.64 (m, 2 H), 1.61 (s, 3 H), 1.68 (s, 3 H), 1.98–2.07 (m, 2 H), 5.05–5.08 (m, 1 H), 5.10 (d, *J* = 10.6 Hz, 1 H), 5.27 (d, *J* = 17.0 Hz, 1 H), 5.75 (dd, *J* = 17.0, 10.5 Hz, 1 H); ^{13}C NMR (100 MHz, CDCl$_3$) δ: 17.6, 23.1, 25.6, 29.2, 42.8, 62.2, 112.8, 122.9, 123.1, 132.3, 141.9; HRMS (DART): *m/z* calcd for C$_{11}$H$_{18}$NO (M+H)$^{+}$ 180.1388, found 180.1385.

20. ^{1}H NMR analysis of the crude oil showed a small signal corresponding to the bromoform proton appearing at δ 6.83. Bromoform does not affect the subsequent step.

21. The submitters purchased dry tetrahydrofuran (anhydrous, 99+%) from Kanto Chemical Co., Inc. and used as received. The water content was 1 ppm by Karl-Fischer titration. The checkers purchased dry tetrahydrofuran from Kanto Chemicals Co., Inc., and purified by a Glass Contour Solvent Dispensing System. The water content was 9.5 ppm by Karl-Fischer titration.

22. *t*-Butyl alcohol, purchased from Wako Pure Chemical Industries, Ltd., was distilled from CaH$_2$ and stored over 4Å molecular sieves.

23. Lithium hexamethyldisilazide (1 M solution in tetrahydrofuran) was purchased from Sigma-Aldrich Chemical Company and used as received.

24. Initially, the submitters employed lithium *t*-butoxide prepared by the reaction of *n*-butyllithium with *t*-butanol in THF. In large-scale experiments, problems were sometimes encountered with the formation of a side product having a similar R$_f$ value to that of Boc-carbamate **6**. Further exploration showed that the side product was the *n*-butyl carbamate as shown below.

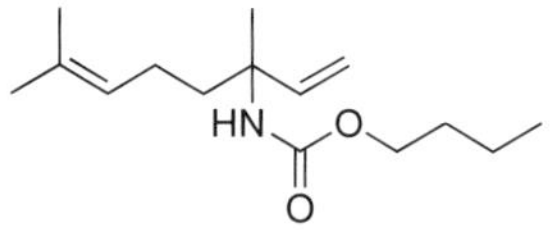

Although this problem could be avoided by using freshly opened *n*-butyllithium coupled with rigorous degassing of *t*-butanol by freeze-thaw cycles immediately before use, lithium hexamethyldisilazide was ultimately

276

selected as the base to obtain reproducible product purity and yields.

25. Reaction progress was monitored by TLC. The isocyanate starting material **5** has an $R_f = 0.40$ (*n*-hexane) and the Boc-carbamate product **6** has an $R_f = 0.47$ (v/v 7:1 *n*-hexane/ethyl acetate).

26. It is important to remove bromoform at this stage because it coelutes with Boc-carbamate **6** during chromatographic purification.

27. The checkers purchased 'Silca Gel 60 (spherical)' (particle size 63–210 μm) from Kanto Chemical Co., Inc. The crude product was transferred to a column (8 × 30 cm, 100 g of silica gel) with a minimum amount of *n*-hexane and eluted with 400 mL of *n*-hexane. Elution was continued with 1.5 L of *n*-hexane:ethyl acetate (v/v 20:1) and fraction collection (100-mL) was begun. Fractions 4–10 were combined and concentrated by rotary evaporation.

28. The submitters reported purity of 97% by GC analysis using an Agilent 6890N instrument equipped with an Agilent J&W DB-5 column (30.0 m × 0.25 mm) and a flame ionization detector using a method of 160 °C isotherm for 10 min, then ramp 20 °C/min to 300 °C, then 300 °C isothermal for 8 min with flow rate 30 cm/sec He carrier gas. The retention time for the product was 6.9 min with unidentified small peaks having retention times 1.7–6.4 min.

29. Kugelrohr distillation of carbamate **6** (oven temperature 130–135 °C at 1.0 mmHg) afforded an analytically pure sample. Spectral data for *t*-butyl 3,7-dimethylocta-1,6-dien-3-ylcarbamate (**6**): IR (NaCl): 3277, 2975, 2929, 1717, 1698 cm^{-1}; ^{1}H NMR (400 MHz, CDCl$_3$) δ: 1.37 (s, 3 H), 1.43 (s, 9 H), 1.60 (s, 3 H), 1.65–1.69 (m, 2 H), 1.68 (s, 3 H), 1.94 (q, $J = 7.3$ Hz, 2 H), 4.61 (br s, 1 H), 5.05–5.12 (m, 3 H), 5.89 (dd, $J = 17.4$, 10.5 Hz, 1 H); ^{13}C NMR (100 MHz, CDCl$_3$) δ: 17.5, 22.5, 24.7, 25.6, 28.4, 39.5, 56.1, 78.8, 111.9, 124.0, 131.8, 143.5, 154.2; Anal. calcd for C$_{15}$H$_{27}$NO$_2$: C, 71.10; H, 10.74; N, 5.53. Found C, 70.71; H, 10.76; N, 5.53; HRMS (DART): *m/z* calcd for C$_{15}$H$_{28}$NO$_2$ (M+H)$^+$ 254.2120, found 254.2117.

Handling and Disposal of Hazardous Chemicals

The procedures in this article are intended for use only by persons with prior training in experimental organic chemistry. All hazardous materials should be handled using the standard procedures for work with chemicals described in references such as "Prudent Practices in the Laboratory" (The National Academies Press, Washington, D.C., 2011 www.nap.edu). All chemical waste should be disposed of in accordance with local regulations. For general guidelines for the management of chemical waste, see Chapter 8 of Prudent Practices.

In the development and checking of these procedures, every effort has been made to identify and minimize potentially hazardous steps. The Editors believe that the procedures described in this article can be carried out with minimal risk if performed with the materials and equipment specified, and in careful accordance with the instructions provided. However, these procedures must be conducted at one's own risk. *Organic Syntheses, Inc.,* its Editors, and its Board of Directors do not warrant or guarantee the safety of individuals using these procedures and hereby disclaim any liability for any injuries or damages claimed to have resulted from or related in any way to the procedures herein.

3. Discussion

Allyl cyanates generated under several reaction conditions are considered to be possible transient intermediates, which undergo instantaneous rearrangement below ambient temperature to afford allyl isocyanates.[2] Historically, the first attempts to synthesize allyl cyanates dates back to a report in 1970 by Holm, who employed a cheletropic reaction of allyl thiatriazoles and found that only rearranged allyl isocyanates were isolated.[3] In 1978, Overman examined the synthesis of allyl cyanates by using the reaction of cyanogen chloride with the lithium salts of allyl alcohols and observed formation of a mixture of allylically rearranged isocyanates and dimeric carbamates.[4] Although these two pioneering reports support the belief that rearrangement of presumed allyl cyanates occurs below room temperature, development of an allyl cyanate-to-isocyanate rearrangement as a synthetic method for preparation of allyl amines is hampered by obstacles associated with preparation of cyanates (esters of cyanic acid: R–OCN).[5] Moreover, no mechanistic investigation had been undertaken to determine if the reaction proceeds via an ionic or concerted mechanism.[6]

In 1991, Ichikawa introduced a new synthetic method for the preparation of allyl cyanates which involves *dehydration reactions of allyl carbamates*.[7] This approach is quite reasonable considering the analogous methods to form carbon-nitrogen triple bonds, such as dehydration reactions of amides or formamides to produce nitriles or isonitriles, respectively (Scheme 1).[8]

278

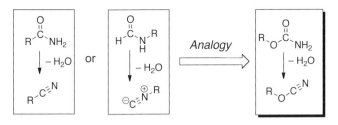

Scheme 1. Retrosynthetic analysis of cyanates by analogy

Ichikawa examined several dehydration conditions of allyl carbamates and found that trifluoromethanesulfonic anhydride (Tf_2O) and diisopropylethylamine (DIPEA) in dichloromethane at -78 °C (method A)[9] and triphenylphosphine, carbon tetrabromide and triethylamine (TEA) in dichloromethane at -20 °C (method B)[10] accomplish the transformation to form allyl cyanates (Scheme 2). It should be noted that these two dehydration methods were originally reported as procedures for the preparation of isonitriles from formamides. Although method B produces triphenylphosphine oxide as co-product, which is often difficult to remove, we preferred method B over method A owing to its experimental simplicity. Furthermore, the reaction conditions employed in method B are sufficiently mild to tolerate a wide range of functional groups. One of the important features of this rearrangement reaction is that the allyl isocyanate products can be flexibly transformed to a variety of allyl amine derivatives. For example, addition of amines, trimethylaluminum or alcohols to the reaction mixture containing the allyl isocyanate affords the respective ureas, acetamides[11] or carbamates.[12]

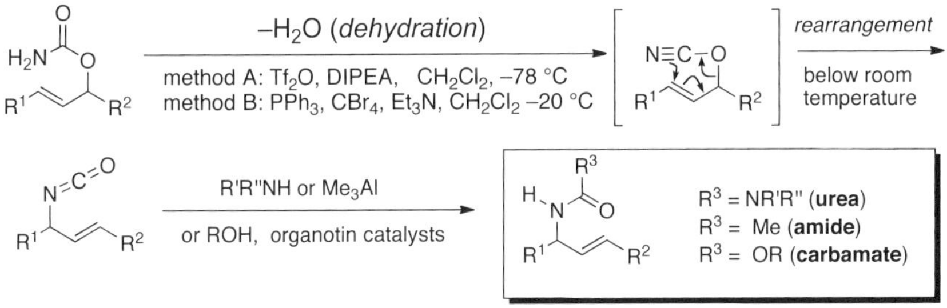

Scheme 2. Allyl cyanate-to-isocyanate rearrangement for the synthesis of allyl amine derivatives

The advantageous feature of the transformation of allyl isocyanates to allyl amine derivatives is demonstrated in synthetic studies of the nucleoside antibiotic, blasticidin S (**11**) (Scheme 3).[13] In the route to this target, transformation of hex-3-enopyranose **7** to 4-amino-hex-2-enopyranose (**10a–c**) was carried out by utilizing an allyl cyanate-to-isocyanate rearrangement (**8**→**9**). Treatment of the produced allyl isocyanate **9** with 2,2,2-trichloroethanol, benzyl alcohol or allyl alcohol furnished the respective Troc (**10a**), Cbz (**10b**) and Alloc carbamate (**10c**) derivatives. After considerable experimentation with these three carbamates to determine the best choice of protecting group, the Troc carbamate (**10a**) was selected as a key intermediate, which led to the first total synthesis of blasticidin S. In addition, it is noteworthy that the intermediate allyl cyanate **8**, having the cyanate group at the pseudo-equatorial position, underwent smooth rearrangement even at 0 °C.[14]

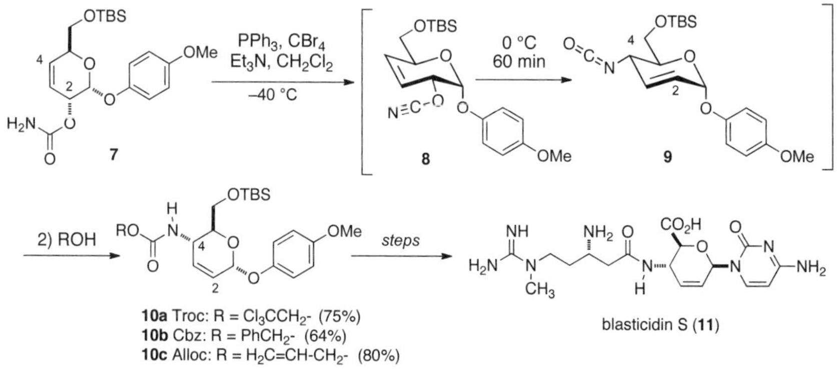

Scheme 3. The key step in the total synthesis of blasticidin S.

The stereochemistry of allyl cyanate-to-isocyanate rearrangement was examined using chiral allyl carbamate **13**, which was prepared from ethyl (*S*)-lactate (**12**) (Scheme 4).[15] Dehydration of **13** followed by treatment of the resulting allyl isocyanate **15** with trimethylaluminum furnished amide **16** in a good yield. The enantiomeric purity of the rearranged product **16** was determined to be 98% by analysis of the corresponding MTPA esters **17**. The absolute stereochemistry of the formed stereogenic center in **16** was determined by analysis of the MTPA amides **18** using the Kusumi method for elucidation of the absolute configuration of primary amines.[16] The results of these studies led to the

Org. Synth. **2013**, *90*, 271-286

conclusion that the rearrangement of allyl cyanates proceeds via a concerted mechanism resulting in [1,3]-chirality transfer with a high level of stereoselectivity.

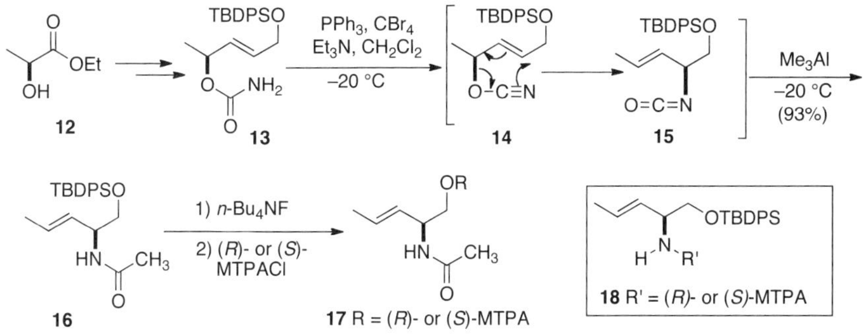

Scheme 4 Examination of stereochemistry in allyl cyanate-to-isocyanate rearrangement.

The availability of a reliable method for the preparation of allyl cyanates through dehydration of allyl carbamates coupled with the stereoselective nature of the rearrangement established that allyl cyanate-to-isocyanate rearrangement is a useful method for the enantioselective synthesis of allyl amine derivatives. As a result, a number of organic chemists have employed this reaction for the syntheses of nitrogen-containing bioactive compounds.[17]

The procedure described herein, highlighting the allyl cyanate-to-isocyanate rearrangement, serves as an efficient method for the preparation of allyl amines as Boc-carbamate derivatives starting from allyl alcohols. This is exemplified by conversion of geraniol (**1**) to the Boc-carbamate **6**. Carbamoylation of **1** is carried out by treatment with trichloroacetyl isocyanate followed by hydrolysis of the trichloroacetyl group of **2** with potassium carbonate in aqueous methanol to provide allyl carbamate **3** in excellent yield (Step A).[18] Although trichloroacetyl isocyanate is an expensive and moisture-sensitive liquid that causes difficulties in handling, it is an ideal reagent for this transformation due to its high reactivity, clean reaction, and excellent yields.[19]

Dehydration of allyl carbamate **3** generates allyl cyanate **4**, which undergoes spontaneous rearrangement to afford allyl isocyanate **5** (Step B). Although the dehydration conditions are mild enough to tolerate a wide range of functional groups, one drawback is the formation of the co-product triphenylphosphine oxide. In the present case, dissolution of the crude reaction mixture in hexane results in precipitation of triphenylphosphine oxide, which is easily removed by filtration. In smaller scale experiments, it is more convenient to purify the product directly by chromatographing the crude mixture rather than by precipitation and filtration. When the product has the same or similar R_f value as triphenylphosphine oxide, tributylphosphine can be used as an alternative.[12] Other dehydration methods, which circumvent problems associated with triphenylphosphine oxide, utilize either Tf$_2$O/DIPEA[7] or trifluoroacetic anhydride (TFAA)/TEA.[20] In particular, in their total synthesis of manzamine A, Fukuyama and co-workers employed TFAA/TEA for dehydrating the functional group-rich carbamate **19** (Scheme 5).[21] It should be noted that use of the TFAA/TEA procedure was originally described as a synthetic method for the preparation of nitriles from amides by Casini.[22]

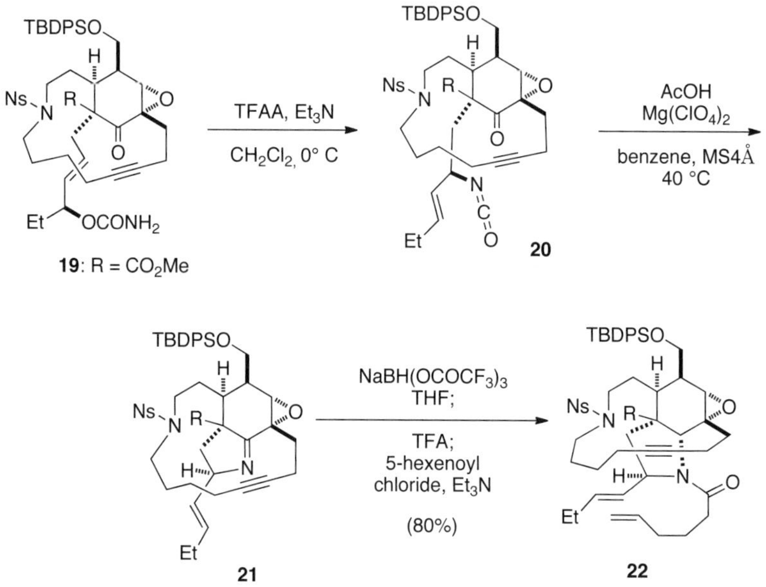

Scheme 5 The key step in the total synthesis of manzamine A

Transformation of the sterically crowded isocyanate **5** to the Boc-carbamate **6** was performed by reaction with lithium *t*-butoxide. Since lithium *t*-butoxide is a strong base, we cannot employ these conditions with isocyanates containing base-sensitive functional groups or epimerization-prone stereogenic centers. In these cases, it is recommended that related procedures be used to transform hindered isocyanates into carbamates under Lewis or Brønsted acid conditions using titanium tetra-*t*-butoxide[23] or trimethylsilyl chloride[24] or molybdenum (VI) dichloride dioxide.[25]

In summary, the transformation of geraniol (**1**) to Boc-carbamate **6** exemplifies a useful synthetic method for the construction of quaternary carbons bearing nitrogen substituents.[26]

1. Faculty of Science, Kochi University, Akebono-cho, Kochi 780-8520 Japan; E-mail: ichikawa@kochi-u.ac.jp.
2. (a) Banert, K.; Groth, S. *Angew. Chem. Int. Ed. Engl.* **1992**, *31*, 866. (b) Banert, K.; Melzer, A. *Tetrahedron Lett.* **2001**, *42*, 6133.
3. Christophersen, C.; Holm, A. *Acta Chem. Scand.* **1970**, *24*, 1512.
4. Overman, L. E.; Kakimoto, M. *J. Org. Chem.* **1978**, *43*, 4564.
5. (a) Grigat, E.; Pütter, R. *Angew. Chem. Int. Ed. Engl.* **1967**, *6*, 206. (b) Jensen, K. A.; Holm, A. In *The Chemistry of Cyanates and Their Thio Derivatives Part 1*, Chapter 16, p. 569; Patai, S., Ed.; Wiley-Interscience: New York, **1977**. (c) Grashey, R. In *Comprehensive Organic Synthesis*, Vol. 6, p. 225; Trost, B. M.; Fleming, I., Eds.; Pergamon: Oxford, **1991**.
6. Alkyl cyanate-to-isocyanate isomerization reactions are known to proceed via an ionization-recombination pathway. See, Martin, D.; Niclas, H-J.; Habisch, D. *Liebigs Ann. Chem.* **1969**, *727*, 10.
7. Ichikawa, Y. *Synlett* **1991**, 238.
8. Y. I. gives thanks to the late Professor Toshio Goto at Nagoya University for his suggestion of this simple and brilliant idea of synthesizing cyanates, described in the following reference: Ichikawa, Y. *Synlett* **2007**, 2927.
9. Baldwin, J. E.; O'Neil, I. A. *Synlett* **1990**, 603.
10. Ichikawa, Y. *J. Chem. Soc. Perkin Trans. 1* **1992**, 2135.
11. Ichikawa, Y.; Yamazaki, M.; Isobe, M. *J. Chem. Soc., Perkin Trans. 1* **1993**, 2429.
12. Ichikawa, Y.; Osada, M.; Ohtani, I. I.; Isobe, M. *J. Chem. Soc., Perkin Trans. 1* **1997**, 1449.

13. (a) Ichikawa, Y.; Hirara, K.; Ohbayashi, M.; Isobe, M. *Chem.–Eur. J.* **2004**, 3241. (b) Ichikawa, Y.; Kobayashi, C.; Isobe, M. *J. Chem. Soc., Perkin Trans. 1* **1996**, 377.

14. For a related synthesis of 4-amino-hex-2-enopyranose using the Overman rearrangement, a high temperature (165 °C in *o*-dichlorobenzene) was necessary to promote the rearrangement. (a) Dyong, I.; Weigand, J.; Merten, H. *Tetrahedron Lett.* **1981**, *22*, 2965. (b) Dyong, I.; Weigand J.; Thiem, J. *Liebigs Ann. Chem.* **1986**, 577.

15. Ichikawa, Y.; Tsuboi, K.; Isobe, M. *J. Chem. Soc., Perkin Trans. 1* **1994**, 2791.

16. Kusumi, T. Fukushima, T.; Ohtani, I.; Kakisawa, K. *Tetrahedron Lett.* **1991**, *32*, 2939.

17. (a) Kapferer, P.; Sarabia, F.; Vasella, A. *Helv. Chim. Acta.* **1999**, *82*, 645. (b) Roulland, E.; Monneret, C.; Florent, J-C.; Bennejean, C.; Renard, P.; Leonce, S. *J. Org. Chem.* **2002**, *67*, 4399. (c) Yokoyama, T.; Yokoyama, R.; Nomura, S.; Matsumoto, S.; Fujiyama, R.; Kiyooka, S.-i. *Bull. Chem. Soc, Jpn.* **2009**, *82*, 1528. (d) Arbour, M.; Roy, S.; Godbout, C.; Spino, C. *J. Org. Chem.* **2009**, *74*, 3806. (e) Yokoyama, R.; Matsumoto, S.; Nomura, S.; Higaki, T.; Yokoyama, T.; Kiyooka, S.-i. *Tetrahedron* **2009**, *65*, 5181. (f) Gagnon, D.; Spino, C. *J. Org. Chem.* **2009**, *74*, 6035. (g) Yamashita, M.; Yamashita, T.; Aoyagi, S. *Org. Lett.* **2011**, *13*, 2204. (h) Liu, Z.; Bittman, R. *Org. Lett.* **2012**, *14*, 620. (i) Boyd, D. R.; Sharma, N. D.; Kaik, M.; McIntyre, P. B. A.; Stevenson, P. J.; Allen, C. C. R. *Org. Biomol. Chem.* **2012**, *10*, 2774.

18. (a) Minami, N.; Ko, S. S.; Kishi, Y. *J. Am. Chem. Soc.* **1982**, *104*, 1109. (b) Hirama, M.; Uei, M. *Tetrahedron Lett.* **1982**, *23*, 5307.

19. For other carbamoylation processes, see: Ichikawa, Y.; Morishita, Y.; Kusaba, S.; Sakiyama, N.; Matsuda, Y.; Nakano, K.; Kotsuki, H. *Synlett* **2010**, 1815, and references therein.

20. Roy, S.; Spino, C. *Org. Lett.* **2006**, *8*, 939.

21. Toma, T.; Kita, Y.; Fukuyama, T. *J. Am. Chem. Soc.* **2010**, *132*, 10233.

22. Campagna, F.; Carotti, A.; Casini, G. *Tetrahedron Lett.* **1977**, *18*, 1813.

23. C. Spino, C.; Joly, M-A.; Godbout. C.; Arbour, M. *J. Org. Chem.* **2005**, *70*, 6118.

24. (a) Ichikawa, Y.; Okumura, K.; Matsuda, Y.; Hasegawa, T.; Nakamura, M.; Fujimoto, A.; Masuda, T.; Nakano, K.; Kotsuki, H. *Org. Biomol. Chem.* **2012**, *10*, 614. (b) Benalil, A.; Roby, P.; Carboni, B.; Vaultier, M. *Synthesis* **1991**, 787.

25. Stock, C.; Brückner, R. *Synlett* **2010**, 2429.

26. (a) Ichikawa, Y.; Yamauchi, E.; Isobe, M. *Biosci. Biotech. Biochem.* **2005**, *69* , 939. (b) Clayden, J.; Donnard, M.; Lefranc, J.; Tetlow, D. J. *Chem. Commun.* **2011**, *47*, 4624.

Appendix
Chemical Abstracts Nomenclature; (Registry Number)

(*E*)-3,7-Dimethylocta-2,6-dien-1-ylcarbamate; (16930-44-2)
Geraniol; (106-24-1)
Trichloroacetyl isocyanate; (3019-71-4)
tert-Butyl 3,7-dimethylocta-1,6-dien-3-ylcarbamate; (1354913-05-5)
Triphenylphosphine; (603-35-0)
Triethylamine; (121-44-8)
Carbon tetrabromide; (558-13-4)
Lithium hexamethyldisilazide; (4039-32-1)

Yoshiyasu Ichikawa was born in Gamagori, Aichi in 1958. He completed his undergraduate studies and PhD at Nagoya University. After postdoctoral studies in the Dyson Perrins Laboratory at Oxford, UK he joined the faculty of Mie University where he pursued his interests in the synthesis of marine natural products based upon the sigmatropic rearrangements. He moved to Nagoya University in 1992 and emigrated to Kochi University in 2004. His research interests are in the area of synthesis of natural products and carbohydrate chemistry.

Noriko Kariya was born in 1986 in Kochi, Japan. She carried out her undergraduate research with Professor Yoshiyasu Ichikawa at Kochi University where she received her MS degree in 2012.

Tomoyuki Hasegawa was born in 1988 in Shizuoka, Japan. He pursued his undergraduate studies with Professor Yoshiyasu Ichikawa at Kochi University where he received his MS degree in 2013.

Eiji Yoshida was born in Ann Arbor, Michigan in 1989. He received his B.S. in 2012 from the University of Tokyo. In the same year, he began his graduate studies at the Graduate School of Pharmaceutical Sciences, the University of Tokyo, under the guidance of Professor Tohru Fukuyama. His research interests are in the area of the total synthesis of natural products.

Palladium-Catalyzed Triazolopyridine Synthesis: Synthesis of 7-Chloro-3-(2-Chlorophenyl)-1,2,4-Triazolo[4,3-a]Pyridine

Submitted by Oliver R. Thiel and Michal M. Achmatowicz.[1]
Checked by Songchuan Tian and Dawei Ma.

1. Procedure

A. *(E)-(2-Chlorobenzylidene)hydrazine (1)*. A 1000-mL 3-necked, round-bottomed flask (24/40 joints) equipped with an overhead mechanical stirrer (Teflon paddle, 6 × 2 × 0.2 cm), water-cooled reflux condenser with

Published on the Web 5/21/2013
© 2013 Organic Syntheses, Inc.

nitrogen inlet and a rubber septum with temperature probe is evacuated and refilled with nitrogen. Hydrazine hydrate ($H_2NNH_2 \cdot 1.5H_2O$) (100 mL, 1.75 mol, 6.8 equiv) (Notes 1 and 2) and ethanol (105 mL) (Note 3) are charged and efficient stirring is established (Note 4). 2-Chlorobenzaldehyde (36.8 g, 0.259 mol, 1.00 equiv) (Note 5) is added using a 60 mL disposable syringe (Note 6) and the mixture (Note 7) is heated with a heating mantle to an internal temperature of 60 °C. The mixture is heated until a colorless solution is obtained (1 h) (Note 8) at which point complete transformation can be ascertained by ^1H NMR analysis (Note 9). The reaction mixture is cooled to room temperature, transferred into a 500-mL round-bottomed flask (24/40 joint) and concentrated by rotary evaporation (Note 10) until approx. 100 g of the distillate is removed (Note 11). The resulting biphasic mixture is transferred into a 500 mL separatory funnel. The product is extracted using two portions of methyl *tert*-butyl ether (0.10 L, 0.05 L) (Note 12). The aqueous residue is discarded (Note 13). The combined organic extracts (Note 14) are transferred into a 500-mL round-bottomed flask (24/40 joint) and concentrated using rotary evaporation (Note 15) to dryness. The resulting liquid is further vacuum-dried (Note 16) for 16 h to afford **1** as a colorless to white solid (Note 17) (41.3–42.1 g, > 99% (Note 18).

    B. *(E)-4-Chloro-2-(2-(2-chlorobenzylidene)hydrazinyl)pyridine (2)*. A 1000-mL 3-necked, round-bottomed flask (2 × 24/40 side joints, 1 × 29/42 middle joint) equipped with an overhead mechanical stirrer (Teflon paddle, 8 × 2 × 0.2 cm), water-cooled reflux condenser with nitrogen inlet and a rubber septum with temperature probe is evacuated and refilled with nitrogen. (*E*)-(2-Chlorobenzylidene)hydrazine (**1**) (37.3 g, 0.241 mol, 1.00 equiv), toluene (300 mL) (Note 19), and potassium carbonate (50.3 g, 0.357 mol, 1.48 equiv) (Note 20) are charged and efficient stirring is established (Note 21). 2,4-Dichloropyridine (36.0 g, 0.243 mol, 1.01 equiv) (Note 22) is added (Note 23) and the resulting suspension is deoxygenated by performing vacuum-nitrogen refill (three cycles). Pd(dppf)Cl$_2 \cdot$CH$_2$Cl$_2$ (0.946 g, 1.16 mmol, 0.05 equiv) (Note 24) is pre-weighed in a 8 mL glass vial equipped with a septum-cap and flushed with nitrogen. The contents of the vial are rapidly transferred into the 1000-mL flask against a gentle positive nitrogen flow (Note 25) and the addition neck is sealed with a rubber septum (Note 26). The mixture (Note 27) is heated with a heating mantle to an internal temperature of 100 °C (Note 28) for 20-22 h, at which time the reaction mixture is sampled for analysis by ^1H NMR (Note 29). The reaction mixture is cooled to room temperature. Water (150 mL) is added and the resulting mixture is stirred (Note 30) for at least 1 h at ambient temperature. The suspension is transferred onto an 800 mL sintered-glass funnel (medium porosity) connected to a 1000 mL suction flask. The filter cake is sequentially triturated on the filter with water (150 mL) and toluene (300 mL in two portions) (Note 31). The wet cake is air-dried overnight (Note 32) to afford crude **2** as fine off-white to yellow needles (61.4 g). The crude product is charged into a 500-mL 3-necked, round-bottomed flask (24/40 joints) equipped with an overhead mechanical stirrer (Teflon paddle, 6 × 2 × 0.2 cm), water-cooled reflux condenser with nitrogen inlet and a rubber septum with temperature probe. DMSO (0.16 L) (Note 33) is added, efficient stirring is established (Note 34), and the headspace is purged with nitrogen. The suspension is heated to an internal temperature of 80 °C for 1.5 h and then is allowed to cool to room temperature. The stirred mixture is aged for at least 1 h at ambient temperature. The suspension is transferred onto an 800 mL sintered-glass funnel (medium porosity) connected to a 1000 mL suction flask. The filter cake is triturated with DMSO (100 mL) (Note 35) and water (400 mL in two portions). The wet cake is air-dried (Note 32) to afford **2** as fine off-white to yellow needles (53.3–56.4 g, 83–88%) (Note 36).

C. *7-Chloro-3-(2-chlorophenyl)-[1,2,4]triazolo[4,3-a]pyridine (3).* A 1000-mL 3-necked, round-bottomed flask (24/40 joints) equipped with an overhead mechanical stirrer (Teflon paddle, 6 × 2 × 0.2 cm), water-cooled reflux condenser with nitrogen inlet and a rubber septum with temperature probe is evacuated and refilled with nitrogen (three cycles). (*E*)-4-Chloro-2-(2-(2-chlorobenzylidene)hydrazinyl)pyridine **(2)** (20.0 g, 0.075 mol, 1.00 equiv) and 2-methyltetrahydrofuran (200 mL) (Note 37) are charged, and efficient stirring is established (Note 38). Chloramine-T trihydrate (25.4 g, 0.090 mol, 1.20 equiv) (Note 39) is added and the mixture is heated with an oil bath to an internal temperature of 60 °C (Note 40). The mixture is heated for 2 h, at which time it is sampled for analysis by TLC (Note 41). Upon complete conversion of starting material the reaction mixture is cooled to room temperature. The reflux condenser is swapped for a 125 mL addition funnel. Sodium sulfite (10.0 g, 0.079 mol, 1.05 equiv) (Note 42) and water (90 mL) are added to a 250 mL Erlenmeyer flask and complete dissolution of the solids is achieved. The aqueous sodium sulfite solution is added via addition funnel to the reaction mixture over 10 min, while maintaining an internal temperature of 15–25 °C (Note 43). The mixture is stirred for at least 10 min and stirring rate is increased to aid in dissolution of solids from the walls of the flask (Note 44). The reaction mixture is transferred to a 1000 mL separation funnel. The reaction flask is rinsed twice with 2-methyltetrahydrofuran (200 mL each) and the rinse is transferred to the separatory funnel (Note 45). The phases are split and the bottom aqueous layer is drained (Note 46). The organic phase is washed twice with aqueous 1 M NaOH (100 mL each) (Notes 46 and 47). The organic phase is washed with aqueous 5 M sodium chloride solution (100 mL) (Note 48). The solution is then transferred to a nitrogen-blanketed 1000 mL 3-necked, round-bottomed flask (24/40 joints) equipped with an overhead mechanical stirrer (Teflon paddle, 6 × 2 × 0.2 cm), water-cooled reflux condenser with nitrogen inlet and a rubber septum with temperature probe. Darco G 60 (100 mesh powder) (4.0 g) (Note 49) is added and the mixture is heated with efficient stirring (Note 50) in an oil bath to an internal temperature of 60 °C for 12–14 h (Note 40). The mixture is cooled to room temperature and vacuum-filtered into a 1000-mL 3-necked, round-bottomed flask (24/40 joints) (Note 51) through a 150 mL sintered-glass funnel (medium porosity), packed with Celite 521 (15 g) (Note 52). The flask is equipped with an overhead mechanical stirrer (Teflon paddle, 6 × 2 × 0.2 cm), water-cooled distillation head with 1000 mL receiving flask and a 250 mL addition funnel with nitrogen-inlet. Distillation is performed under efficient stirring in an oil bath heated to 100–105 °C (Note 53). Solvent is removed until the premarked 120 mL solvent line is reached (Note 51), upon which the reaction mixture is cooled to room temperature. Crystallization of the target compound is observed. The mixture is aged for at least 1 h at room temperature and then heptane (300 mL) (Note 54) is added via addition funnel over 90 min. The mixture is stirred for at least 1 h and then the product is collected by vacuum filtration. The filter cake is rinsed with a mixture of heptane/2-methyltetrahydrofuran (3:1 v/v, 2 portions of 50 mL). The solids are air-dried to afford analytically pure **3** as a yellow solid (17.4–18.0 g, 87–90%) (Note 55).

2. Notes

1.   Hydrazine hydrate ($H_2NNH_2 \cdot 1.5H_2O$, 85%) was purchased from Lingfeng Shanghai by the checkers and used as received. The submitters purchased hydrazine hydrate ($H_2NNH_2 \cdot 1.5H_2O$, 50-60%) from Aldrich and it was used as received.

Org. Synth. **2013**, *90*, 287-300

2. Use of excess hydrazine hydrate is required to ensure selective formation of mono-hydrazone **1**.

3. Ethanol ($\geqslant$99.5%) was purchased from Zhenxing Shanghai and used as received.

4. The stirring was set to 400 rpm.

5. 2-Chlorobenzaldehyde (97%) was purchased from Alfa by the checkers and used as received. 2-Chlorobenzaldehyde (99%) was purchased by the submitters from Acros and used as received

6. Addition is mildly exothermic. Adiabatic temperature increase from 20 °C to 32 °C was observed.

7. The reaction mixture was a colorless to pale yellow solution with a small amount of yellow precipitate (bis-hydrazone **1a**).

8. During the warming up period the bis-hydrazone **1a** dissolved and the reaction mixture became a yellow solution. Complete decolorization was observed as the yellow colored bis-hydrazone gradually converted into the colorless mono-hydrazone **1** at 60 °C.

9. Approx. 0.1 mL sample of the reaction mixture was diluted with 0.6 mL of d_6-DMSO. Typically nearly pure mono-hydrazone **1** is observed, no bis-hydrazone **1a** is detected.

10. Bath was set to 50 °C and distillation was carried out at 30-60 mmHg.

11. During the distillation product **1** separated as a liquid resulting in a milky emulsion.

12. Methyl *tert*-butyl ether (99.9%) was purchased from Aldrich and used as received.

13. The extracted aqueous layer consists predominantly of hydrazine hydrate and should be treated appropriately. *Combining with waste containing heavy metal impurities must be avoided.* Dilution with copious amount of water prior disposal is advised.

14. Combined MTBE extracts (150 g) were colorless and opalescent.

15. Bath was set to 50 °C and distillation was carried out at 40–300 mmHg.

16. Room temperature, 2 mmHg.

17. Neat product is typically a supercooled yellowish liquid. Brief cooling of the flask results in a rapid crystallization affording a waxy solid.

18. Yield is calculated on the basis of 97% purity of the starting aldehyde. Analytical data for compound **1**: mp 35.5–36.0 °C (neat); IR 3392, 3199, 1594, 1472, 1441, 1391, 1215, 1127, 1048, 1032, 914, 753, 707,

629 cm^{-1}; ^{1}H NMR (400 MHz, DMSO-d_6) δ: 7.22–7.31 (m, 4 H), 7.40 (dd, J = 7.8, 1.6 Hz, 1 H), 7.83 (dd, J = 7.7, 1.9 Hz, 1 H), 8.06 (s, 1 H); ^{13}C NMR (100 MHz, DMSO-d_6) δ: 125.4, 127.1, 128.4, 129.5, 130.6, 133.1, 133.5; HRMS (m/z): [M+H$^+$] calcd for $C_7H_8ClN_2{}^+$: 155.03705; Found 155.03738. The crude product was utilized directly in the next step without further purification.

19. Toluene (99.8%) was purchased from Tianlian Shanghai and used as received.

20. Potassium carbonate (98%) was purchased from Sinopharm Chemical reagent and used as received.

21. The stirring was set to 400 rpm.

22. 2,4-Dichloropyridine (99%) was purchased from Energy Shanghai by the checkers and used as received. 2,4-Dichloropyridine (99%) was purchased by the submitters from Oakwood and used as received.

23. Minor bubbling was observed immediately following 2,4-dichloropyridine charge.

24. Pd(dppf)Cl$_2$·CH$_2$Cl$_2$ was purchased from Aldrich by the checkers and used as received. Pd(dppf)Cl$_2$·CH$_2$Cl$_2$ was purchased from Strem by the submitters and used as received.

25. With the addition neck capped the nitrogen flow was set to high using bubbler as an indicator. The cap was removed and the nitrogen flow was adjusted as necessary to maintain minimal nitrogen flow through the bubbler.

26. Nitrogen pressure was adjusted to maintain a visible flow through the bubbler (approx. 1 bubble per sec.).

27. Orange-pink suspension of K$_2$CO$_3$ and pre-catalyst.

28. Upon reaching approximately 30 °C the suspension rapidly changed color from orange-pink to yellow. Gradual thickening of the suspension was observed as the reaction progresses. Gradual precipitation of palladium black was occasionally observed.

29. Approx. 0.1 mL sample of the supernatant was diluted with 0.6 mL of d$_6$-DMSO. The conversion was calculated by comparing the integration of diagnostic signals of 2,4-dichloropyridine (7.78 ppm, d, 1H) and/or 1 (8.02 ppm, s, 1H) with the integration of toluene signal at (2.30 ppm, s, 3H). Typical conversion is ≥90%.

30. Efficient stirring (400-600 rpm) facilitates dissolution of inorganic by-products in the aqueous layer. Fine uniform suspension was obtained.

31. Combined filtrates consisting of dark-orange organic layer and colorless aqueous layer were discarded.

32. Ambient air was passed through the wet cake overnight.

33. DMSO ($\geq$99%) was purchased from J&K and used as received.

34. The stirring was set to 250-450 rpm.

35. The dark supernatant was allowed to fully drain using suction before DMSO wash was introduced on the filter. The suction was cut-off and the filtercake was triturated with DMSO using spatula and again fully drained using suction.

36. Analytical data for compound **2**: mp 250–251 °C (DMSO); IR 1582, 1431, 849, 750, 704 cm^{-1}; ^{1}H NMR (400 MHz, DMSO-d_6, 80 °C) δ: 6.88–6.89 (m, 1 H), 7.28 (s, 1 H), 7.35–7.41 (m, 2 H), 7.48 (d, J = 7.3 Hz, 1 H), 8.08 (d, J = 7.7 Hz, 1 H), 8.11 (d, J = 5.1 Hz, 1 H), 8.43 (s, 1 H), 11.42 (s, 1 H); ^{13}C NMR (100 MHz, DMSO-d_6, 80 °C) δ: 105.1, 114.6, 125.8, 126.8, 129.1, 129.5, 131.3, 131.5, 135.4, 143.5, 148.8, 157.3; HRMS (*m/z*): [M+H$^+$] calcd for $C_{12}H_{10}Cl_2N_3^+$: 266.02463; Found 266.02474.

37. 2-Methyltetrahydrofuran (98%) was purchased from Aldrich and used as received.

38. The stirring was set to 300 rpm and an off-white to grey slurry formed.

39. Chloramine-T trihydrate (ACS reagent, 98%) was purchased from Alfa by the checkers and used as received. Chloramine-T trihydrate (ACS reagent, 98%) was purchased by the submitters from Aldrich or Acros and used as received.

40. The internal temperature was maintained in a range from 55 to 65 °C.

41. The reaction was monitored via TLC using the following method: The R$_f$ values are 0.66 (EtOAc/petroleum ether=1/4) for product and 0.4 (EtOAc/petroleum ether=1/1) for starting material. The submitters monitored the reaction by HPLC by the following method: column: XBridge C18, 100x3 mm, 3.5 11m; flow rate: 0.8 mL/min; solvent A: 0.1% trifluoroacetic acid in water; Solvent B: 0.1% trifluoroacetic acid in acetonitrile; gradient: 5% B to 100% B over 12 min; wavelength: 235 nm; Retention times: **(2):** 6.28 min; **(3):** 6.12 min; Chloramine-**T:** 6.45 min; toluenesulfonamide: 4.34 min.

42. Sodium sulfite (ACS reagent, $\geq$98.0%) was purchased from Aldrich and used as received.

43. The addition is mildly exothermic; an ice-bath may be used to control the temperature.

44. The stirring was set to 600 rpm and a homogenous biphasic mixture was obtained.

45. The additional amount of solvent was required to avoid supersaturation and crystallization of the product during aqueous work-up and charcoal-treatment.

46. A small amount of rag layer was removed with the aqueous layer.

47. The 1M sodium hydroxide solution was prepared by dissolution of sodium hydroxide (Sinopharm Chemical reagent, 40 g) in water (1 L).

48. The 5M sodium chloride solution was prepared by dissolution of sodium chloride (Sinopharm Chemical reagent, 292 g) in water (1 L)

49. Darco G 60 (10 mesh) was purchased from Aldrich and used as received.

50. Stirring was set a 400 rpm.

51. 2-Methyltetrahydrofuran (120 mL) was added to the flask and the solvent line was marked with a pen. This line was used as reference mark during the distillation. Subsequently the flask is emptied.

52. Celite 521 was purchased from Sinopharm Chemical reagent and used as received.

53. Depending on solvent loss during vacuum filtration and efficiency of distillate condensation solvent amounts between 460 and 580 mL were collected.

54. Heptane (Chromasolv, $\geq$99.0%) was purchased from Tianlian Shanghai and used as received.

55. mp 140–142 °C (MeTHF/heptane); IR 1630, 1522, 1435, 1371, 1257, 1051, 982, 936, 864, 747, 714 cm^{-1}; ^{1}H NMR (400 MHz, CDCl$_3$) δ: 6.85 (d, J = 7.2 Hz, 1 H), 7.49–7.51 (m, 1 H), 7.56–7.61 (m, 2 H), 7.68 (d, J = 7.6, 1 H), 7.75 (d, J = 7.4, 1 H), 7.85 (s, 1 H). ^{13}C NMR (101 MHz, CDCl$_3$) δ: 115.1, 115.9, 124.0, 125.4, 127.6, 130.3, 132.3, 133.3, 134.0, 134.3, 145.2, 150.2; HRMS (m/z): [M+H$^+$] calcd for (C$_{12}$H$_8$Cl$_2$N$_3$): 264.0090; Found 264.0100. Anal. Calcd. for C$_{12}$H$_7$Cl$_2$N$_3$: C, 54.57; H, 2.67; N, 15.91. Found: C, 54.57; H, 2.71; N, 15.87.

Handling and Disposal of Hazardous Chemicals

The procedures in this article are intended for use only by persons with prior training in experimental organic chemistry. All hazardous

294

materials should be handled using the standard procedures for work with chemicals described in references such as "Prudent Practices in the Laboratory" (The National Academies Press, Washington, D.C., 2011 www.nap.edu). All chemical waste should be disposed of in accordance with local regulations. For general guidelines for the management of chemical waste, see Chapter 8 of Prudent Practices.

These procedures must be conducted at one's own risk. *Organic Syntheses, Inc.*, its Editors, and its Board of Directors do not warrant or guarantee the safety of individuals using these procedures and hereby disclaim any liability for any injuries or damages claimed to have resulted from or related in any way to the procedures herein.

3. Discussion

Triazolopyridines constitute an important class of heteroaromatic compounds. The [1,2,4]triazolo[4,3-*a*]pyridine moiety[2] can be found in a variety of biologically active compounds, including antibacterial, antithrombotic, antiinflammatory, antiproliferative, and herbicidal agents.[3] Traditional approaches to this class of compounds rely on oxidative or dehydrative cyclizations of a linear precursor.[2] These intermediates are usually obtained through reaction of 2-hydrazinopyridines with aldehydes or acid chlorides. Depending on the underlying heterocyclic core the access to the required 2-hydrazinopyridines can be challenging.

To overcome this issue we recently described a palladium-catalyzed coupling reaction of 2-chloropyridines with aldehyde derived hydrazones (Scheme 1).[4] Aldehyde-derived mono-hydrazones can be obtained in a straightforward fashion by the reaction with an excess of hydrazine.[5] The initial mixture of kinetically favored bis-hydrazone and mono-hydrazone can be readily equilibrated in the presence of an excess of hydrazine to afford the desired mono-hydrazone in nearly quantitative yields. The reaction system for the coupling is very simple and the catalyst Pd(dppf)Cl$_2$ is stable and readily available. Procedurally the reaction is simple and the products can be isolated by direct filtration at the end of the reaction. The oxidative cyclization step makes use of Chloramine-T as a clean oxidant.[6] The isolation involves a simple basic aqueous work-up to remove the toluene-sulfonamide byproduct and a charcoal treatment to remove colored trace-impurities. Crystallization is achieved from 2-methyltetrahydrofuran and heptane.

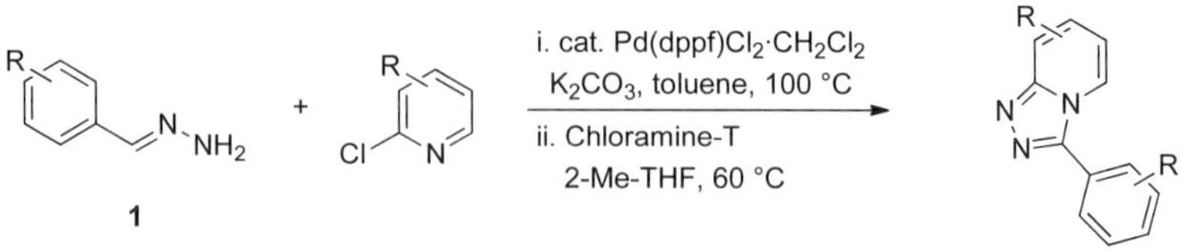

Scheme 1. Synthesis of triazolopyridines.

The reaction has a broad scope with regards to the pyridine and aldehyde component (Table 1). Additional examples have been disclosed in the original communication.[4] Satisfyingly the reaction is also suitable for other chloroazines, with a partial scope shown in Table 2, with additional examples in the original manuscript.[4]

Table 1. Synthesis of triazolopyridines.

Entry	Product	Yield % - Step 2	Yield % - Step 3
1	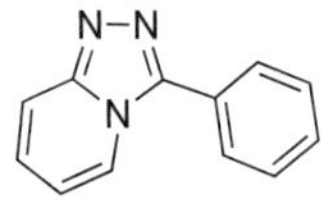	89	91
2	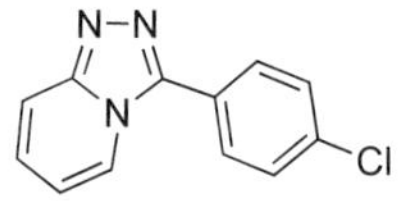	85	83
3	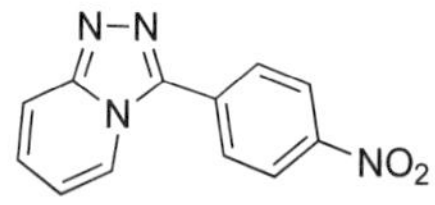	84	75
4	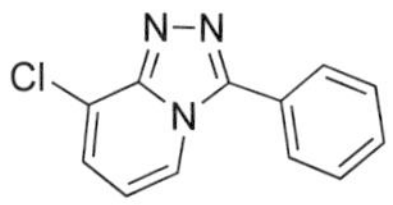	42	95
5	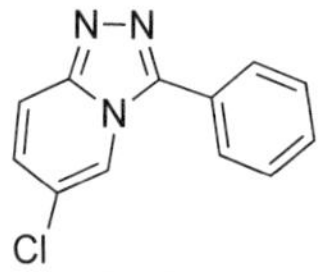	58	93

Table 2. Synthesis of related heterocycles.

Entry	Product	Yield % - Step 2	Yield % - Step 3
1	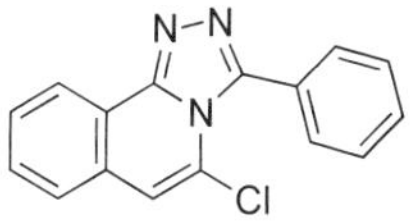	81	69
2	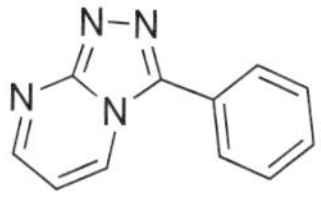	63	99
3	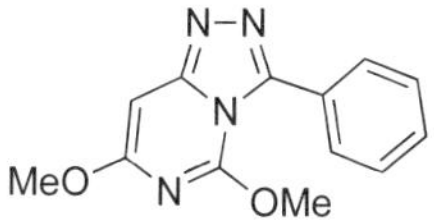	90	79
4	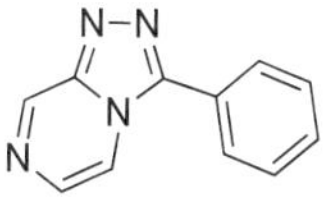	87	75
5	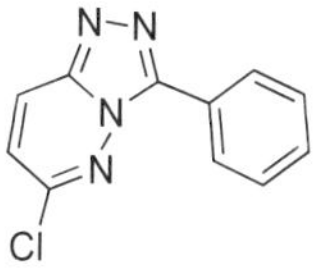	79	82

The compound prepared in this procedure can be a useful building block for further functionalization. This was demonstrated by conducting a second metal-catalyzed coupling reaction on the more activated chloride substituent. Both palladium-catalyzed Suzuki-coupling[7] and iron-catalyzed coupling with a Grignard-reagent[8] afforded the desired products in good yields (Scheme 2).

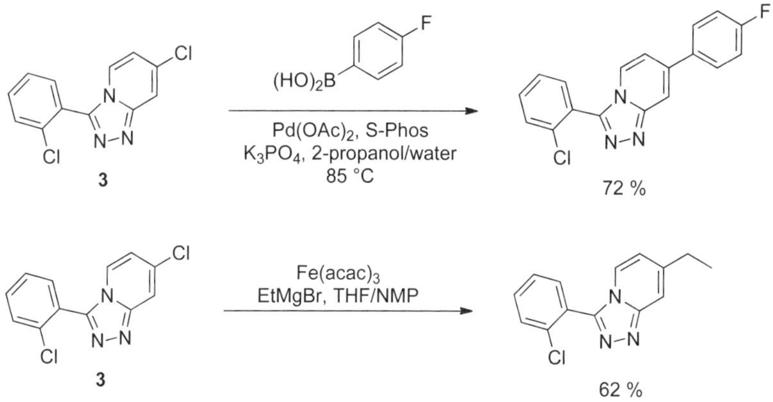

Scheme 2. Further functionalization by metal-mediated coupling reactions.

1. Amgen, Chemical Process Research and Development, One Amgen Center Drive, Thousand Oaks, CA 91320-1799.
2. Jones, G. *Adv. Heterocyclic Chem.* **2002**, *83*, 1.
3. (a) Yoshimura, Y.; Tomimatsu, K.; Nishimura, T.; Miyake, A.; Hashimoto, N. *J. Antibiot.* **1992**, *45*, 721; (b) Sadana, A. K.; Mirza, Y.; Aneja, K. R.; Prakash, O. *Eur. J. Med. Chem.* **2003**, *38*, 533; (c) Lawson, E. C.; Hoekstra, W. J.; Addo, M. F.; Andrade-Gordon, P.; Damiano, B. P.; Kauffman, J. A.; Mitchell, J. A.; Maryanoff, B. E. *Bioorg. Med. Chem. Lett.* **2001**, *11*, 2619; (d) Kalgutkar, A. S.; Hatch, H. L.; Kosea, F.; Nguyen, H. T.; Choo, E. F.; McClure, K. F.; Taylor, T. J.; Henne, K. R.; Kuperman, A. V.; Dombroski, M. A. Letavic, M. A. *Biopharm. Drug Dispos.* **2006**, *27*, 371; (e) McClure, K. F.; Abramov, Y. A.; Laird, E. R.; Barberia, J. T.; Cai, W.; Carty, T. J.; Cotina, S. R.; Danley, D. E.; Dipesa, A. J.; Donahue, K. M.; Dombroski, M. A.; Elliott, N. C.; Gabel, C. A.; Han, S.; Hynes, T. R.; LeMotte, P. K.; Mansour, M. N.; Marr, E. S.; Letavic, M. A.; Pandit, J.; Ripin, D. B.; Sweeney, F. J.; Tan, D.; Tao, Y. *J. Med. Chem.* **2005**, *48*, 5728.
4. Thiel, O. R.; Achmatowicz, M. M.; Reichelt, A.; Larsen, R. D. *Angew. Chem. Int. Ed.* **2010**, *49*, 8395.
5. Shastin, A. V.; Korotchenko, V. N.; Nenajdenko, V. G.; Balenkova, E. S. *Tetrahedron* **2000**, *56*, 6557.
6. Bourgeois, P.; Cantegril, R.; Chene, A.; Gelin, J.; Mortier, J.; Moyroud, J. *Synth. Commun.* **1993**, *23*, 3195.
7. Barder, T. E.; Walker, S.D.; Martinelli, J.R.; Buchwald, S. L. *J. Am. Chem. Soc.* **2005**, *127*, 4685.
8. Fürstner, A.; Leitner, M.; Mèndez, H.; Krause, H. *J. Am. Chem. Soc.* **2002**, *124*, 13856–13863.

Appendix
Chemical Abstracts Nomenclature; (Registry Number)

Hydrazine hydrate; (10217-52-4)
Benzaldehyde, 2-chloro-; (35913-09-8)
Benzaldehyde, 2-chloro-, hydrazone; (52372-78-8)
Potassium carbonate; (584-08-7)

Pyridine, 2,4-dichloro; (26452-80-2)

Dichloro[1,1'-bis(diphenylphosphino)ferrocene]palladium (II) dichloromethane adduct; (72287-26-4)

Benzaldehyde, 2-chloro-: 2-(4-chloro-2-pyridinyl)hydrazone; (1258542-95-8)

2-Methyltetrahydrofuran: Furan, tetrahydro-2-methyl-; (96-47-9)

Chloramine-T trihydrate: Benzenesulfonamide, N-chloro-4-methyl-, sodium salt, hydrate (1:1:3); (7080-50-4)

Sodium sulfite; (7757-83-7)

1,2,4-Triazolo[4,3-a]pyridine, 7-chloro-3-(2-chlorophenyl)-; (1019918-88-7)

Oliver R. Thiel studied chemistry at the Technical University Munich, Germany, and completed a Diploma thesis on rhodium-catalyzed hydroaminations under supervision of Professor Matthias Beller. He then pursued a Ph.D. (1998-2001) at the Max-Planck-Institut für Kohlenforschung Mülheim, Germany, under guidance of Professor Alois Fürstner, exploring RCM-reactions in natural product synthesis. After a postdoctoral appointment (2001-2003) at Stanford University with Professor Barry M. Trost, Oliver joined the Chemical Process Research & Development department at Amgen in Thousand Oaks, where he has been involved in the development of synthetic processes of numerous clinical candidates.

Michal Achmatowicz received his undergraduate education at the University of Warsaw, Poland where he completed his undergraduate thesis in chemistry in 1997. He then joined Prof. Janusz Jurczak's research group at the Institute of Organic Chemistry of Polish Academy of Sciences in Warsaw to pursue his Ph.D. in organic chemistry. From 2001 to 2003 he was a postdoctoral research fellow with Prof. Louis S. Hegedus at the Colorado State University. Subsequently he joined the Chemical Process Research and Development group at Amgen in Thousand Oaks, California, where he has been developing robust processes toward active pharmaceutical ingredients, co-authoring several publications, and enjoying rock-climbing in the spare time.

Songchuan Tian received his undergraduate education in China Pharmaceutical University, China where he completed his undergraduate thesis in traditional medicine in 2007. He then pursued a Ph. D. in organic chemistry (2007-2012) in Prof. Dawei Ma's research group at Shanghai Institute of Organic Chemistry of Chinese Academy of Sciences in Shanghai. Now he is an assistant researcher in Prof. Ma's group.

Preparation of DABSO from Karl-Fischer reagent

Submitted by Ludovic Martial, and Laurent Bischoff.*[1]
Checked by Changming Qin and Huw M. L. Davies.

Caution: Karl-Fischer reagent A, which consists of a pyridine-sulfur dioxide solution in methanol, should be used under a well-ventilated fume hood.

1. Procedure

1,4-Diazabicyclo[2,2,2]octane bis (sulfur dioxide) adduct. An oven-dried 500-mL round-bottomed flask containing a 5 cm Teflon-coated oval stir bar is fitted with a rubber septum and allowed to cool to room temperature under vacuum (Note 1). At room temperature (22 °C), 1,4-diazabicyclo[2,2,2]octane (DABCO) (15.0 g, 134 mmol, 1 equiv) (Note 2) is added and the flask is connected to a vacuum line *via* a needle inserted through the septum. The flask is evacuated then back-filled with argon. This process is repeated three times. Dry THF (180 mL) (Note 3) is introduced *via* syringe under a positive argon pressure. After complete dissolution, the reaction medium is ice-cooled and Karl-Fischer reagent (solution A, 120 mL, ~280 – 375 mmol SO_2) (Note 4) is added dropwise *via* syringe at 0 °C (external temperature) over 30 min. Precipitation of DABSO is observed shortly after the addition is initiated. After stirring 30 min at 0 °C, the ice-bath is removed and the suspension is stirred at room temperature for an additional 3 h. The suspension is filtered through a 250-mL sintered-glass funnel under reduced pressure (vacuum pump, 15 mmHg). The white solid is washed on the funnel by triturating a few seconds with diethyl ether (50 mL) (Note 5) without suction. The collected solid is transferred back to the original 500-mL flask. Diethyl ether (200 mL) is added and the suspension is stirred at room temperature for 15 min under a nitrogen atmosphere. The solid is collected on the same sintered-glass funnel. The re-slurry procedure is repeated twice in the same manner. Since DABSO is a hygroscopic solid, it is immediately transferred into a 250-mL round-bottomed flask and dried

Published on the Web 5/28/2013

in a desiccator under vacuum for 12 h (Note 6) to afford DABSO (30.9–31.0 g, 96–97%) as a colorless powder (Note 7).

2. Notes

1.   The apparatus is dried in an oven at > 70 °C overnight and cooled to room temperature under vacuum (0.2 mmHg).

2.   DABCO (95 %) was purchased from Sigma-Aldrich and used as received from a freshly opened bottle. The submitters report that older samples of DABCO can be purified by recrystallization (Armarego, W. L.F.; Chai, C. L. L. *Purification of Laboratory Chemicals* (5th Edition), **2003**, Elsevier, p 191).

3.   Anhydrous THF (99.9%) was obtained from Acros and used as received.

4.   Karl-Fischer reagent (reagent for volumetric two-component KF titration, pyridine-containing working medium, solution A: pyridine-sulfur dioxide) was purchased from Fluka (Sigma-Aldrich) and used as received. This reagent contains approximately15-20% SO_2.

5.   Anhydrous diethyl ether (99.0%) was obtained from Fisher Scientific and used as received.

6.   Phosphorus pentoxide (15 g) was placed in the desiccator and the product was dried under high vacuum (~ 0.05 mmHg).

7.   1,4-Diazabicyclo[2,2,2]octane bis (sulfur dioxide) adduct has the following physical and spectroscopic properties: mp 141–143 °C; IR: ν (cm^{-1}): 3400, 2962, 2892, 2802, 1651, 1465, 1324, 1178, 1053, 999, 949, 842, 804, 785, 718; ^{1}H NMR (400 MHz, CD_3OD) δ: 3.21 (s) ; ^{13}C NMR (100 MHz, CD_3OD) δ: 45.6; Anal calcd: C 29.99, H 5.03, N 11.66; found C 30.38, H 5.06, N 11.58.

Handling and Disposal of Hazardous Chemicals

The procedures in this article are intended for use only by persons with prior training in experimental organic chemistry. All hazardous materials should be handled using the standard procedures for work with chemicals described in references such as "Prudent Practices in the Laboratory" (The National Academies Press, Washington, D.C., 2011 www.nap.edu). All chemical waste should be disposed of in accordance with local regulations. For general guidelines for the management of chemical

302

waste, see Chapter 8 of Prudent Practices.

These procedures must be conducted at one's own risk. *Organic Syntheses, Inc.*, its Editors, and its Board of Directors do not warrant or guarantee the safety of individuals using these procedures and hereby disclaim any liability for any injuries or damages claimed to have resulted from or related in any way to the procedures herein.

3. Discussion

DABSO consists of a 1:2 complex between DABCO and sulfur dioxide. This reagent has recently found an increasing number of applications, such as sulfonamide formation *via* palladium coupling, synthesis of sulfonylureas, etc.[2,3,4] It consists of a white, stable, however hygroscopic powder, and its handling is much more convenient and safer than gaseous sulfur dioxide. In addition, the regulations concerning the transportation and storage of gaseous sulfur dioxide render its replacement with DABSO even more useful. However, DABSO is an expensive commercially-available reagent, generally sold in small gram-scale amounts, and is currently sold by few chemical suppliers. Its preparation has been described in the literature,[5] either as isolated DABSO itself, or prepared *in situ*. In each case gaseous sulfur dioxide was either condensed or bubbled through a solution of DABCO. These procedures required a large excess of toxic sulfur dioxide, which should be trapped to avoid discarding in the atmosphere.

We herein report a safe and cheap alternative for the preparation of DABSO, using commercially-available Karl-Fischer reagent as the source of sulfur dioxide. Care must be taken when purchasing the solution, since both volumetric and coulometric titration solutions exist. The coulometric solution, containing iodide, SO_2, pyridine in methanol should not be used for DABSO preparation in order to avoid iodide-contamination of the resulting DABSO. When a SO_2-pyridine solution in methanol is used (sold as 'solution A'), no trace of pyridine nor methanol is observed.

Since the SO_2 concentration (15-20%) is not accurately provided by the supplier, DABCO is used as the limiting reagent. With the use of a higher DABCO:SO_2 molar ratio, yields of DABSO might be improved; however, we preferred to ensure an excess SO_2 is used to avoid contamination of DABSO with residual DABCO.

Thus, when a THF solution of DABCO is treated with a Karl-Fischer solution, precipitation of DABSO occurs almost immediately. Following filtration and washing of the solid with ether to remove pyridine, methanol and THF, clean DABSO is obtained. The reagent is stable enough to afford drying under vacuum without loss of SO_2.

1. IRCOF – INSA Rouen – UMR CNRS 6014 COBRA – Université de Rouen, place Emile Blondel 76130 Mont-Saint-Aignan, France ; e-mail: laurent.bischoff@univ-rouen.fr. This work was supported by the Région Haute Normandie, CRUNCH network.
2. Nguyen, B.; Emmett, E. J.; Willis, M. C. *J. Am. Chem. Soc.* **2010**, *132*, 16372–16373.
3. Woolven, H.; González-Rodríguez, C.; Marco, I.; Thompson, A. L.; Willis, M. C. *Org. Lett.* **2011**, *13*, 4876–4878.
4. Emmett, E. J.; Richards-Taylor, C. S.; Nguyen, B.; Garcia-Rubia, A.; Hayter, B. R.; Willis, M. C. *Org. Biomol. Chem.* **2012**, *10*, 4007–4014.
5. Santos, P. S.; Mello, M. T. S. *J. Mol. Struct.* **1988**, *178*, 121–133.

Appendix
Chemical Abstracts Nomenclature; (Registry Number)

1,4-Diazabicyclo[2,2,2]octane bis (sulfur dioxide) adduct (119752-83-9)
1,4-Diazabicyclo[2.2.2]octane (280-57-9)

Laurent Bischoff was born in France in 1969. After a Ph-D thesis in J.P. Genet laboratory (E.N.S.C.P. – Université Paris 6), he worked at a post-doctoral fellow in the laboratory pf Pr. J.E. Baldwin (Oxford, U.K., total synthesis of Manzamine fragments). After 8 years as an associate professor at the Faculty of Pharmacy of Paris (group of Pr. Roques then Pr. C. Garbay, working on metalloenzyme then phosphatase inhibitors), he moved in 2003 to Rouen for a professor position, now studying new tools for heterocyclic chemistry, and some extensions of the Mitsunobu reaction.

Ludovic Martial was born in France in 1986. He received his BSC in Molecular Chemistry at the University of Aix-Marseille III, and his MSC in Physics and Chemistry of Medicines at the University of Rouen. He's now ongoing his Ph-D research at the University of Rouen under the guidance of Professor Laurent Bischoff.

Changming Qin was born in Shandong, China in 1982. He did his undergraduate work at Ludong University on the preparation of polymer nanocomposites under the guidance of Prof. Yucai Hu. He obtained his Masters degree in organic chemistry in 2008 at Wenzhou University under the supervision of Prof. Huayue Wu, working on palladium-catalyzed transformations of aryl boronic acids. After graduation, he worked at the University of Hong Kong with Prof. Chi-Ming Che in 2008-2009, and then joined Prof. Huw Davies' group at Emory University in 2010. His current research is focused on design and synthesis of chiral dirhodium catalysts and their application in novel asymmetric carbeniod transformations.

Low-epimerization Peptide Bond Formation with Oxyma Pure: Preparation of Z-*L*-Phg-Val-OMe

Submitted by Ramon Subirós-Funosas,* Ayman El-Faham*, and Fernando Albericio.*[1]
Checked by Asher Lower and Margaret Faul.

1. Procedure

Z-L-Phg-Val-OMe. A 500-mL, three-necked, round-bottomed flask equipped with a nitrogen gas inlet, thermocouple and Teflon-coated cylindrical magnetic stirrer bar (50 x 8 mm) (Note 1), is charged with EDC·HCl (4.72 g, 25 mmol, 1 equiv) and dissolved in DCM/DMF (1:1) (110 mL) (Notes 2 and 3). The colorless solution of the carbodiimide is then stirred (700 rpm) at room temperature for approximately 20 min to allow dissolution, followed by immersion in an ice bath at 0 °C. Z-*L*-Phg-OH (7.26 g, 25 mmol, 1 equiv) and Oxyma Pure (3.67 g, 25 mmol, 1 equiv) are added to the cold solution of EDC·HCl as solids (Notes 4, 5 and 6). Two minutes after the addition of the reactants, H-Val-OMe·HCl (4.27 g, 25 mmol, 1 equiv) is added as a solid to the preactivated cocktail, followed by addition of DIEA (4.29 mL, 25 mmol, 1 equiv) using a 5-mL syringe (Note 7). The flask is then flushed with nitrogen and sealed with a polyethylene cap. The coupling cocktail is allowed to stir at 700 rpm in the ice bath for 1 h and then the reaction progresses at room temperature overnight. After 14–15 hours, the extent of peptide bond formation is monitored by TLC, showing trace amounts of starting acid and amine (Note 8). The solvent is removed by rotary evaporation (40 °C, 28 mmHg) (Note 9). The crude oily residue is diluted in AcOEt (500 mL) and transferred to a 1-L graduated separatory funnel (Note 10). The organic solution is extracted and washed with 1 N HCl (3 × 300 mL), 1 N Na$_2$CO$_3$

Org. Synth. **2013**, *90*, 306-315
Published on the Web 5/28/13

(3 × 350 mL), and saturated NaCl (3 × 350 mL). The resulting pale yellow organic fraction is dried over approximately 25 g of anhydrous MgSO$_4$, filtered to a 1-L round-bottomed flask, and concentrated by rotary evaporation (25 °C, 28 mmHg) to afford a white solid, with partial yellow tone, which is dried overnight in a vacuum oven (50 °C, ≤10 mmHg) to yield crude product (9.57 g, 97.5% yield) (Note 11). Ethanol (100 mL) is added to the peptidic material at room temperature and the resulting slurry is stirred at 200 rpm on the rotary evaporator. After complete dissolution of the solid at 60 °C bath temperature, the stirring and heating of the solution is stopped and the solution is allowed to cool slowly to room temperature over 2 h, resulting in crystallization of the dipeptide as white needles within the yellow solution. The resulting needles are isolated by filtration through a Büchner funnel, washed with ice-cold ethanol (100 mL), collected in a 25 x 100 mm crystallizing dish and dried overnight in a vacuum oven (50 °C, 3 mmHg) to provide Z-L-Phg-Val-OMe dipeptide (7.98–8.25 g, 81–84% yield) as white needles, free of the DL-epimer (Note 12).

2. Notes

1. In order to enhance the efficiency of the coupling process, all glassware was dried overnight in an oven at 70 °C and flushed with a stream of nitrogen after cooling to prevent the presence of air or moisture.

2. Anhydrous dichloromethane (DCM) (99.8%, stabilized with amylene) and *N,N*-dimethylformamide (DMF, anhydrous, 99.8%) were purchased from Sigma Aldrich. Z-L-Phg-OH (99.5%) was obtained from Iris Biotech. H-Val-OMe hydrochloride (99%) was purchased from Sigma Aldrich. 1-Ethyl-3-(3-dimethylaminopropyl)carbodiimide hydrochloride (EDC·HCl) (premium quality, crystalline) was purchased from Fisher Scientific. Ethyl 2-cyano-2-hydroxyimino acetate (Oxyma Pure) (99.9%) was obtained from Bachem. *N,N*-Ethyldiisopropylamine (DIEA) (99.5%) was purchased from Fisher Scientific. All chemicals were used as received and stored at room temperature on the bench top, except for the amino acids and EDC·HCl, which were stored in a freezer at –5°C given their hygroscopic nature.

3. New bottles of DCM and DMF packaged in 100-mL Sure-Seal amber bottles were used.

4. Addition of the reagents should take place no more than 5 minutes after the cooling of the carbodiimide solution, since EDC·HCl tends to

precipitate at 0 °C. Following addition of Z-*L*-Phg-OH and Oxyma Pure any insoluble EDC·HCl quickly dissolves.

5. Z-*L*-Phg-OH and Oxyma Pure are preferably weighed together and added as a solid mixture for an optimized coupling.

6. Immediately after contacting the acid and Oxyma Pure, the carbodiimide solution turned from colorless to bright yellow.

7. After addition of the amine and base, the solution acquired a bright orange color, which slowly decreased with reaction progress to pale yellow.

8. TLC analyses were performed using plates pre-coated with silica Gel 60 F254, purchased from EMD Chemicals Inc., using 10% DCM/MeOH with 1% AcOH as eluent. The R_f value of the title dipeptide product is 0.73, whereas the starting Z-*L*-Phg-OH and H-Val-OMe·HCl have R_f values of 0.41 and 0.14 respectively. Phosphomolybdate dip solution allowed clear visualization of Z-*L*-Phg-OH and Z-*L*-Phg-Val-OMe in the crude mixture, whereas H-Val-OMe·HCl (bright red) could be clearly spotted with alternative ninhydrin staining.

9. DMF was co-evaporated at 40 °C with approximately 200 mL of toluene.

10. Although the dipeptide product is more easily dissolved in dichloromethane, more facile extraction and greater yields are obtained with this methodology.

11. A small sample (10 mg) of the crude mixture obtained after work-up is analyzed by ^{1}H NMR, showing approximately 1.6% of the DL-epimer The most reliable and accurate method to determine the extent of epimerization in dipeptides containing *N*-terminal Phg residues is the integration of methoxy peaks corresponding to the different epimers in the ^{1}H NMR spectra, taking advantage of the methyl ester nonequivalence.[2] With regard to the dipeptide herein described, LL and DL epimers displayed OMe signals at δ 3.63 and 3.73 ppm, respectively, in CDCl₃.[3] The integration method described by Davies can be applied, and the result is in good agreement with simple normalization of the methyl resonance.[4] The limit of detection for NMR is considered ≤0.2% and limit of quantification is ≤0.5%. Analysis of the degree of epimerization was also attempted by reverse-phase HPLC, but complete separation of the LL and DL epimers could not be achieved.

12. Diastereomeric purity of the recrystallized solid was again examined by ^{1}H NMR spectroscopy, showing complete disappearance of the OMe signal of the DL-epimer at δ 3.73 ppm. The recrystallized material

showed also excellent overall purity (>99%) according to HPLC and NMR techniques. The submitter reported the following HPLC conditions could be used to assess purity. HPLC was conducted on a X-Bridge BEH column (C_{18} 3.5 µm, 4.6 x 100 mm), using a linear gradient of 5 to 100% CH_3CN in H_2O/0.1% TFA over 8 min, with a flow rate 1.0 mL/min and detection at 220 nm, Retention time of (Z-*L*-Phg-Val-OMe) = 7.25 min. The Z-*L*-Phg-Val-OMe obtained using this procedure, which is stable after long storage in a freezer at –5 °C, has the following physical and spectral data: mp: 147–148 °C; FTIR (neat, cm^{-1}): 3338 (m), 3281 (s), 3061 (w), 3033 (w), 2967 (w), 2874 (w), 1740 (s), 1706 (m), 1659 (s), 1531 (s), 1497 (w), 1456 (w), 1433 (w), 1390 (w), 1363 (m), 1341 (w), 1305 (w), 1278 (w), 1250 (m), 1213 (m), 1141 (m), 1114 (w), 1079 (w), 1050 (w), 1028 (w), 1012 (w), 964 (w), 915 (w), 898 (w), 753 (m), 733 (m), 699 (s), 629 (s). ^{1}H-NMR (400 MHz, $CDCl_3$) δ: 0.86 (d, *J* = 6.7 Hz, 3 H, CH_3), 0.92 (d, *J* = 6.9 Hz, 3 H, CH_3), 2.15 (m, 1 H, CH), 3.63 (s, 3 H, OCH_3), 4.49 (dd, *J* = 4.9, 8.6 Hz, 1 H), 5.09 (d, *J* = 12.3 Hz, 2 H, OCH_2), 5.27 (d, *J* = 3.6 Hz, 1 H, NH), 6.06 (d, *J* = 7.4 Hz, 1 H, NH), 6.18 (d, *J* = 8.1 Hz, 1 H, CH), 7.27–7.40 (m, 10 H, CH_{ar}); ^{13}C NMR (100 MHz, $CDCl_3$) δ: 17.8, 18.9, 31.3, 52.1, 57.5, 59.0, 67.1, 127.3, 128.1, 128.1, 128.5, 128.7, 129.1, 136.2, 137.5, 155.7, 169.7, 171.7; $[\alpha]^{20}_{D}$= +77.4 ($CHCl_3$, c = 1.00); HRMS (*m/z*) (ESI): calcd. for $C_{21}H_{27}N_2O_5$ [M+H] 399.19145, found 399.19124; TLC (Hexane/AcOEt = 1:1) R_f = 0.69;

Handling and Disposal of Hazardous Chemicals

The procedures in this article are intended for use only by persons with prior training in experimental organic chemistry. All hazardous materials should be handled using the standard procedures for work with chemicals described in references such as "Prudent Practices in the Laboratory" (The National Academies Press, Washington, D.C., 2011 www.nap.edu). All chemical waste should be disposed of in accordance with local regulations. For general guidelines for the management of chemical waste, see Chapter 8 of Prudent Practices.

These procedures must be conducted at one's own risk. *Organic Syntheses, Inc.*, its Editors, and its Board of Directors do not warrant or guarantee the safety of individuals using these procedures and hereby disclaim any liability for any injuries or damages claimed to have resulted from or related in any way to the procedures herein.

3. Discussion

Peptides have emerged in the last decade as promising therapeutic scaffolds in the pharmaceutical industry, given their high potency, specificity and low toxicity.[5] Retention of optical purity throughout the chemical process is pivotal for these New Chemical Entities and consequently, methodological tools have been developed to minimize C_α-epimerization.[6,7] In this regard, the inclusion of N-hydroxylamines in the coupling cocktail helps prevent the loss of stereochemical integrity in the activated intermediate species.[8] The N-hydroxybenzotriazole family, which comprises HOBt, HOAt and 6-Cl-HOBt as the most renowned members, have dominated the field in the preceding decades over succinimides (HOSu), benzotriazines (HODhb), pyridinones (HOPy) and triazoles (HOCt), as a result of their enhanced reactivity, absence of side reactions, and overall moderate prices.[7] However, the dangerous safety profile of N-hydroxybenzotriazoles and derived coupling reagents prompted a search for alternative templates.[7-9] Recently, acidic ketoximes featuring electron-withdrawing substituents have been proposed as additives to carbodiimides, leading to the widespread reception of ethyl 2-cyano-2-hydroxyimino acetate (Oxyma Pure) as a reliable and safe coupling choice.[7,8,10] The ability of Oxyma Pure to preserve stereochemical configuration during activation of peptide fragments and Cys proved much higher than that of HOBt and comparable

Oxyma Pure

to that of HOAt.[10] Best results have been obtained during activation of the highly sensitive α-phenylglycine (Phg) residue towards epimerization, with Oxyma Pure exceeding the performance of HOAt in minimizing the level of Z-D-Phg-Pro-NH$_2$ (Table 1).[2,11] Moreover, the use of Oxyma Pure in combination with formamidinium salts results in further reduction of epimerization.[12] In addition, Oxyma Pure provides considerably broader scope than HOBt by mediating difficult junctions, such as those containing N-Me or Aib residues, and is fully compatible with microwave irradiation.[10]

310

Recently, other applications such as ester and amide bond formation, and the minimization of base-catalyzed side reactions, have been discovered.[13,14]

This procedure describes the application of Oxyma Pure in the preparation of the Z-*L*-Phg-Val-OMe dipeptide, which combines the epimerization-prone α-phenylglycine amino acid and the sterically hindered valine residue.[3,15,16] Therefore, this dipeptide (which has been reported a few times in the literature) is an excellent platform to test both the capability of Oxyma Pure to retain stereochemical purity and to promote acylation of bulky residues.[3,15,16] An advantage of this peptide model over other Phg-containing dipeptides such as Z-Phg-Pro-NH$_2$ is the ability to monitor the degree of epimerization by ^{1}H-NMR (see Note 11) and the ease of spectral analysis, given the absence of cis/trans isomerism.[2,3,16] In addition, EDC·HCl is employed as carbodiimide because of its more facile removal than DIC during aqueous work-up. Although the basic center contained in its structure could promote epimerization, this effect is not observed in solution-phase; moreover, EDC·HCl accelerates active ester formation to a greater extent than DIC.[17] A similar enhancement in the activation process is observed when DCM is used as coupling solvent, with simultaneous reduction of loss of chirality.[11,17] Using this combination of reagents and solvents, an optimized method for Oxyma Pure-mediated peptide bond formation is presented. All reagents are commercially available at low to moderate cost and furthermore, Oxyma Pure is comparatively less expensive than HOAt.

By means of the abovementioned procedure, the suitability of the EDC·HCl/Oxyma Pure coupling system in solution-phase is unambiguously demonstrated (Table 2). In comparison to previous procedures on the same peptide platform using *N*-hydroxybenzotriazoles, Oxyma Pure afforded a substantially lower content of DL epimer than HOBt (0.1 vs 3.7%) with concomitant increase of yield.[15] The extent of epimerization was also considerably lower than that induced by HOAt (0.1 vs. <1-2%).[15] Similar to the tendency observed with the *N*-hydroxylamines, Oxyma Pure also outperformed the corresponding benzotriazole-based aminium salts HATU and HBTU, and formamidinium salt TFFH, even when the low epimerization-inducing DCM is used as solvent (Table 2).[3,15,16] Further, not only yield and stereochemical integrity, but also purity, is enhanced with Oxyma Pure, based on the melting point of the target dipeptide obtained using the present (147-148°C) vs. previous (136-141°C) procedures.[3,16]

Moreover, the procedure showed that after recrystallization the level of the epimer is negligible.

Table 1. Extent of epimerization during stepwise formation of Z-*L*-Phg-Pro-NH$_2$ in solution phase.

entry	Coupling Reagent	Yield (%)	DL (%)
1	DIC/HOAt	81.4	3.3
2	DIC/HOBt	81.9	9.3
3	DIC/Oxyma Pure	89.9	1.0
4	DIC/HOPy	88.2	17.4
5[a]	TFFH/Oxyma Pure	85.0	0.54

[a] 2 min. preactivation was performed with 2 equiv of DIEA.

Table 2. Extent of epimerization during stepwise formation of Z-*L*-Phg-Val-OMe in solution phase.

entry	Coupling Reagent	Base	Solvent	Yield (%)	DL (%)
1	EDC/HOAt	Proton Sponge	DMF	70-90	<1-2
2	EDC/HOBt	Proton Sponge	DMF	70-90	3.7
3	HATU	Proton Sponge	DMF	70-90	<1-2
4	HBTU	Proton Sponge	DMF	70-90	4.0
5[a]	EDC/Oxyma Pure	DIEA	DMF/DCM 1:1	81-84	0.1
6	TFFH	DIEA	DMF	80	3.6
7	TFFH	DIEA	DCM	82	1.3

[a] This work, after recrystallization

1. Chemistry and Molecular Pharmacology Program, Institute for Research in Biomedicine, Barcelona Science Park, Baldiri Reixac 10, 08028-Barcelona, Spain; ramon.subiros@gmail.com,

aymanel_faham@hotmail.com, albericio@irbbarcelona.org; Research in the laboratory of the authors was partially funded by the Secretaría de Estado de Cooperación Internacional (AECI), the CICYT (CTQ2009-07758), the Generalitat de Catalunya (2009SGR 1024), the Institute for Research in Biomedicine Barcelona (IRB Barcelona), and the Barcelona Science Park.

2. Carpino, L. A. *J. Org. Chem.* **1988**, *53*, 875–878
3. Carpino, L. A.; El-Faham, A. *US Patent, 5712418A,* **1998.**
4. Claridge, T. D. W.; Davies, S. G.; Polywka, M. E. C.; Roberts, P. M.; Russell, A. J.; Savory, E. D.; Smith, A. D. *Org. Lett.* **2008**, *10*, 5433.
5. a) Peptides as Drugs. Discovery and Development; Groner, B., Ed.; WILEY-VCH: KGaA, Weinheim, **2009**; pp 1–219; b) Edwards, C. M. B.; Cohen, M. A.; Bloom, S. R. *Q. J. Med.* **1999**, *92*, 1-4; c) Loffet, A. *J. Peptide Sci.* **2002**, *8*, 1–7.
6. Albericio, F., Chinchilla, R., Dodsworth, D. J., Najera, C. *Org. Prep. Proced. Int.* **2001**, *33*, 203.
7. El-Faham, A.; Albericio, F. *Chem. Rev.* **2011**, *111*, 6557–6602.
8. Subirós-Funosas, R.; El-Faham, A.; Albericio, F. *"N-Hydroxylamines for Peptide Synthesis,"* in *Patai's Chemistry of Functional Groups*, John Wiley & Sons, **2011***, 2*, 623–730.
9. Wehrstedt, K. D.; Wandrey, P. A.; Heitkamp, D. *J. Hazard. Mater.* **2005**, *126*, 1–7.
10. Subirós-Funosas, R.; Prohens, R.; Barbas, R.; El-Faham, A.; Albericio, F. *Chem. Eur. J.* **2009**, *15*, 9394–9403.
11. Wenschuh, H.; Beyermann, M.; Haber, H.; Seydel, J. K.; Krause, E.; Bienert, M. *J. Org. Chem.* **1995**, *60*, 405–410.
12. Khattab, S. N. *Bull. Chem. Soc. Jpn.* **2010,** *83*, 1374–1379.
13. Subirós-Funosas, R.; Khattab, S. N.; Nieto-Rodriguez, L.; El-Faham, A.; Albericio, F. *Aldrichimica Acta* **2013**, *46,* 21–40
14. Subirós-Funosas, R.; El-Faham, A.; Albericio, F. *Biopolymers* **2011**, *98,* 89–97.
15. Carpino, L. A. *J. Am. Chem. Soc.* **1993**, *115*, 4397–4398.
16. Carpino, Louis A.; El-Faham, A. *PCT Int. Appl., WO 9604297,* **1996**
17. Carpino, L. A.; El-Faham, A. *Tetrahedron* **1999**, *55*, 6813–6830.

Appendix
Chemical Abstracts Nomenclature; (Registry Number)

Z-*L*-Phg-OH: Benzeneacetic acid, α-[[(phenylmethoxy)carbonyl]amino]-, (α*S*)-; (53990-33-3)

H-Val-OMe·HCl: L-Valine, methyl ester, hydrochloride (1:1); (6306-52-1)

Z-*L*-Phg-Val-OMe: L-Valine, (2*S*)-2-phenyl-*N*-[(phenylmethoxy)carbonyl]glycyl-, methyl ester; (159487-14-6)

EDC·HCl: 1,3-Propanediamine, N^3-(ethylcarbonimidoyl)-N^1,N^1-dimethyl-, hydrochloride (1:1); (25952-53-8)

Oxyma Pure: Acetic acid, 2-cyano-2-(hydroxyimino)-, ethyl ester (3849-21-6)

DIEA: 2-Propanamine, *N*-ethyl-*N*-(1-methylethyl)-; (7087-68-5)

DCM: Methane, dichloro-; (75-09-2)

DMF: Formamide, *N*,*N*-dimethyl-; (68-12-2)

Professor Ayman El-Faham received his B.Sc. and M.Sc. in Physical Organic Chemistry from the University of Alexandria, Egypt. In 1991 he received his Ph.D. in organic chemistry through a joint project between the University of Alexandria and the University of Massachusetts. From 1992 to 1999, he worked on new coupling reagents at Professor Carpino's Lab. Following a position as Head of the Chemistry Department, Beirut Arab University (2000 to 2004), and as Direct Manager of both the NMR and Central Lab at Alexandria University (2004 to 2008), he worked at King Saud University, as a Professor of Organic Chemistry (2008 to 2010). Currently he is working back at King Saud University. His research interests include synthesis of peptides, natural products, heterocycles, and biologically active targets.

Ramon Subirós-Funosas received his B.Sc. in Chemistry from the University of Barcelona in 2007. Previously, he moved for one year to GlaxoSmithKline, in Stevenage, UK, where he joined the Medicinal Chemistry Department in 2005. Ramon obtained his Ph.D. in Organic Chemistry in 2011 from the University of Barcelona, under the supervision of Prof. Fernando Albericio, developing a new family of coupling reagents based on Oxyma Pure. In 2010, he visited the group of Professor Dawson at The Scripps Research Institute, La Jolla, USA, for a 4-month internship, acquiring knowledge in the synthesis of small proteins. His major research interests include peptide synthesis and methodology of native chemical ligation. He is currently a Marie Curie IEF fellow at Humboldt Universität-zu-Berlin with Prof. Oliver Seitz.

Professor Fernando Albericio obtained his Ph.D. in Chemistry from the University of Barcelona. After postdoctoral work at Tufts University, at the Université d'Aix-Marseille, and at the University of Minnesota, he joined the University of Barcelona as an Associate Professor. In 1992-1994, he was appointed Director of Peptide Research at Milligen/Biosearch, Boston, USA, then returned to the University of Barcelona. Currently, he is holding a triple appointment as Professor at the University of Barcelona and Group Leader at the Institute for Research in Biomedicine, and Research Professor at the University of KwaZuluNatal (Durban, South Africa). From 2005 to 2012, he has been Executive Director in the Barcelona Science Park. His major research interests cover practically all aspects of peptide synthesis and combinatorial chemistry methodology.

Asher Lower was born in San Francisco, CA in 1981. He received his Ph.D. in synthetic organic chemistry from the University of California, Santa Barbara in 2006 under the direction of Bruce H. Lipshutz. His doctoral research focused on the development of new stereoselective transformations using inexpensive and environmentially friendly Cu and Ni catalysts. Asher started his career at Ampac Fine Chemicals where he held the position of Sr. Scientist specializing in commercial-scale manufacture of hazardous APIs. Asher joined Amgen in 2009 and currently supports late-stage and commercial process development.

Practical Synthesis of Di-*tert*-Butyl-Phosphinoferrocene

Submitted by Carl A. Busacca, Magnus C. Eriksson,[1] Nizar Haddad, Z. Steve Han, Jon C. Lorenz, Bo Qu, Xingzhong Zeng, and Chris H. Senanayake.
Checked by Corey M. Reeves and Brian M. Stoltz.

Caution! tert-Butyllithium is extremely pyrophoric and must not be allowed to come into contact with the atmosphere. This reagent should only be handled by individuals trained in its proper and safe use. It is recommended that transfers be carried out by using a 20-mL or smaller glass syringe filled to no more than 2/3 capacity or by cannula. For a discussion of procedures for handling air-sensitive reagents, see Aldrich Technical Bulletin AL-134.

1. Procedure

Di-tert-butylphosphinoferrocene. An oven-dried 1000-mL four-necked (one 34/45 joint and three 24/40 joints) round-bottomed flask is allowed to cool in a desiccator over anhydrous calcium sulfate. Once cool, the central joint is equipped with an overhead mechanical stirrer, the glass rod of which is fitted with 7.2 x 2 cm Teflon paddle, coated with lubricant (Note 1) and sheathed by a 34/45 jointed glass stirrer bearing. The remaining three necks are fitted with a thermocouple in a 24/40 adapter, an argon line connected to a 24/40 adapter, and rubber septum. The rubber septum is removed from the fourth neck and the flask is charged with ferrocene (8.0 g, 43.0 mmol, 1 equiv) (Note 2). A 250-mL pressure-equalizing addition funnel with a 24/40 joint is fitted in the fourth neck and the reaction set-up is flushed with argon for 5 min (see Note 3 for an image of the reaction setup). The top of the addition funnel is sealed with a rubber septum and the entire set-up is maintained under a positive pressure of argon

316

throughout the remaining operations. Tetrahydrofuran (480 mL) (Note 4) is added via cannula through the rubber septum and down through the addition funnel resulting in an orange solution that is stirred at room temperature. A solution of potassium *tert*-butoxide in tetrahydrofuran (1.0 M, 5.25 mL, 5.25 mmol, 0.12 equiv) (Note 5) is added to the flask through the rubber septum using a gastight syringe. The addition funnel is rinsed with tetrahydrofuran (20 mL) followed by pentane (10 mL) (Note 6). The reaction solution is cooled to about –73 °C by immersion of the flask in a dry ice-acetone bath. A solution of *tert*-butyllithium (1.7 M, 49.6 mL, 84.3 mmol, 2.0 equiv) (Note 7) is added dropwise via the addition funnel over 10–15 min, keeping the internal temperature around –70 °C. The addition funnel is rinsed with pentane (10 mL) with a gastight syringe. A 2-3 °C exotherm is observed during the addition and an orange slurry forms toward the end or shortly after completion of the addition. The slurry is stirred at about –70 °C for 1 h and then the dry ice-acetone bath is replaced with an ice-water bath and the content of the flask is stirred and allowed to warm. Once the internal temperature reaches ca. 0 °C (Note 8), di-*tert*-butylchlorophosphine (8.8 mL, 46.3 mmol, 1.1 equiv) (Note 9) is charged to the addition funnel via a gastight syringe and added to the reaction mixture drop-wise over 2–3 min. The addition funnel is rinsed with tetrahydrofuran (10 mL). The ice/water bath is removed and the mixture is stirred at ambient temperature for 12 h. Reaction progress is monitored by TLC analysis (Note 10).

Tetrafluoroboric acid-diethyl ether complex (11.5 mL, 84.5 mmol, 2.0 equiv) (Note 11) is added at ambient temperature to the reaction mixture via gastight syringe through the septum over ca. 5 min. A slight exotherm (5–7 °C) is observed as the solution becomes turbid. A solid begins to precipitate within 0.5–1 h. The suspension is stirred at ambient temperature for 2 h to afford a red thin suspension (Note 12). Deionized water (100 mL) and methyl *tert*-butyl ether (100 mL) are added via syringe through the septum and the resulting turbid homogeneous two-layer solution is stirred for 30 min at ambient temperature. The mixture is then vacuum-filtered through a 1 cm pad of Celite (ca. 30 g) in a 300-mL fritted Büchner funnel with a vacuum pressure of ca. 15 mmHg. The filter pad is rinsed with tetrahydrofuran (40 mL) and the rinse combined with the filtrate. The filtrate is transferred to a flask and concentrated via rotary evaporation (Note 13) until all organics are removed and a red solid as a suspension in ca. 100 mL water is obtained. The suspension is stirred at ambient temperature for 1 h and vacuum-filtered using a 100-mL fritted Büchner funnel, fitted with a VWR brand No. 5 filter paper (55 mm), and with a vacuum pressure of ca. 15 mmHg. The filter cake is washed with water (40 mL) and dried on the frit for 1 h using house-vacuum to obtain the crude product as an orange solid (Note 14).

A 250-mL three-necked round-bottomed flask with 24/40 joints is equipped with a 5 cm magnetic stirbar, thermocouple in one neck, a 60 mL 24/40 joint pressure-equalizing addition funnel that is capped with a rubber septum in the center neck and an argon line connected to a 24/40 adapter in the third neck. The flask is charged with crude **1** and a 16-gauge needle is used to puncture the rubber septum attached to the addition funnel, allowing the flask to be flushed with argon for 5 min. Methyl *tert*-butyl ether (50 mL) is added via gastight syringe to the reaction flask via the addition funnel. The suspension is heated with stirring to 50–55 °C via an oil bath and, once the internal temperature reaches 50 °C, methanol (50 mL) is added via gastight syringe to the addition funnel and then to the mixture drop-wise over 5 min. The mixture is stirred at 50 °C for an additional 10–15 min to obtain a dark red solution. The solution is vacuum-filtered (with a vacuum pressure of ca. 15 mmHg) through a Büchner funnel with #1 filter paper into a pre-warmed 500-mL 24/40 joint round-bottomed flask (Note 15) that contains a 7 cm magnetic stirbar. The Büchner funnel is replaced with a 125 mL pressure-equalizing addition funnel that is charged with methyl *tert*-butyl ether (100 mL) and an argon line is attached to the top of the addition funnel. The methyl *tert*-butyl ether in the addition funnel is added dropwise over 5–10 min, which results in the formation of a slurry. The oil bath is removed and the slurry stirred at ambient temperature for 3 h. The slurry is filtered on a Büchner medium-fritted funnel and the cake is washed with *tert*-butyl methyl ether (40 mL). The filter cake is dried using house-vacuum to obtain **1** (11.0 g, 61%) as a light orange solid (Notes 16).

2. Notes

1. "Stir-Lube" is obtained from Ace Glass (Trubore stirrer lubricant, product number 811710).
2. Ferrocene was purchased from Aldrich (98%) and used as received.

3.

4. Anhydrous tetrahydrofuran containing 250 ppm BHT as inhibitor was purchased from Aldrich and used as received.

5. Potassium *tert*-butoxide as a 1.0 M solution in tetrahydrofuran was purchased from Aldrich and used as received. Fresh bottles of potassium *tert*-butoxide solution gave superior results.

6. The purpose of the pentane rinse is to remove tetrahydrofuran from the addition funnel since *tert*-butyllithium reacts instantaneously with tetrahydrofuran at room temperature.

7. The *tert*-butyllithium was purchased from Aldrich as a 1.7M solution in pentane and used as received.

8. It takes about 5–10 minutes for the internal temperature to rise from –70 °C to ca. 0 °C.

9. The di-*tert*-butylchlorophosphine was purchased from Aldrich and used as received.

10. Thin Layer Chromatographic analysis is performed in 10 % methanol in dichloromethane, with the product having an $R_f = 0.4$.

11. Tetrafluoroboric acid diethyl ether complex was purchased from Aldrich and used as received.

12. The suspension may be stirred overnight at ambient temperature before further work-up.

13. House vacuum (10–14 mmHg) and ca. 19 °C water bath were used.

14. The yield of crude **1** is approximately 15–16 g (85–90%).

15. The round-bottomed flask is warmed through immersion in a 50–55 °C oil bath prior to filtration.

16. Purified **1** was isolated in 64% yield (5.8 g) when the reaction was performed on 21.5 mmol scale. Compound **1** exhibits the following physical and spectroscopic properties: $R_f = 0.4$ (10 % methanol in dichloromethane); mp = 175–176 °C; ^{1}H NMR (CDCl$_3$, 300 MHz), δ: 1.52 (d, J_{H-P} = 16.6 Hz, 18 H), 4.43 (s, 5 H), 4.66 (s, 2 H), 4.79 (s, 2 H), 6.86 (d, J_{H-P} = 493 Hz, 1 H); ^{13}C NMR (CDCl$_3$, 126 MHz), δ: 28.4, 34.3 (d, J_{C-P} = 37.8 Hz), 56.5 (d, J_{C-P} = 82.1 Hz), 71.7, 73.4 (d, J_{C-P} = 9.1 Hz), 73.9 (d, J_{C-P} = 9.9 Hz); ^{31}P NMR (CDCl$_3$, 121 MHz) δ: 40.8; ^{19}F NMR (CDCl$_3$, 282 MHz) δ: –150.04; ^{11}B NMR (CDCl$_3$, 128 MHz) δ: –0.85; IR (neat film, NaCl): 3109, 2968, 2430, 1709, 1477, 1415, 1375, 1281, 1168, 1070, 1049, 1031, 883, 838, 812 cm^{-1}; Elem. Anal. calc'd for C$_{18}$H$_{28}$BF$_4$FeP: C, 51.72, H, 6.75; Found: C, 51.70; H, 6.96; HRMS for free phosphine C$_{18}$H$_{27}$FeP [M + H]$^+$: m/z calc'd, 331.1272; found, 331.1281.

Handling and Disposal of Hazardous Chemicals

The procedures in this article are intended for use only by persons with prior training in experimental organic chemistry. All hazardous materials should be handled using the standard procedures for work with chemicals described in references such as "Prudent Practices in the Laboratory" (The National Academies Press, Washington, D.C., 2011 www.nap.edu). All chemical waste should be disposed of in accordance with local regulations. For general guidelines for the management of chemical waste, see Chapter 8 of Prudent Practices.

These procedures must be conducted at one's own risk. *Organic Syntheses, Inc.*, its Editors, and its Board of Directors do not warrant or guarantee the safety of individuals using these procedures and hereby disclaim any liability for any injuries or damages claimed to have resulted from or related in any way to the procedures herein.

3. Discussion

Electron-rich dialkylphosphines have emerged in recent years as highly useful ligands for a variety of cross-coupling reactions. Specifically, di-alkylphosphinoferrocenes are very useful ligands that have found applications in metal-catalyzed C-C[2], C-N and C-S[3] bond formation. The syntheses of these ligands typically proceed through deprotonation/metallation of ferrocene followed by reaction of the metallated ferrocene with the appropriate chlorophosphine. Di-*tert*-butylphosphinoferrocene is a particularly challenging case due to its electron-rich nature and ease of oxidation to the phosphine oxide.[4] Furthermore, mono-metallation of ferrocene is not a trivial task and requires carefully optimized reaction conditions to avoid incomplete reactions and formation of di-metallated species.[5]

The existing procedures for the synthesis of di-*tert*-butylphosphinoferrocene give medium-low to low yields[6], require reactions and purifications in dry-boxes,[7] or start from expensive bromo-ferrocene.[8] The synthesis of di-*tert*-butylphosphinoferrocene was reported in three different publications with some variations in the procedure. The first procedure[6] in 28% isolated yield used a large excess of ferrocene and *tert*-BuLi for the deprotonation/metallation and required dry-box work and sublimations of the product for purification. The second modified procedure[7] used an excess of ferrocene as well and relied on a well-defined ratio of THF/pentane for a successful mono-metallation. It also required purification on silica gel and dry-box work. The third most recent procedure[8] used expensive bromo-ferrocene and metal-halogen exchange at –78 °C. This procedure also involved silica gel chromatography and dry-box work.

Based on the work by Mueller-Westerhoff on metallation of ferrocene[5], we developed a practical scaleable synthesis of di-*tert*-butylphosphinoferrocene. To avoid the issues associated with its air-sensitivity we isolated the compound as its HBF$_4$ salt[9] or the corresponding phosphine-borane.[10] The protocol of Mueller-Westerhoff that served as the starting point for our work suggested that it is important to conduct the metallation at –78 °C for best results. We have found that while it is important to perform the deprotonation cold, it is not necessary to be at –78 °C. In our case, the metallation has been conducted at up to –45 °C with successful results. In addition, we found it to be important to allow the reaction mixture to warm up to between –20 °C and 0 °C before addition of the chloro-phosphine. When the chloro-phosphine was added at –78 °C following the metallation we obtained a complex mixture of the desired product along with several unknown impurities. The ligand synthesis was demonstrated on multigram scale and 200 gram scale. The HBF$_4$-salt (**1**) is an air-stable solid that can be stored and used without further purification in Pd-catalyzed aminocarbonylation reactions as indicated by two examples shown in Scheme 1.[11]

Scheme 1. Aminocarbonylation of aryl halides.

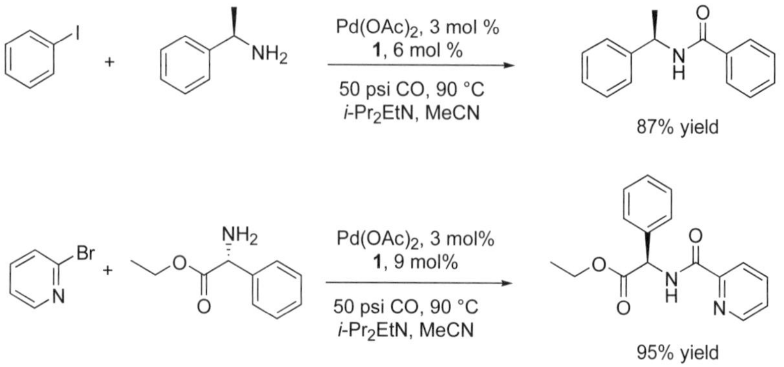

1. Department of Chemical Development, Boehringer Ingelheim Pharmaceuticals, Inc., P. O. Box 368, Ridgefield, CT 06877. E-mail: magnus.eriksson@boehringer-ingelheim.com
2. Baillie, C.; Zhang, L.; Xiao, J. *J. Org. Chem.* **2004**, *69*, 7779–7782.
3. Hartwig, J. F. *Acc. Chem. Res.* **2008**, *41*, 1534–1544.
4. The ^{31}P NMR chemical shift for the phosphine oxide in CDCl$_3$ was 58.9 ppm in agreement with literature data from reference **2**.
5. Sanders, R.; Mueller-Westerhoff, U. T. *J. Organometal. Chem.* **1996**, *512*, 219–224.
6. Mann, G.; Incarvito, C.; Rheingold, A. L.; Hartwig, J. F. *J. Am. Chem. Soc.* **1999**, *121*, 3224–3225.
7. Shelby, Q.; Kataoka, N.; Mann, G.; Hartwig, J. *J. Am. Chem. Soc.* **2000**, *122*, 10718–10719.
8. Fujita, K.; Yamashita, M.; Puschmann, F.; Alvarez-Falcon, M. M.; Incarvito, C. D.; Hartwig , J. F. *J. Am. Chem. Soc.* **2006**, *128*, 9044–9045. The cost of bromoferrocene is $27-44/gram.
9. A synthesis of the HBF$_4$ salt **1** reported recently: Oms, O.; Jarrosson, T.; Tong, L. H.; Vaccaro, A.; Bernardinelli, G.; Williams, A. F. *Chem. Eur. J.* **2009**, *15*, 5012–5022. The ligand itself was prepared following the procedure by Kataoka, N.; Shelby, Q.; Stambuli, J. P.; Hartwig, J. F. *J. Org. Chem.* **2002**, *67*, 5553–5566.
10. The procedure originally submitted to *Organic Syntheses* contained a protocol for isolation of the phosphine-borane complex of di-*tert*-

butylphosphinoferrocene. Although the crude phosphine borane complex was isolated in good yield by the checkers per the submitted procedure, the protocol for purification by recrystallization failed to provide product of the expected quality. The protocol for isolation of the phosphine-borane complex of di-*tert*-butylphosphinoferrocene was therefore omitted from the published procedure. Pure phosphine borane complex of di-*tert*-butylphosphinoferrocene can be isolated by silica gel chromatography using ethyl acetate/hexanes. For a leading reference on the synthesis of phosphine boranes, see Imamoto, T.; Oshiki, T.; Onozawa, T.; Kusumoto, T.; Sato, K. *J. Am. Chem. Soc.* **1990**, *112*, 5244–5252.

11. Qu, B.; Haddad, N.; Han, Z. S.; Rodriguez, S.; Lorenz, J. C.; Grinberg, N.; Lee, H.; Busacca, C. A.; Krishnamurthy, D.; Senanayake, C. *Tetrahedron Lett.* **2009**, *50*, 6126–6129.

Appendix
Chemical Abstracts Nomenclature; (Registry Number)

Di-*tert*-butylphosphinoferrocene: Ferrocene, [bis(1,1-dimethylethyl)phosphino]-; (223655-16-1)

Ferrocene; (102-54-5)

Potassium *tert*-butoxide: 2-Propanol, 2-methyl-, potassium salt (1:1); (865-47-4)

tert-Butyllithium: Lithium, (1,1-dimethylethyl)-; (594-19-4)

Di-*tert*-butylchlorophosphine: Phosphinous chloride, *P,P*-bis(1,1-dimethylethyl)- ; (13716-10-4)

Tetrafluoroboric acid diethyl ether complex: Borate(1-), tetrafluoro-, hydrogen (1:1) Ethane, 1,1'-oxybis-; (67969-82-8)

Dr. Carl Busacca received his BS in Chemistry from North Carolina State University, and did undergraduate research in Raman spectroscopy and ^{60}Co radiolyses. After three years with Union Carbide, he moved to the labs of A.I. Meyers at Colorado State University, earning his Ph.D. in 1989 studying asymmetric cycloadditions. He worked first for Sterling Winthrop before joining Boehringer-Ingelheim in 1994. He has worked extensively with anti-virals, and done research in organopalladium chemistry, ligand design, organophosphorus chemistry, asymmetric catalysis, NMR spectroscopy, and the design of efficient chemical processes. He is deeply interested in the nucleosynthesis of transition metals in supernovae.

Dr. Magnus Eriksson was born in Stockholm, Sweden. He received his undergraduate degree in Chemical Engineering and his PhD in Organic Chemistry from Chalmers University of Technology in Gothenburg in 1995 under the guidance of Professor Martin Nilsson working on copper-promoted 1,4-additions to carbonyl compounds. After post-doctoral work at Boehringer Ingelheim Pharmaceuticals and at MIT with Professor Stephen Buchwald, he joined Boehringer Ingelheim Pharmaceuticals in 2000 where he is currently a Principal Scientist. His research interests include Process Research, catalytic transformations and synthetic methodology.

Dr. Nizar Haddad received his B.A. (1984) and D.Sc. in chemistry (1988, Prof. Dan Becker) from the Technion, Israel. After postdoctoral research at the University of Chicago (Prof. J. D. Winkler) then Harvard University (Prof. Y. Kishi), he joined the faculty in the Chemistry Department at the Technion in 1991. Following sabbatical leave with Prof. K. C. Nicolaou at Scripps, he joined Boehringer Ingelheim Pharmacuticals in 1998 where he is currently Senior Principal Scientist heading the Automation/Catalysis group. His research interests include the development of catalytic asymmetric reactions and their applications in devising new and economical processes.

Zhengxu (Steve) Han was born in China. He received his B.S degree and Ph.D. in organic chemistry from Lanzhou University with Professor You-Cheng Liu. In 1991, he moved to Tübingen University, Germany as a fellow of the Alexander von Humboldt Foundation with Professor Anton Rieker. After four years of postdoctoral research with Professor Stuart Linn at University California, Berkeley and Professor Lee Magid at the University of Tennessee, Knoxville, he joined the process research group at Sepracor, Inc., in 1998 and then Boehringer Ingelheim in 2005. His research interests center on efficient process and methodology development for asymmetric synthesis.

Dr. Jon C. Lorenz obtained his undergraduate degree from Whitman College in Walla Walla Washington. After teaching science in the northwest province of Cameroon as a Peace Corps Volunteer he attended graduate school at Colorado State University where he received his Ph.D. under the guidance of Prof. Yian Shi in 2002. Subsequently he joined Chemical Development at Boehringer Ingelheim Phamaceuticals where he is currently a Principal Scientist. His research interests include development and application of catalytic asymmetric reactions, and the many facets of practical process development.

Dr. Bo Qu was born in China, where she received a B.S. degree in chemistry. She then completed her M.S. at University of Science and Technology of China. She obtained her Ph.D. from the University of South Carolina in 2002 under the guidance of Prof. Richard Adams. After 3 years of postdoctoral studies at Cornell University with Prof. David Collum, she joined the Department of Chemical Development at Boehringer Ingelheim Pharmaceuticals in Ridgefield, CT, where she is currently a Senior Scientist. Dr. Qu's research interests focus on development of new catalytic transformations for efficient chemical processes, organometallic chemistry, and automated parallel syntheses.

Xingzhong Zeng was born in China in 1970. He received his Ph.D. in 2004 from Purdue University under the supervision of Prof. Ei-ichi Negishi and then spent 2 years (2004-2006) as a Post Doctoral Fellow in Chemical Development, Boehringer Ingelheim Pharmaceuticals, Inc (BIPI) under the supervision of Dr. Vittorio Farina, working on mechanistic studies and methodology development of RCM reaction and its application to large scale synthesis of BILN 2061. He joined BIPI as a Senior Scientist in 2007 and his current research interests are process research and scale-ups, new methodology development and metal-catalyzed transformations.

Dr. Chris H. Senanayake obtained his Ph.D. with Professor James H. Rigby at Wayne State University followed by postdoctoral fellow with Professor Carl R. Johnson. In 1989, he joined Process Development at Dow Chemical Co. In 1990, he joined the Merck Process Research Group. After Merck, he accepted a position at Sepracor, Inc. in 1996 where he was appointed to Executive Director of Chemical Process Research. In 2002, he joined Boehringer Ingelheim Pharmaceuticals. Currently, he is the Vice President of Chemical Development. He is the co-author more than 250 papers and patents in many areas of synthetic organic chemistry.

Corey M. Reeves was born in Santa Monica, CA, USA. Corey obtained a BS in Chemistry and BA in Sociology from Columbia University in New York City in 2009. During this time, he completed undergraduate research under the guidance of Prof. Tristan Lambert. In 2010, he began doctoral studies at the California Institute of Technology, working in the laboratory of Prof. Brian Stoltz.

Preparation of 1-Benzyl-2-methyl-3-(ethoxycarbonyl)-4-(2-phenylethyl)-1H-pyrrole from 4-Phenyl-1,2-butadiene

A.

B.

C.

Submitted by Shengjun Ni, Can Zhu, Jie Chen and Shengming Ma.[*1]
Check by Mélanie Charpenay and Kay Brummond.

1. Procedure

Caution! Ethyl α-diazoacetoacetate is a potentially explosive material. Transformations involving this compound should be performed behind a blast shield in a well-ventilated fume hood.

A. *4-Phenyl-1,2-butadiene* (*1*). A 3-necked 250-mL round-bottomed flask containing a magnetic stir bar (3 cm length x 1.5 cm diameter) is equipped with a reflux condenser and two rubber septa (Note 1). The top of the condenser is fitted with a stopcock connected to a vacuum line with an argon flow. Magnesium turnings (7.2 g, 0.3 mol, 1.2 equiv) are added and the apparatus is flame-dried under vacuum. After cooling to room temperature under an argon purge, the flask is charged with diethyl ether (100 mL) and equipped with a thermometer. 1,2-Dibromoethane (0.5 mL) (Note 2) is added to initiate the reaction with a slight increase of the internal temperature (2–3 °C). Benzyl bromide (30 mL, 0.25 mol) (Note 2) is injected into the mixture with a syringe pump at a rate so as to maintain a gentle reflux (15 mL/h). After the addition, the mixture is heated under reflux with an oil bath (oil bath temperature, 55 °C) for another 2 h. Another 3-necked 250-mL round-bottomed flask equipped with a magnetic stir bar (3 cm length x 1.5 cm diameter) (Note 1) is fitted with two rubber stoppers and a 250-mL pressure equalizing addition funnel. The flask and funnel are flame-dried under vacuum. After cooling to room temperature under an argon purge, propargyl bromide (33.4 mL of an 80 wt. % solution in toluene, 0.3 mol, 1.2 equiv) (Note 2) and diethyl ether (50 mL) (Note 2) are added and the flask is fitted with a thermometer. The mixture is cooled to 0 °C (internal temperature) with an ice-salt bath. The prepared Grignard reagent is transferred to the addition funnel using an 18-gauge cannula and added to the propargyl bromide over 3 h. After the addition, the resulting mixture is stirred for an additional 2 h at 0 °C (Note 3). Then the mixture is poured into a 1-L Erlenmeyer flask containing a cold saturated aqueous solution of NH$_4$Cl (200 mL). The mixture is transferred to a 500 mL separatory funnel, the phases are separated, and the aqueous phase is back-extracted with diethyl ether (100 mL x 2) (Note 2). The combined organic phases are washed with brine (100 mL), dried over anhydrous Na$_2$SO$_4$, and filtered. The solvent is removed by rotary evaporation (20 mmHg, 30 °C water bath). Purification is performed by vacuum distillation (70–80 °C/23 mmHg, fractionating column: 2 cm diameter x 15 cm length) to afford 4-phenyl-1,2-butadiene (**1**) (18.6 g, 57% by the checker; 24.9 g, 77% by the submitter, see discussion) as a colorless liquid (Note 4).

B. *1-Acetyl-2-phenethylidenecyclopropanecarboxylic acid ethyl ester (2).* A 3-necked 250-mL round-bottomed flask is equipped with a reflux condenser, the top of which is fitted with a stopcock connected to a vacuum line with an argon flow. The other two necks are fitted with rubber septa. After evacuating and backfilling with argon the flask is charged with a magnetic stir bar (3 cm length x 1.5 cm diameter), Rh$_2$(OAc)$_4$ (180 mg, 0.4 mmol, 0.008 equiv) (Note 5), 4-phenyl-1,2-butadiene (**1**) (19.5 g, 150 mmol, 3.00 equiv) (Note 6) and CH$_2$Cl$_2$ (40 mL) (Note 2) under argon. The mixture is heated to reflux using an oil bath (55 °C, external temperature) and after 5 min at reflux, a solution of ethyl α-diazoacetoacetate[2] (7.8 g, 50 mmol, 1 equiv) in CH$_2$Cl$_2$ (10 mL) is added using a syringe pump (10 mL/h). Upon completion of the addition, an additional quantity of Rh$_2$(OAc)$_4$ (40 mg, 0.1 mmol, 0.002 equiv) (Note 7) is added and the resulting mixture is refluxed for 2 h; the progress of the reaction is monitored by TLC (Note 3). Upon completion, the reaction is cooled to

328

room temperature and concentrated by rotary evaporation (20 mmHg, 35 °C water bath). The residue is purified by column chromatography (column ø = 80 mm with 300 g of silica gel (230-400 mesh); ~150 mL fractions; *n*-Hexane (1.5 L) (Note 2) is used to recover 4-phenyl-1,2-butadiene **1** (fractions 5-12, 7.64 g, 39%); then a mixture of 3:1 *n*-hexane:ethyl acetate (Note 2) (2 L) is used to elute 1-acetyl-2-phenethylidene-cyclopropanecarboxylic acid ethyl ester **2** (fractions 20-25, 15.1 g) (Note 8).

C. *1-Benzyl-2-methyl-3-(ethoxycarbonyl)-4-(2-phenylethyl)-1H-pyrrole (3)*. To a 100-mL round-bottomed flask is added, in succession, a magnetic stir bar (3 cm length x 1.5 cm diameter) (Note 1), crude product **2** (15.0 g), CH₃CN (15 mL, Note 2), benzylamine (9.8 mL, 90 mmol) (Note 2) and CH₃CN (15 mL). The flask is equipped with a reflux condenser and the top of the condenser left open to the air. After being heated with an oil bath (80 °C) for 19 h the reaction is judged complete as evidenced by TLC (Note 3). The resulting mixture is concentrated by rotary evaporation (20 mmHg, 35 °C water bath) and the residue is purified by column chromatography (column ø = 80 mm, 300 g of silica gel, 230-400 mesh). A mixture of petroleum ether (8.0 L) (Note 2) and ethyl acetate (0.2 L) (Note 2) are used as eluent, collecting ~150 mL fractions. Concentration of fractions 21-50 by rotary evaporation (20 mmHg, 40 °C water bath) affords 1-benzyl-2-methyl-3-(ethoxycarbonyl)-4-(2-phenylethyl)-1*H*-pyrrole **3** (9.05 g, 52% combined yield for steps B and C) as a light yellow solid (Notes 9 and 10).

2. Notes

1.   All glassware was thoroughly washed and dried in an oven (150 °C). Magnetic stir bars were washed with acetone and dried.

2.   The submitters prepared their own magnesium turnings from magnesium ingot (>99%) purchased from Sinopharm Chemical Reagent Co., Ltd and obtained a higher yield for the reaction (77% vs 57%). The checkers used magnesium turnings purchased from Sigma Aldrich. The submitters used anhydrous diethyl ether (≥99%) purchased from Shanghai Experimental Reagent Co., Ltd and distilled over sodium wire with diphenyl ketone as the indicator and the checkers used column-dried diethyl ether. The submitters used dichloromethane distilled from calcium hydride and the checkers used column-dried dichloromethane. Benzyl bromide (≥98%) was purchased from Sigma Aldrich and distilled from MgSO₄ before use. The

submitters purchased propargyl bromide (≥98%) from Shanghai Darui Finechemical Co., Ltd and distilled before use. The checkers purchased an 80 wt. % solution of propargyl bromide in toluene (Sigma-Aldrich) and used as received. 1,2-Dibromoethane (≥98%) was purchased from Sigma-Aldrich, petroleum ether from Fisher Scientific, and all the other chemicals from Sigma Aldrich. All chemicals were used as received unless specified above. Ethyl diazoacetoacetate was prepared using an *Organic Syntheses* method.[2]

3.   TLC analysis: Step A: R_f of compound **1** = 0.68 (eluent: petroleum ether); Step B: R_f of ethyl α-diazoacetoacetate = 0.39 (eluent: petroleum ether/EtOAc (10/1)) and visualized using UV (254 nm); R_f of compound **2** = 0.42 (eluent: petroleum ether/EtOAc (10/1)) and visualized with $KMnO_4$. Step C: R_f of compound **3** = 0.48 (eluent: petroleum ether/EtOAc (10/1)) and visualized using UV (254 nm) and an aqueous solution of $KMnO_4$.

4.   4-Phenyl-1,2-butadiene[6] was colorless after distillation under reduced pressure, but turned yellow after a few hours at room temperature. GC purity: 94.6% (conditions: Rtx-5MS column (30×0.25×0.25); oven: 50 °C, then 10 °C/min, 270 °C for 5 min; injector: 230 °C; EI, interphase temperature: 280 °C; split: 42:1; He: 3.0 mL/min). ^{1}H NMR (400 MHz, CDCl$_3$) δ: 3.36–3.40 (m, 2 H), 4.72–4.75 (m, 2 H), 5.30 (quint, J = 7.0 Hz, 1 H), 7.21-7.34 (m, 5 H); ^{13}C NMR (100 MHz, CDCl$_3$) δ: 35.3, 75.2, 89.7, 126.4, 128.5, 128.6, 140.4, 209.1; IR (neat) ν 3085, 3062, 3027, 2978, 2909, 2859, 1936, 1686, 1603, 1519, 1496, 1427, 1336, 1271, 1180, 1085, 1030, 968 cm^{-1}; MS (EI) m/z 131 (89, M+H$^+$), 103 (100), 91 (98), 77 (88), 65 (80).

5.   The submitters prepared rhodium(II) acetate dimer (Rh$_2$(OAc)$_4$) according to the procedure in *Inorganic Synthesis*.[3] The checkers purchased Rh$_2$(OAc)$_4$ from Strem Chemicals.

6.   The submitters reported an increase in the yield of this reaction with increasing quantities of 4-phenyl-1,2-butadiene: 2.0 equiv gave 41% yield; 3.0 equiv gave 53% yield, and 4 equiv gave 60% yield of **2**. After the reaction, excess 4-phenyl-1,2-butadiene was recovered by column chromatography.

7.   Rh$_2$(OAc)$_4$ should be added in two portions because ethyl α-diazoacetoacetate did not fully react when the same quantity of Rh$_2$(OAc)$_4$ was added in one portion.

8.   The crude 1-acetyl-2-phenethylidenecyclopropanecarboxylic acid ethyl ester **2** (a mixture of Z/E isomers,[9] ratio = 1.7:1) was used directly in the next step without further purification. A portion of this crude material (60.8 mg) was purified by chromatographic separation on silica gel with an

330

eluent of *n*-hexane:ethyl acetate = 30:1 for characterization data. ^{1}H NMR (300 MHz, CDCl$_3$) δ: 1.23–1.28 (m, 3 H), 2.07–2.32 (m, 2 H), 2.32–2.35 (m, 3 H), 3.50–3.56 (m, 2 H), 4.15–4.23 (m, 2 H), 6.01 (t, *J* = 7.2 Hz, 0.34 H, =CH), 6.05–6.12 (m, 0.66 H, =CH)], 7.13–7.31 (m, 5 H); ^{13}C NMR (100 MHz, CDCl$_3$) δ: 14.2 (2 C), 17.7, 19.0, 28.5, 28.7, 37.5, 37.9, 39.9 (2 C), 61.6 (2 C), 118.6 (2 C), 125.1, 125.2, 126.5 (2 C), 128.6 (2 C), 128.7 (2 C), 139.3, 139.4, 169.0, 169.1, 200.4, 200.8; IR (neat) ν 1686, 1496, 1427, 1356, 1273, 1235, 1207, 1180, 1129, 1084, 1030 cm^{-1}; MS *m/z* 259 (M+H$^+$, 100), 258 (39), 213 (13), 212 (7); HRMS calcd. for C$_{16}$H$_{19}$O$_3$: 259.1334. Found: 259.1332.

9.   *1-Benzyl-2-methyl-3-(ethoxycarbonyl)-4-(2-phenylethyl)-1H-pyrrole* (**3**): mp = 42 °C; ^{1}H NMR (400 MHz, CDCl$_3$) δ: 1.37 (t, *J* = 7.1 Hz, 3 H), 2.44 (s, 3 H), 2.88–2.92 (m, 2 H), 2.99–3.03 (m, 2 H), 4.32 (q, *J* = 7.1 Hz, 2 H), 4.97 (s, 2 H), 6.30 (s, 1 H), 6.96 (d, *J* = 6.7 Hz, 2 H), 7.16–7.34 (m, 8 H); ^{13}C NMR (100 MHz, CDCl$_3$) *δ*: 11.8, 14.7, 29.1, 37.2, 50.4, 59.3, 111.3, 119.7, 124.9, 125.7, 126.5, 127.7, 128.2, 128.7, 128.9, 136.6, 137.3, 142.8, 166.2; MS (EI) *m/z* 348 (M+H$^+$, 100), 347 (4), 302 (2), 256 (3); IR (neat) ν 1950, 1686, 1642, 1564, 1506, 1496, 1445, 1428, 1355, 1339, 1273, 1235, 1180, 1129, 1084, 1030 cm^{-1}; HRMS (EI) calcd for C$_{23}$H$_{26}$NO$_2$$^+$ (M$^+$): 348.1964; Found: 348.1954; Anal. calcd. for C$_{23}$H$_{25}$NO$_2$: C, 79.51; H, 7.25; N, 4.03; Found: C, 79.67; H, 7.15; N, 3.89.

10. This compound was recrystallized from petroleum ether and EtOAc for melting point determination. For recrystallization, **3** (0.53 g) was dissolved in EtOAc (1 mL) with heating. Then petroleum ether (20 mL) was added. The solution was sealed and kept at –8 °C for one day. The crystals (65 mg, 12%) were collected via filtration and 465 mg (88%) was recovered by evaporation of the mother liquor.

Handling and Disposal of Hazardous Chemicals

The procedures in this article are intended for use only by persons with prior training in experimental organic chemistry.   All hazardous materials should be handled using the standard procedures for work with chemicals described in references such as "Prudent Practices in the Laboratory" (The National Academies Press, Washington, D.C., 2011 www.nap.edu).   All chemical waste should be disposed of in accordance with local regulations. For general guidelines for the management of chemical waste, see Chapter 8 of Prudent Practices.

These procedures must be conducted at one's own risk. *Organic Syntheses, Inc.*, its Editors, and its Board of Directors do not warrant or guarantee the safety of individuals using these procedures and hereby disclaim any liability for any injuries or damages claimed to have resulted from or related in any way to the procedures herein.

3. Discussion

Pyrroles are one of the most prevalent heterocyclic compounds. In recent years, polysubstituted pyrroles with various substituents have been prepared.[4] However, there are very limited reports on syntheses of 2,3,4-trisubstituted pyrroles.[5] In this contribution, we have demonstrated an intermolecular cyclization reaction of alkylidenecyclopropyl ketones and amines which provides an efficient route to 2,3,4-trisubstituted pyrroles.

4-Phenyl-1,2-butadiene was first prepared by the group of Hirao et al. by first generating an organovanadium compound from reaction of benzylmagnesium bromide with VCl$_3$, then reacting the resulting species with propargyl bromide to afford the allene in 31% yield with 4% of 4-phenyl-1-butyne as a byproduct.[6] Other methods of generating alkyl- and aryl-substituted allenes have been mediated by CuX.[7] However, we found 4-phenyl-1,2-butadiene could be prepared without CuX or any additional metal as catalyst. New glassware was used to ensure no residual Cu(I) was present as a contaminant. Initial efforts using this procedure resulted in the formation of >10% of 4-phenyl-1-butyne.[7] However, its formation was diminished by slow addition of the Grignard reagent into propargyl bromide. The preparation of benzyl magnesium bromide is very important to ensure a higher yield. The checkers did not see the formation of 4-phenyl-1-butyne but instead observed a 16% recovery of propargyl bromide and an 11% yield of 1,2-diphenylethane as estimated by crude NMR, due to the less efficient formation of the benzyl magnesium bromide forming the benzyl bromide-homocoupling product 1,2-diphenylethane leading to the recovery of the propargyl bromide in the subsequent step.

We have prepared various 2,3,4-trisubstituted pyrroles via reaction of alkylidenecyclopropyl ketones with amines (Table 1).[8] In step B, the alkylidenecyclopropylketone products are contaminated with some minor by-products, but the crude product obtained by simple filtration through a column of silica gel can be used directly for the next step. In the multi-gram scale reaction of step C, moisture is easily controlled; thus, the reaction can be conducted without MgSO₄.

Table 1. Intermolecular Cyclization of Alkylidenecyclopropyl Ketones with Amines Affording 2,3,4-Trisubstitued Pyrroles

entry	R¹	R²	R³	R⁴	yield (%)
1	C_7H_{15}	COOEt	CH_3	Bn	78
2	C_4H_9	COOEt	CH_3	Bn	78
3	Bn	COOEt	CH_3	Bn	75
4	C_7H_{15}	$COCH_3$	CH_3	Bn	64
5	C_7H_{15}	SO_2Ph	CH_3	Bn	75
6	C_4H_9	SO_2Ph	CH_3	Bn	82
7	C_7H_{15}	COOEt	CF_3	Bn	40
8	C_7H_{15}	COOEt	Ph	p-MeOBn	67
9	C_7H_{15}	COOEt	CH_3	p-MeOBn	86
10	Bn	COOEt	CH_3	p-MeOBn	77
11	C_7H_{15}	COOEt	CH_3	n-C_4H_9	50
12	C_7H_{15}	COOEt	CH_3	t-C_4H_9	50

Two possible mechanistic pathways can be envisioned to afford the corresponding pyrroles. One pathway starts from nucleophilic attack of the amine at the less sterically hindered carbon atom of the 3-membered ring to afford intermediate **4** (path a). The nucleophilic nitrogen then reacts with the carbonyl group leading to 3-alkylidene-5-hydroxy tetrahydropyrrole intermediate **5**. Subsequent dehydration and aromatization generates product **3** via the intermediacy of **6**. An alternative pathway starts with the intermolecular condensation of **1** and **2** to afford the cyclopropylimine intermediate **7** (path b). Cloke-type rearrangement of intermediate **7** readily leads to ring expansion via the subsequent nucleophilic attack of the nitrogen atom at the less sterically hindered carbon atom in the cyclopropane ring, which causes the distal cleavage to form **6**.[10] Subsequent aromatization affords product **3**.

Scheme 1. Proposed reaction mechanism

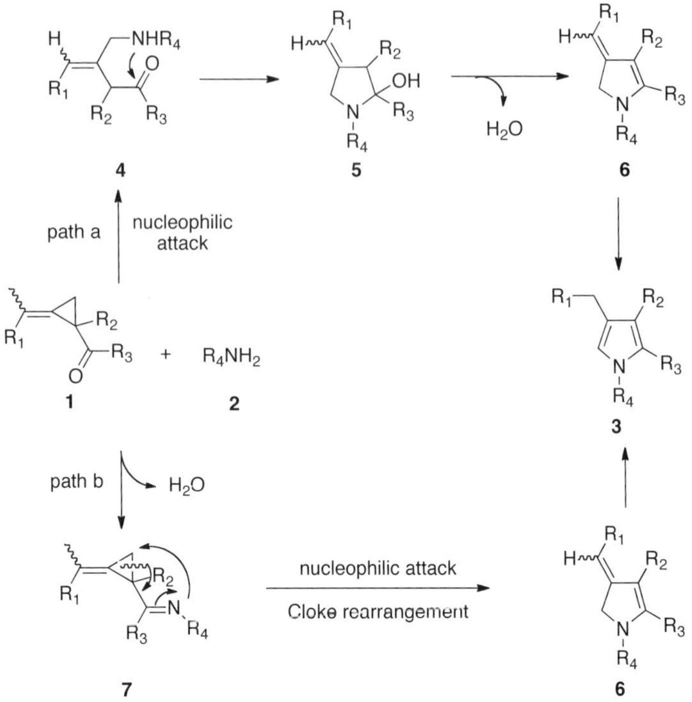

1. State Key Laboratory of Organometallic Chemistry, Shanghai Institute of Organic Chemistry, Chinese Academy of Sciences, 345 Lingling Road, Shanghai, China. E-mail: masm@sioc.ac.cn. We thank the National Natural Science Foundation of China (21232006) and the National Basic Research Program of China (2009CB825300) for financial support.

2. Davies, H. M. L.; Cantrell, Jr. W. R.; Romines, K. R.; Baum, J. S. *Org. Synth.* **1992**, *70*, 93.
3. Rempel, G. A.; Legzdius, P.; Smith, H.; Wilkinson, G. *Inorg. Synth.* **1972**, *13*, 90.
4. For recent examples, see: (a) Ngwerume, S.; Camp, J. *J. Org. Chem.* **2010**, *75*, 6271. (b) Saito, A.; Konishi, T.; Hanzawa, Y. *Org. Lett.* **2010**, *12*, 372. (c) Ackermann, L.; Sandmann, R.; Kaspar, L. T. *Org. Lett.* **2009**, *11*, 2031. (d) Herath, A.; Cosford, N. D. P. *Org. Lett.* **2010**, *12*, 5182. (e) Morin, M. S. T.; St-Cyr, D. J.; Arndtsen, B. A. *Org. Lett.*

2010, *12*, 4916. (f) Donohoe, T. J.; Race, N. J.; Bower, J. F.;   Callens, C. K. A. *Org. Lett.* **2010**, *12*, 4094. (g) Rakshit, S.; Patureau, F. W.; Glorius, F. *J. Am. Chem. Soc.* **2010**, *132*, 9585.

5.   For recent examples, see: (a) St-Cyr, D. J.; Morin, M. S. T.; Bélanger-Gariépy, F.; Arndtsen, B. A.; Krenske, E. H.; Houk, K. N. *J. Org. Chem.* **2010**, *75*, 4261. (b) Liao, Q.; Zhang, L.; Wang, F.; Li, S.; Xi, C. *Eur. J. Org. Chem.* **2010**, *28*, 5426.

6.   Hirao, T.; Misu, D.; Yao, K.; Agawa, T. *Tetrahedron Lett.* **1986**, *27*, 929.

7.   VenKruijsse, H. D.; Brandsma, L. *Synthesis of Acetylenes, Allenes and Cummulenes. A Laboratory Manual;* Elsevier, Amsterdam, **1981**.

8.   Lu, L.; Chen, G.; Ma, S. *Org. Lett.* **2006**, *8*, 835.

9.   Ma, S.; Lu, L.; Zhang, J. *J. Am. Chem. Soc.* **2004**, *126*, 9645.

10. Cloke, J. B. *J. Am. Chem. Soc.* **1929**, *51*, 1174.

Appendix
Chemical Abstracts Nomenclature; (Registry Number)

Magnesium; (7439-95-4)

Benzene, (bromomethyl)-; (100-39-0)

1-Propyne, 3-bromo-; (106-96-7)

Magnesium, bromo(phenylmethyl)-; (1589-82-8)

Benzenemethanamine; (100-46-9)

Butanoic acid, 2-diazo-3-oxo-, ethyl ester; (2009-97-4)

Rhodium, tetrakis[μ-(acetato-κO:κO')]di-, (Rh-Rh); (15956-28-2)

Ethane, 1,2-dibromo-; (106-93-4)

Prof. Shengming Ma was born in 1965 in Zhejiang, China. He graduated from Hangzhou University (1986) and received his Ph.D. degree from Shanghai Institute of Organic Chemistry (1990). He became an assistant professor of SIOC in 1991. After his postdoctoral appointments at ETH with Prof. Venanzi and Purdue University with Prof. Negishi from 1992–1997, he joined the faculty of SIOC in 1997. From February 2003 to September 2007, he was jointly appointed by SIOC and Zhejiang University. He works for East China Normal University and SIOC from October 2007.

Shengjun Ni was born in Jining, China. He received his B.S. degree in applied chemistry from China University of Petroleum (CUP) in 2007. He is now pursuing his M.S. degree in Shanghai Institute of Organic Chemistry (SIOC), Chinese Academy of Sciences under the supervision of Prof. Shengming Ma. His research interest is transition metal catalyst in modern organic synthesis.

Can Zhu was born in Nantong, China. He received his B.Sc. in chemistry from University of Science and Technology of China (USTC) in 2009. He is currently pursuing a Ph.D. in Shanghai Institute of Organic Chemistry (SIOC) under the supervision of Prof. Shengming Ma, working on the application of transition metal catalysis in modern organic synthesis.

Jie Chen was born in Nanyang, China. She received her B.Sc. in chemistry from Henan University in 2005. In 2010, she obtained her Ph.D. under the supervision of Prof. Shengming Ma from Shanghai Institute of Organic Chemistry (SIOC), Chinese Academy of Sciences. In 2011, she joined the laboratory of Prof. Joseph M. Ready as a postdoctoral fellow at the University of Texas Southwestern Medical Center (UTSW).

Mélanie Charpenay was born in 1986 in Voiron, France. She studied organic chemistry at the European Engineering School of Chemistry, Polymers and Materials Science in Strasbourg and graduated in 2009. The same year, she obtained her Masters Degree at the University of Strasbourg. Then she carried out Ph.D. studies at the same university under the supervision of Dr. Jean Suffert, focusing on the development of new synthetic methodologies to reach highly strained polycyclic structures such as fenestradienes and cyclooctatrienes using palladium catalyzed cascade reactions. She is currently working as a postdoctoral fellow at the University of Pittsburgh with Professor Kay Brummond and group on the synthesis of biologically relevant fluorescent probes.

Synthesis of Highly Enantiomerically Enriched Amines by Asymmetric Transfer Hydrogenation of N-(tert-Butylsulfinyl)Imines

A.

B.

C.

Submitted by David Guijarro, Óscar Pablo and Miguel Yus.[1]
Checked by Chintan S. Sumaria and Viresh H. Rawal.

1. Procedure

A. (R_S)-N-[1-Phenylethylidene]-2-methylpropane-2-sulfinamide   (3).
An oven-dried 100-mL two-necked round-bottomed flask containing an octagonal Teflon-coated magnetic stirring bar (3.2 × 0.8 cm), fitted with a vacuum adapter and a rubber septum, is charged with acetophenone (1) (6.01 g, 50.0 mmol, 1.0 equiv) and (R_S)-2-methylpropane-2-sulfinamide (2) (6.06 g, 50.0 mmol, 1.0 equiv) (Note 1). After applying vacuum followed by flushing with argon, the stirred mixture is treated with Ti(OEt)$_4$ (21.3 mL, 22.8 g, 100 mmol, 2.0 equiv) (Note 2), added by syringe. At this point, the reaction shows an intense yellow color. The rubber septum is replaced with a glass stopper, and the stirred (700 rpm) mixture is heated so as to maintain the temperature of the oil bath at 72 °C for 4 h (Note 3), during which period

338

Org Synth. 2013, 90, 338-349
Published on the Web 7/31/2013
© 2013 Organic Syntheses, Inc.

its color changes to orange. The reaction mixture is cooled to room temperature using a water bath, diluted with ethyl acetate (40 mL), and then transferred to a 250-mL Erlenmeyer flask containing a mixture of ethyl acetate:brine (25 mL:25 mL). The resulting pale yellow suspension is filtered through a short plug of Celite (Note 4), which is washed with ethyl acetate (about 1000 mL total) until no more product elutes (as checked by TLC). The organic layer is concentrated by rotary evaporation (40–80 mmHg, 28 °C bath temperature) to afford crude imine **3** as a yellow oil (Note 5). The crude material is purified by flash column chromatography (24 × 4.5 cm column containing 160 g of silica gel) (Note 6). A mixture of hexanes:ethyl acetate is used as the eluent, with a gradient of polarity as follows: hexanes:ethyl acetate 100:0 (about 500 mL), 20:1 (about 300 mL), 15:1 (about 300 mL), 9:1 (about 600 mL), 3:2 (about 300 mL), and finally 1:1 until no more product elutes (about 300 mL) (Notes 7 and 8). Fractions containing pure product are concentrated by rotary evaporation (40-80 mmHg, 28 °C bath temperature) and the resultant product is then subjected to high vacuum (5 mmHg, 40 °C bath temperature) to afford imine **3** (9.54–9.78 g, 85–88% yield) as a yellow solid (Notes 9 and 10).

B. (R,R$_S$)-N-(1-Phenylethyl)-2-methylpropane-2-sulfinamide (**4**). An oven-dried 500-mL two-necked round-bottomed flask is equipped with an octagonal Teflon-coated magnetic stirring bar (3.8 × 0.9 cm), a condenser topped with a vacuum adapter with an argon inlet, and a rubber septum. After applying vacuum and flushing with argon, the flask is charged with [RuCl$_2$(p-cymene)]$_2$ (490 mg, 0.8 mmol, 0.025 equiv), 2-amino-2-methylpropan-1-ol (143 mg, 1.6 mmol, 0.05 equiv), and 4 Å MS (3 g) while maintaining a positive pressure of argon (Notes 11 and 12). The system is again evacuated and filled with argon [5 cycles vacuum (6 seconds), then argon], and a positive pressure of argon is maintained throughout the remainder of the experiment. Dry isopropyl alcohol (25 mL) is added using a syringe (Note 13) and the rubber septum is replaced by a glass stopper. The stirred (450 rpm) reaction mixture is then placed in a preheated oil bath (bath temperature 86 °C) for 25 min, resulting in a color change to cherry-red (Note 14). The oil-bath is replaced with a water-bath (preheated to 50 °C) and heating is adjusted to maintain the temperature at about 50 °C (Note 15). After replacing the glass stopper with a rubber septum, a previously prepared solution of imine **3** (7.147 g, 32.0 mmol, 1.0 equiv) in dry isopropyl alcohol (160 mL) under argon is added to the reaction mixture using a cannula (over about 10 min), and the stirring speed is increased to

600 rpm. The temperature of the system is allowed to stabilize at 50 °C over approximately 5 minutes, then a previously prepared solution of potassium *tert*-butoxide (449 mg, 4.0 mmol, 0.125 equiv) in dry isopropyl alcohol (40 mL) is added to the reaction mixture using a syringe (Note 16). The resulting brownish mixture is stirred at 50 °C (external temperature) for 2 h (Note 17). The reaction mixture is cooled to room temperature using a water bath for 15 min, then filtered through a short plug of silica gel (Note 18), which is washed with ethyl acetate (about 600 mL total) until no more product elutes (as checked by TLC). In order to ensure the removal of the ruthenium complex, solvents are removed by rotary evaporation (40–80 mmHg, 28 °C bath temperature) and the residue is dissolved in 40 mL of ethyl acetate and passed through a short plug of silica gel (Note 19), which is washed with ethyl acetate (about 500 mL total) until no more product elutes (as checked by TLC). The filtrate is concentrated by rotary evaporation (40–80 mmHg, 28 °C bath temperature) and then subjected to high vacuum (5 mmHg, 40 °C bath temperature) to afford pure sulfinamide **4** (7.20–7.22 g, >99% yield) as a brown oil (Notes 20, 21, and 22).

C. *(R)-1-Phenylethanamine* (**5**). A 250-mL one-necked round-bottomed flask is equipped with an octagonal Teflon-coated magnetic stirring bar (3.2 × 0.8 cm) and sealed with a septum fitted with needle connected to a vacuum/nitrogen line. The flask is purged using a vacuum followed by nitrogen introduction (3 times) and maintained under a positive pressure of nitrogen. It is then charged with **4** (6.76 g, 30.0 mmol, 1.0 equiv) and dry methanol (60 mL) (Note 23) using a syringe. The solution is cooled to 0−5 °C using an ice-water bath. While stirring, a 2M solution of HCl in diethyl ether (75 mL, 150 mmol, 5.0 equiv) is added by syringe (Note 24). After 5 min, the cooling bath is removed and the reaction mixture is stirred (600 rpm) at room temperature for 3 h. The solvent is removed by rotary evaporation (10–80 mmHg, 28 °C bath temperature). The residue is dissolved in aqueous 2M HCl solution (100 mL) then extracted with ethyl acetate (2 × 80 mL). While the combined organic layers are discarded, the acidic aqueous layer is cooled to 0−5 °C using an ice-water bath and treated with aqueous 6M NaOH solution (100 mL), which is added in portions over 5 min while stirring, resulting in a pH > 12. The resulting mixture is extracted with ethyl acetate (3 × 50 mL) and the combined organic layers are washed with H_2O (25 mL) and brine (25 mL). The organic layer is dried over anhydrous magnesium sulfate (4 g), filtered, then concentrated by rotary evaporation (40–80 mmHg, 28 °C bath temperature) (Note 25) to afford pure amine **5** (3.09–3.22 g, 85–88% yield, 95% *ee* of the (*R*)-enantiomer) as a pale yellow oil (Notes 26 and 27).

2. Notes

1. Acetophenone **1** (98%) was purchased from Acros Organics (Ref. No. 102412500). (*R*$_S$)-2-Methylpropane-2-sulfinamide **2** (99%) was purchased from AK Scientific, Inc. (Ref. No. 70308). Both compounds were used as received.

2. Titanium tetraethoxide (33-35% TiO$_2$) was purchased from Alfa Aesar (Ref. 77142) and used as received.

3. The reaction progress can be monitored by TLC analysis on Merck Silica Gel 60 F$_{254}$ glass plates (Ref. 1057150001) and visualization with a UV lamp. Using hexanes:ethyl acetate (7:3) as eluent, acetophenone **1** has an R$_f$ = 0.64, sulfinamide **2** has an R$_f$ = 0 and *N*-(*tert*-butylsulfinyl)imine **3** has an R$_f$ = 0.30.

4. The dimensions of the plug of Celite: 8 cm diameter × 2 cm height.

5. ^{1}H NMR analysis of the reaction crude showed that the conversion to imine **3** was around 90%.

6. Silica gel 60, 230-400 mesh (pore size 60 Å, particle size 40-63 µm) purchased from Sorbent Technologies was used.

7. Imine **3** has an intense yellow color while the other reagents are colorless, allowing visual monitoring of product elution. When the yellow band reaches the bottom of the column (after about 500 mL of 9:1 hexanes:ethyl acetate) fractions of 20 mL are collected using 30 mL test-tubes until no more product is eluted.

8. The column fractions were checked by TLC analysis as described in Note 3.

9. The submitters note that if **3** does not solidify after removing the solvent under reduced pressure, the resulting oil solidifies to a yellow solid when kept in a freezer (under argon atmosphere, ca. −25 °C) overnight.

10. The physical and spectroscopic properties of imine **3** are as follows:[2] mp 36 °C; ^{1}H NMR (500 MHz, CDCl$_3$) δ: 1.33 (s, 9 H), 2.77 (s, 3 H), 7.38–7.52 (m, 3 H), 7.89 (d, *J* = 7.5 Hz, 2 H); ^{13}C NMR (125 MHz, CDCl$_3$) δ: 19.7, 22.4 (3 C), 57.3, 127.1 (2 C), 128.3 (2 C), 131.6, 138.6, 176.3; IR (neat, cm^{-1}) 2979, 1606, 1594, 1572, 1276, 1089, 1067; HRMS (ES+) m/z calcd for C$_{12}$H$_{17}$NOSNa (M$^+$ + Na): 246.0929 Found: 246.0927. [α]^{20}$_D$ −14.4 (c 1.09, CH$_2$Cl$_2$), {Submitters: [α]^{20}$_D$ −13.0 (c 2.22, CH$_2$Cl$_2$)}.

Anal. calcd. for C$_{12}$H$_{17}$NOS: C, 64.53; H, 7.67; N, 6.27. Found: C, 64.76; H, 7.57; N, 6.30.

11. [RuCl$_2$(*p*-cymene)]$_2$ (98%) was purchased from Strem Chemicals (Ref. 44-0190). 2-Amino-2-methylpropan-1-ol (≥95%) was purchased from Sigma-Aldrich Chemical Co. (Ref. A9199). Both were used as received.

12. Molecular sieves (4 Å; 8-12 mesh beads), purchased from Fischer Scientific (Ref. M514-500), were activated before use by heating at 190 °C for 3 h in a Kugelrohr apparatus under reduced pressure (5 mmHg).

13. Isopropyl alcohol (puriss, absolute over MS, ≥99.5%) (H$_2$O ≤ 0.005%) was purchased from Sigma-Aldrich Chemical Co. (Ref. 59309) and used as received.

14. The submitters report that ruthenium complex formation was incomplete if the oil-bath temperature was lower than 82 °C.

15. The change from an oil-bath to a water-bath speeds up cooling of the reaction mixture to 50 °C.

16. Potassium *tert*-butoxide (≥98%) was purchased from Sigma-Aldrich Chemical Co. (Ref. 156671). It was activated before use by heating for 4 h at 175 °C in a Kugelrohr apparatus under reduced pressure (5 mmHg).

17. The reaction progress can be monitored by TLC analysis on Merck Silica Gel 60 F$_{254}$ glass plates (Ref. 1057150001) and visualized with phosphomolibdic acid (5% in EtOH). Using hexanes:ethyl acetate (7:3) as eluent, imine **3** has an R$_f$ = 0.30 and the reduction product **4** has an R$_f$ = 0.15.

18. Silica gel 60 plug (230-400 mesh, pore size 60 Å, particle size 40-63 μm): 6 cm diameter × 7 cm height.

19. Silica gel 60 plug (230-400 mesh, pore size 60 Å, particle size 40-63 μm): 5 cm diameter × 5 cm height.

20. ^{1}H NMR and ^{13}C NMR analysis of the isolated crude material indicated that imine **3** had cleanly and completely converted to sulfinamide **4**, making further purification unnecessary.

21 The physical and spectroscopic properties of compound **4** are as follows:[3] ^{1}H NMR (500 MHz, CDCl$_3$) δ: 1.24 (s, 9 H), 1.51 (d, J = 6.5 Hz, 3 H), 3.42 (br s, 1 H), 4.55 (qd, J = 6.5, 3.0 Hz, 1 H), 7.27–7.38 (m, 5 H); ^{13}C NMR (125 MHz, CDCl$_3$) δ: 22.6 (3 C), 22.7, 53.9, 55.4, 126.5 (2 C), 127.8, 128.7 (2 C), 144.0; IR (neat, cm^{-1}) 3214, 2974, 2927, 1455, 1363, 1055, 938, 762, 700; HRMS (ES+) *m/z* calcd for C$_{12}$H$_{19}$NOSNa (M$^+$ + Na): 248.1085 Found: 248.1087. Checkers: [α]$^{20}_D$ −40.6 (c 1.2, EtOAc); Submitters: [α]$^{20}_D$

−42.5 (c 1.9, EtOAc). Anal. Calcd. For $C_{12}H_{19}NOS$: C, 63.96; H, 8.50; N, 6.22. Found: C, 63.67; H, 8.34; N, 6.29.

22. A 300 mg sample was purified by column chromatography (6 × 1.5 cm column containing 6 g of silica gel). A mixture of hexanes:ethyl acetate was used as eluent, with a gradient of polarity as follows: 7:3 (10 mL), 3:2 (20 mL), 1:1 (20 mL), and 2:3 until no more product elutes (about 50 mL). Compound **4** was isolated by rotary evaporation (276 mg, 92 % recovery). ^{1}H and ^{13}C NMR spectra, as well as $[\alpha]_D^{20}$, were identical to the crude product (Note 21). Anal. Calcd. For $C_{12}H_{19}NOS$: C, 63.96; H, 8.50; N, 6.22. Found: C, 64.03; H, 8.49; N, 6.30.

23. Anhydrous methanol (99.8%) was purchased from Sigma-Aldrich Chemical Co. (Ref. 322415) and used as received.

24. The 2M solution of HCl in diethyl ether was purchased from Alfa Aesar (Ref. H26914) and used as received.

25. **Important:** The final product <u>should not</u> be concentrated to dryness. The concentration should be halted with about 5 mL of liquid remaining. Final solvent removal was carried out by gently blowing a stream of nitrogen or argon over the top of the liquid at atmospheric pressure for 5-10 min, until NMR showed the clean product.

26. The physical and spectroscopic properties of amine **5** are as follows:[4] ^{1}H NMR (300 MHz, $CDCl_3$) δ: 1.39 (d, J = 6.5 Hz, 3 H), 1.56 (br s, 2 H), 4.11 (q, J = 6.5 Hz, 1 H), 7.20–7.27 (m, 1 H), 7.30–7.39 (m, 4 H); ^{13}C NMR (75 MHz, $CDCl_3$) δ: 25.6, 51.3, 125.6 (2 C), 126.7, 128.4 (2 C), 147.8; IR (neat, cm^{-1}) 3364, 3029, 2969, 1617, 1559, 1474, 1450, 1376, 1027, 764, 696; HRMS (ES+) m/z calcd for $C_8H_{11}NNa$ (M$^+$ + Na): 144.0789 Found: 144.0780. Submitters: $[\alpha]_D^{20}$ +31.5 (c 2.7, CHCl$_3$); Aldrich {Ref. 77879, for (R)-**5** (99% ee)}: $[\alpha]_D^{20}$ +33.8 (c 1.2, CHCl$_3$).

27. The enantiomeric excess of amine **5** was evaluated as follows. A solution of **5** (24 mg, 0.2 mmol) in dichloromethane (5 mL) was cooled to 0−5 °C using an ice-water bath. An aqueous 2M NaOH solution (5 mL) was added followed by benzoyl chloride (46 µL, 0.4 mmol), and the resulting mixture was stirred for 4 h at room temperature. The organic layer was separated and washed with aqueous 2M NaOH solution (2 × 5 mL). The organic layer was dried over anhydrous sodium sulfate, filtered, and evaporated to dryness under reduced pressure (40–80 mmHg), to afford the corresponding benzamide **6** (40 mg, 89% yield) as a white solid. Compound **6** was analyzed by HPLC on an Agilent 1100-series instrument equipped with a Chiracel OD-H column (25 × 0.46 cm), using 10% isopropyl alcohol

in *n*-hexane as eluent, a flow rate of 0.5 mL/min and UV detection at 240 nm. The retention times of the two enantiomers were 27.2 min (*R*, 97.4%) and 39.9 min (*S*, 2.6%). The racemic amine *rac*-**5** was prepared by reductive alkylation of ammonia by acetophenone following a literature procedure[5] and was benzoylated as described above.

6

Handling and Disposal of Hazardous Chemicals

The procedures in this article are intended for use only by persons with prior training in experimental organic chemistry. All hazardous materials should be handled using the standard procedures for work with chemicals described in references such as "Prudent Practices in the Laboratory" (The National Academies Press, Washington, D.C., 2011 www.nap.edu). All chemical waste should be disposed of in accordance with local regulations. For general guidelines for the management of chemical waste, see Chapter 8 of Prudent Practices.

These procedures must be conducted at one's own risk. *Organic Syntheses, Inc.*, its Editors, and its Board of Directors do not warrant or guarantee the safety of individuals using these procedures and hereby disclaim any liability for any injuries or damages claimed to have resulted from or related in any way to the procedures herein.

3. Discussion

The diastereoselective reduction of *N*-(*tert*-butylsulfinyl)ketimines is among the most useful methods for the preparation of enantiomerically enriched amines. The asymmetric reduction of this type of imine has been performed using boranes,[6a,6b,7] sodium or lithium borohydrides,[6,7,8] aluminium hydrides,[6b,7,8] and diethylzinc in the presence of Ni(acac)$_2$.[9] The present procedure utilizes a methodology based on asymmetric transfer hydrogenation, which is a very convenient reduction protocol because it avoids the handling of molecular hydrogen or metallic hydrides, among other reasons. Thus, highly enantiomerically enriched α-branched primary

344

amines are obtained in excellent yields by the diastereoselective reduction of a variety of sulfinylimines followed by desulfinylation of the nitrogen atom.[10] The *N*-(*tert*-butylsulfinyl)imines are prepared by reaction of stoichiometric amounts of the corresponding ketones and 2-methylpropane-2-sulfinamide in the presence of two equivalents of Ti(OEt)$_4$ under solvent-free conditions.[11] The obtained imines **3** are reduced by asymmetric transfer hydrogenation using isopropyl alcohol as hydrogen source and, as a catalyst, a ruthenium complex bearing a commercially available simple achiral β-aminoalcohol: 2-amino-2-methylpropan-1-ol. This catalyst has been shown to be very versatile, giving good results for the reduction of a variety of aromatic, heteroaromatic and aliphatic sulfinylketimines, including several examples of sterically congested imines. The reduction products **4** are treated with a solution of HCl in a mixture of Et$_2$O and MeOH to afford the

Table 1. Other primary amines prepared by this procedure

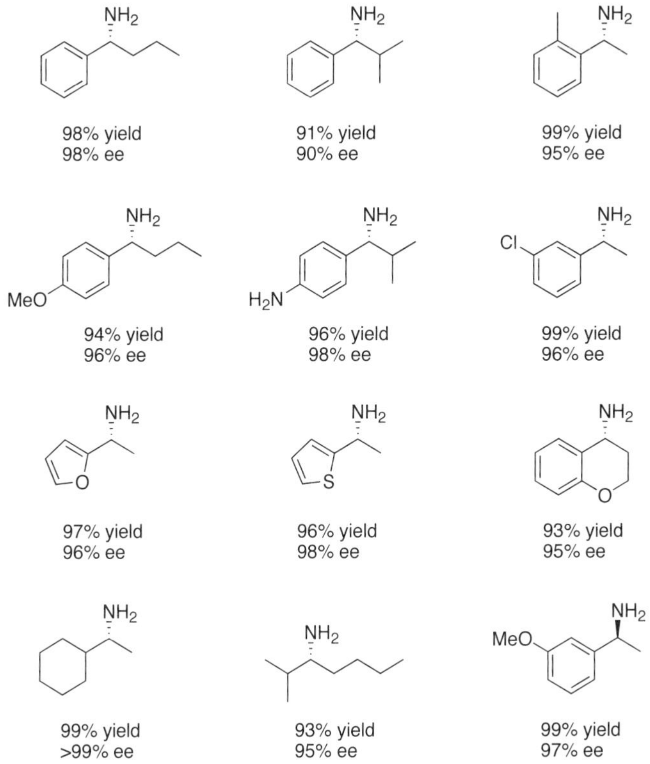

98% yield
98% ee

91% yield
90% ee

99% yield
95% ee

94% yield
96% ee

96% yield
98% ee

99% yield
96% ee

97% yield
96% ee

96% yield
98% ee

93% yield
95% ee

99% yield
>99% ee

93% yield
95% ee

99% yield
97% ee

corresponding free primary amines **5** in excellent yields and enantiomeric excesses (Table 1).

It is worth noting that very good results are obtained for the reduction of the less reactive aliphatic imines. Our procedure represents a general method for the synthesis of highly enantiomerically enriched aromatic and aliphatic primary amines. The enantiomeric (*S*)-amines are also readily accessible with the same optical purity using the same ruthenium catalyst by changing the absolute configuration of the sulphur atom in the starting sulfinamide (see the last example in Table 1).

The synthetic utility of some of the obtained α-branched primary amines has already been shown. For instance, (*S*)-1-(3-methoxyphenyl)ethanamine has been used as a key intermediate in a short total synthesis of (*S*)-Rivastigmine, a cholinesterase inhibitor used for the treatment of moderate dementia in patients with Alzheimer or Parkinson diseases.[12]

(*S*)-Rivastigmine

1. Departamento de Química Orgánica, Facultad de Ciencias and Instituto de Síntesis Orgánica (ISO), Universidad de Alicante, Apdo. 99, 03080 Alicante, Spain. E-mail: dguijarro@ua.es; oscar.pablo@ua.es; yus@ua.es. We thank the Spanish Ministerio de Ciencia e Innovación (MICINN; grant no. CONSOLIDER INGENIO 2010-CSD2007-00006, CTQ2007-65218 and CTQ11-24151), the Generalitat Valenciana (Grant No. PROMETEO/2009/039 and FEDER), and the University of Alicante for generous and continuous financial support, as well as MEDALCHEMY S.L. for a gift of chemicals. O. P. thanks the Spanish Ministerio de Educación for a predoctoral fellowship (Grant no. AP-2008-00989).

2. Imine **3** is a known compound. See, for instance: Colyer, J. T.; Andersen, N. G.; Tedrow, J. S.; Soukup, T. S.; Faul, M. M. *J. Org. Chem.* **2006**, *71*, 6859.

3. Sulfinamide **4** is a known compound. See, for instance: Cano, R.; Ramón, D. J.; Yus, M. *J. Org. Chem.* **2011**, *76*, 5547.

346

4. The physical and spectroscopic properties of the obtained amine **5** matched those of a commercially available sample (99% *ee*) purchased from Aldrich (Ref. 77879).
5. Miriyala, B.; Bhattacharyya, S.; Williamson, J. S. *Tetrahedron* **2004**, *60*, 1463.
6. See, for instance: (a) Ellman, J. A. *Pure & Appl. Chem.* **2003**, *75*, 39. (b) Zhou, P.; Chen, B.-C.; Davis, F. A. *Tetrahedron* **2004**, *60*, 8003. (c) Ferreira, F.; Botuha, C.; Chemla, F.; Pérez-Luna, A. *Chem. Soc. Rev.* **2009**, *38*, 1162.
7. See, for instance: (a) Dutheuil, G.; Couve-Bonnaire, S.; Pannecoucke, X. *Angew. Chem. Int. Ed.* **2007**, *46*, 1290. (b) Liu, Z.-J.; Liu, J.-T. *Chem. Commun.* **2008**, 5233. (c) Martjuga, M.; Shabashov, D.; Belyakov, S.; Liepinsh, E.; Suna, E. *J. Org. Chem.* **2010**, *75*, 2357.
8. See, for instance: (a) Peltier, H. M.; Ellman, J. A. *J. Org. Chem.* **2005**, *70*, 7342. (b) Tanuwidjaja, J.; Peltier, H. M.; Ellman, J. A. *J. Org. Chem.* **2007**, *72*, 626. (c) Denolf, B.; Leemans, E.; De Kimpe, N. *J. Org. Chem.* **2007**, *72*, 3211.
9. Xiao, X.; Wang, H.; Huang, Z.; Yang, J.; Bian, X.; Qin, Y. *Org. Lett.* **2006**, *8*, 139.
10. Pablo, O.; Guijarro, D.; Kovács, G.; Lledós, A.; Ujaque, G.; Yus, M. *Chem. Eur. J.* **2012**, *18*, 1969.
11. Collados, J. F.; Toledano, E.; Guijarro, D.; Yus, M. *J. Org. Chem.* **2012**, *77*, 5744.
12. Mangas-Sánchez, J.; Rodríguez-Mata, M.; Busto, E.; Gotor-Fernández, V.; Gotor, V. *J. Org. Chem.* **2009**, *74*, 5304.

Appendix
Chemical Abstracts Nomenclature; (Registry Number)

Acetophenone: 1-phenyl-ethanone; (98-86-2)

(R_S)-2-Methylpropane-2-sulfinamide: [$S(R)$]-2-methyl-2-propanesulfinamide; (196929-78-9)

Titanium(IV) ethoxide: tetraethyl titanate; (3087-36-3)

[RuCl$_2$(*p*-cymene)]$_2$: Di-μ-chlorodichlorobis[(1,2,3,4,5,6-μ)-1-methyl-4-(1-methylethyl)benzene]di-Ruthenium; (52462-29-0)

2-Amino-2-methylpropan-1-ol: 2-amino-2-methyl-1-propanol; (124-68-5)

(R)-1-Phenylethanamine: (R)-α-methyl-benzenemethanamine; (3886-69-9)

Miguel Yus was born in Zaragoza (Spain) in 1947, and received his BSc (1969), MSc (1971) and PhD (1973) degrees from the University of Zaragoza. After spending two years as a postdoctoral fellow at the Max Planck Institut für Kohlenforschung in Mülheim, Ruhr he returned to the University of Oviedo where he became associate professor in 1977, being promoted to full professor in 1987. In 1988 he moved to the University of Alicante where he is currently the head of the Organic Synthesis Institute (ISO). His current research interest is focused on the preparation of reactive functionalized organometallic compounds and their use in synthetic organic chemistry and asymmetric catalysis.

David Guijarro was born in Alicante in 1967. He received his BSc (1990), MSc (1991) and PhD (1994) degrees in Chemistry from the University of Alicante. He did a postdoctoral stay at the University of Uppsala (Sweden) (1995-1997), after which he went back to the University of Alicante, where he has occupied the following academic positions: University College Professor (2000-2003), Associate Professor (2003-2011) and Full Professor (currently). He has been visiting scientist at the University of Warwick (Coventry, UK) (2009). His current research interests are focused on asymmetric synthesis, especially in stereoselective additions of organozinc reagents to carbonyl compounds and imines and in reduction processes of the same type of compounds by hydrogen transfer reactions.

Óscar Pablo Rodríguez was born in 1985 in Alicante (Spain). He obtained his BSc in Chemistry in 2008 from the University of Alicante with an honorary mention for his academic achievements. In 2008 he started his PhD studies at the same university under the supervision of Professors Miguel Yus and David Guijarro. His current research activities are focused on the asymmetric transfer hydrogenation of activated imines. He is interested in asymmetric synthesis and the development of new synthetic methodologies. He is co-author of 5 publications in international journals.

Chintan S. Sumaria was born in 1987 in Mumbai, India. He obtained his undergraduate education at the Indian Institute of Technology – Bombay (IIT Bombay), Mumbai, where he carried out undergraduate research in the areas of chemical biology and synthetic methodology. In 2011, he joined the University of Chicago Chemistry Graduate Program and has been working on expanding the scope the Inverse-Electron Demand Diels Alder reactions of azadienes in the research group of Professor Viresh H. Rawal. He has been the recipient of the KVPY scholarship funded by the Department of Science and Technology of the Government of India and the Burjor-Godrej scholarship offered by the Department of Chemistry, IIT Bombay.

Simplified Preparation of Dimethyldioxirane (DMDO)

Submitted by Douglass F. Taber[1,*], Peter W. DeMatteo and Rasha A. Hassan.

Checked by John L. Wood and John A. Enquist, Jr.

Caution! Reactions and subsequent operations involving peracids and peroxy compounds should be run behind a safety shield. For relatively fast reactions, the rate of addition of the peroxy compound should be slow enough so that it reacts rapidly and no significant unreacted excess is allowed to build up. The reaction mixture should be stirred efficiently while the peroxy compound is being added, and cooling should generally be provided since many reactions of peroxy compounds are exothermic. New or unfamiliar reactions, particularly those run at elevated temperatures, should be run first on a small scale. Reaction products should never be recovered from the final reaction mixture by distillation unless all residual active oxygen compounds (including unreacted peroxy compounds) have been destroyed. Decomposition of active oxygen compounds may be accomplished by the procedure described in Korach, M.; Nielsen, D. R.; Rideout, W. H. Org. Synth. 1962, 42, 50 (Org. Synth. 1973, Coll. Vol. 5, 414).

Dimethyldioxirane is a volatile peroxide and should be treated as such. The preparation and all reactions of the dioxirane should be carried out in a hood.

1. Procedure

A. *Dimethyldioxirane.* Distilled H_2O (20 mL), acetone (30 mL), and $NaHCO_3$ (24 g, 0.285 mol) are combined in a 1-L round-bottomed flask and chilled in an ice/water bath with magnetic stirring (oblong stirbar, 3 cm x 1 cm x 1.5 cm) for 20 min (Note 1). After 20 min, stirring is halted and Oxone (25 g, 0.0406 mol) is added in a single portion (Note 2). The flask is loosely covered and the slurry is stirred vigorously for 15 min while still submerged in the ice bath. After 15 min, the stir bar is removed from the reaction flask and rinsed with a small portion of distilled water.

Org. Synth. **2013**, *90*, 350-357
Published on the Web 9/4/2013
© 2013 Organic Syntheses, Inc.

The flask containing the reaction slurry is then attached to a rotary evaporator with the bath at room temperature. The bump bulb (250 mL) (Note 3) is chilled in a dry ice/acetone bath and a vacuum of 155 mmHg is applied via a benchtop diaphragm pump and an accompanied in-line vacuum regulator. During this process, the flask is rotated vigorously (210 rpm) to prevent bumping of the slurry into the bump trap (Note 4). After 15 min, the bath temperature is raised to 40 °C over the course of 10 min. When the bath reaches 40 °C, the distillation is halted immediately via releasing the vacuum and raising the flask from the heated water bath.

The pale yellow acetone solution of DMDO is decanted from the bump bulb directly into a graduated cylinder to measure the total volume of the solution (an average of 25 mL) and then the solution is dried over Na_2SO_4. The Na_2SO_4 is removed by filtration and rinsed with 10 mL of acetone. Titration of the obtained DMDO solution is then performed according to the procedure of Adam, et al.[2] (Note 5). Results consistently show 2.1–2.3 mmol total DMDO in the solution, giving a final concentration typically in the range of 60–65 mM (Note 6). The DMDO solution was used immediately following titration.

2. Notes

1. The checkers used distilled water obtained from Colorado State University's distilled water system. The checkers purchased acetone (purity of 99.7%) and $NaHCO_3$ (purity of 99.9%) from Fisher Scientific and used both directly without further purification.

2. The checkers employed oxone monopersulfate purchased from Aldrich, with purity unspecified, and used directly without further purification.

3. A standard rotary evaporator bump bulb like the one illustrated below was used.

4. The entire setup is illustrated in the accompanying photograph. The checkers found it necessary to constantly refresh the dry ice/acetone bath with additional dry ice and acetone throughout the distillation. Otherwise, cooling of the bump trap diminished significantly, and the desired DMDO solution was recovered in a much lower volume.

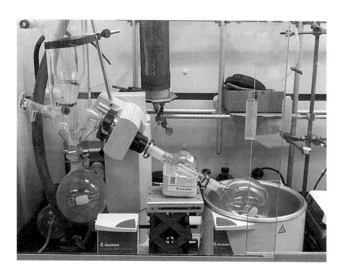

5. The following procedure is a representative example of the method employed by the checkers to assay the concentration of the obtained DMDO solution:[3c] In a 1 mL volumetric test tube, a 0.7 M solution of thioanisole in acetone-d_6 is prepared, to a total volume of 1 mL. A 0.6 mL portion of this solution is transferred to a 20 mL scintillation vial and chilled to ca. 10 °C in a 1,4-dioxane bath chilled with a small quantity of dry ice. Upon reaching 10 °C, 3.0 mL of the obtained DMDO solution is added to the thioanisole solution. The resulting solution is stirred for 10 min and then a portion of the solution is added directly to an NMR tube. Analysis of the ^{1}H NMR via signal integration of the sulfoxide phenyl protons at δ 7.6 – 7.9 ppm against the thioanisole phenyl protons at δ 7.1 – 7.3 ppm allows for determination of the ratio of oxidized product to excess thioanisole. From this data, concentration of the larger DMDO solution could be calculated.

The submitter's original procedure for assaying the concentration of the DMDO sample is as follows: The entire liquid collected in the bump bulb is transferred to a round-bottomed flask with CH_2Cl_2. A weighed excess amount of *trans*-stilbene is added. After 24 h, the solvent is removed under

352

high vacuum, and the ^{1}H NMR spectrum of the residue is recorded. All ten protons of *trans*-stilbene and the arene protons of the epoxide appear around δ 7.0, while the epoxide methine protons of the product appear at δ 3.95. Knowing how much *trans*-stilbene had been added, it is possible to calculate the amount of DMDO that had been generated.

6. In order to determine the efficacy of the prepared DMDO solution, the checkers performed a reaction wherein a sample of *trans*-stilbene was oxidized as follows: To a glass vial containing *trans*-stilbene (97.0 mg, 0.538 mmol, 1.0 equiv) is added 9.3 mL of a previously prepared and assayed solution of DMDO (65.6 mM in acetone, 0.610 mmol, 1.1 equiv). The resulting solution is stirred for 16 h, and then solvent is removed in *vacuo*. The resulting residue is dissolved in CH_2Cl_2 and dried over Na_2SO_4. After filtration and removal of solvent *in vacuo*, *trans*-stilbene oxide (104 mg, 98 %) is isolated in essentially pure form.

Trans-stilbene oxide has the following spectral properties: ^{1}H NMR (400 MHz, CDCl₃) δ: 3.90 (s, 2 H), 7.42–7.37 (m, 10 H); ^{13}C NMR (100 MHz, CDCl₃) δ: 62.96, 125.6, 128.4, 128.7, 137.3; IR (solid) cm^{-1}: 3036, 2989, 1603, 1493, 1461, 1285, 1072, 846, 795. Exact mass, Calculated for $C_{14}H_{13}O$ (M+H): 197.0966. Found: 197.0961. Anal. Calcd. for $C_{14}H_{12}O$: C, 85.68. H, 6.16. Found: C, 85.47. H, 6.24.

Handling and Disposal of Hazardous Chemicals

individuals using these procedures and hereby disclaim any liability for any injuries or damages claimed to have resulted from or related in any way to the procedures herein.

3. Discussion

Dimethyldioxirane (DMDO), easily prepared by the reaction of Oxone with acetone, has been used in organic synthesis for many years.[3] In particular, the "acetone free" form, prepared by partitioning the acetone solution of DMDO between CH_2Cl_2 and aqueous phosphate buffer, is prized for its ability to generate even very sensitive epoxides under neutral conditions.[4] This is clearly illustrated by the conversion of **1** to **3** by way of epoxide **2**, as reported by Rainer (Scheme 1).[5a]

Scheme 1. Preparation of a reactive epoxide with DMDO

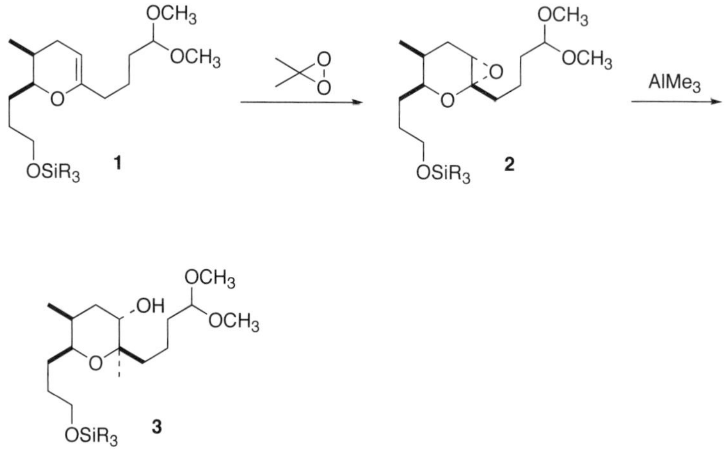

Despite this utility, DMDO has not commonly been used for laboratory scale epoxidations, the commercial reagent MCPBA being widely preferred. The reluctance to employ DMDO may likely be due to the complicated cryogenic purification prescribed[3a] for the preparation of the reagent as a solution in acetone.

We have found that practical quantities of DMDO in acetone can be prepared by simple rotary evaporation of the Oxone/$NaHCO_3$/acetone slurry with collection of the distillate in the bump bulb of the rotary evaporator.

354

Using a 1-L round-bottomed flask, we could routinely prepare 2.1-2.3 mmol DMDO. Titration was accomplished by the addition of a measured amount of the DMDO solution to a known quantity of thioanisole (Note 5), with ^{1}H NMR analysis of the resulting sulfide/sulfoxide ratio affording data on the concentration of the prepared solution. The DMDO solution prepared as described was used in the epoxidation of 100 mg of *trans*-stilbene in near quantitative yield.[6]

1. Department of Chemistry and Biochemistry, University of Delaware, Newark, DE 19716 taberdf@udel.edu

2. Adam, W.; Chan, Y. Y.; Cremer, D.; Gauss, J.; Scheutzow, D.; Scheutzow, D.; Schindler, M. *J. Org. Chem.* **1987**, *52*, 2800–2803.

3. For current methods for the preparation of solutions of DMDO, see: a) Murray, R. W.; Jeyaraman, R. *J. Org. Chem.* **1985**, *50*, 2847–2853. b) Adam. W.; Bialas, J.; Hadjiarapoglou, L. *Chem. Ber.* **1991**, *124*, 2377–2377. c) Murray, R. W.; Singh, M. *Org. Syn.* **1997**, *74*, 91; *Coll. Vol. 9*, **1998**, 288. c) Mikula, H.; Svatunek, D.; Lumpi, D.; Glöcklhofer, F.; Hametner, C.; Fröhlich, J. *Org. Process Res. Dev.* **2013**, *17*, 313–316.

4. For the preparation of "acetone free" DMDO, see Gilbert, M.; Ferrer, M.; Sánchez-Baez, F.; Messeguer, A. *Tetrahedron* **1997**, *53*, 8643-8650.

5. For applications of "acetone free" DMDO, see: a) Zhang, Y.; Rohanna, J.; Zhou, J.; Iyer, K.; Rainier, J. D. *J. Am. Chem. Soc.* **2011**, *133*, 3208–3216. b) Alberch, L.; Cheng, G.; Seo, S.-K.; LI, X.; Boulineau, F., Wei, A. *J. Org. Chem.* **2011**, *76*, 2532--2547. c) Ghosh, P.; Lotesta, S. D.; Williams, L. J. *J. Am. Chem. Soc.* **2007**, *129*, 2438–2439. d) Crimmins, M. T.; McDougall, P. J.; Emmitte, K. A. *Org. Lett.* **2005**, *7*, 4033–4036. e) Bach, R. D.; Dmitrenko, O.; Adam, W.; Schambony, S. *J. Am. Chem. Soc.* **2003**, *125*, 924–934. f) Ndakala, A. J.; Howell, A. R. *J. Org. Chem.* **1998**, *63*, 6098–6099.

6. For a recent preparation and characterization of *trans*-stilbene oxide, see Mercier, E. A.; Smith, C. D., Parvez, M.; Back, T. G. *J. Org. Chem.* **2012**, *77*, 3508–3517.

Appendix
Chemical Abstracts Nomenclature; (Registry Number)

DMDO; dimethyldioxirane (74087-85-7)
Oxone; (70693-62-8)

Douglass F. Taber was born in 1948 in Berkeley, California. He earned a B.S. in Chemistry with Honors from Stanford University in 1970, and a Ph.D. in Organic Chemistry from Columbia University in 1974 (G. Stork). After a postdoctoral year at the University of Wisconsin (B.M. Trost), Taber accepted a faculty position at Vanderbilt University. He moved to the University of Delaware in 1982, where he is currently Professor of Chemistry. Taber is the author of more than 200 research papers on organic synthesis and organometallic chemistry. He is also the author of the weekly Organic Highlights published at http://www.organic-chemistry.org/

Rasha Hassan was born on June 12, 1978 in Egypt. On 1995, she joined the Faculty of Pharmacy, Cairo University for college study. In May 2000, she graduated with an excellent honors grade, ranked 13th out of 1400 student in her class. Based on these accomplishments, Rasha joined the faculty of the Organic Chemistry department as a lecturer assistant, teaching laboratory Organic Chemistry to the pharmacy students. Simultaneously, she carried out research in organic synthesis to earn her Master's degree, awarded in May 2005. Since May 2011 she has continued her research toward a Ph.D. as a visiting scholar in the laboratory of Professor Douglass F. Taber at the University of Delaware.

356

Peter W. DeMatteo received a dual B.S. in Chemical Engineering and Chemistry from Lehigh University in 2002. He continued his chemistry studies under Ned Heindel and received his M.S. In Chemistry in 2004. He then transferred into the natural products lab of Douglass F. Taber, receiving his Ph.D. in 2011. He is currently employed as an IRTA postdoc at the National Institute for Drug Abuse where he synthesizes opioid analogs under the supervision of Dr. Kenner C. Rice.

John A. Enquist, Jr. received a B.S. in Chemistry from the University of California at San Diego in 2004. Later, he received a Ph.D. in Chemistry from the California Institute of Technology under the direction of Professor Brian M. Stoltz. He is currently a post-doctoral fellow in the laboratories of Professor John L. Wood, at the Colorado State University.

Practical and Efficient Synthesis of *N*-Formylbenzotriazole

Submitted by Mhairi Matheson, Adele E. Pasqua, Alan L. Sewell and Rodolfo Marquez.*[1]
Checked by Diptarka Hait, Jonathan Truong, and Mohammad Movassaghi.

1. Procedure

A 500-mL, 3-necked, round-bottomed flask (Note 1) equipped with a 5 cm Teflon-coated cylindrical magnetic stir bar, an internal thermometer, a rubber septum, and a Vigreux condenser capped with a rubber septum with an argon inlet, is charged with acetic anhydride (26.5 mL, 28.6 g, 280 mmol, 1.0 equiv) (Note 2). The contents are cooled to –10 °C by means of a Dewar filled with isopropanol and an immersion cooler (Neslab CC 100) (Note 3), then 100% formic acid (21.1 mL, 25.8 g, 560 mmol, 2.0 equiv) (Note 4) is added in portions to the flask over 3 min via a measuring cylinder [Caution: exotherm from –10 °C to 1 °C]. The resulting colorless reaction mixture is then heated in an oil bath on a hot plate to 43 °C (internal temperature) over 25 min. The reaction mixture is stirred vigorously for 3 h (42–44 °C, internal temperature) and monitored by ^{1}H NMR analysis (Note 5).

^{1}H NMR analysis of crude reaction aliquots is employed to determine the quantity of acetic formic anhydride in the reaction mixture (18.1 g, 206 mmol). Formic anhydride (2.4 g, 32 mmol) was also present (Notes 6 and 7). The crude anhydride mixture is then cooled to 5 °C over 30 min using an ice bath. While maintaining a positive pressure of argon, the Vigreux condenser is removed and replaced with a rubber septum with an argon inlet.

A 500-mL, two-necked round-bottomed flask, equipped with a 5 cm Teflon-coated magnetic stir bar, an internal thermometer, and a rubber septum with an argon inlet, is charged with benzotriazole (23.8 g, 200 mmol, 0.84 equiv relative to formylating agents (Notes 6 and 8). Anhydrous tetrahydrofuran (100 mL) is added to the benzotriazole via a measuring

Org. Synth. **2013** *90*, 358-366
Published on the Web 10/1/2013
© 2013 Organic Syntheses, Inc.

cylinder and the resulting light yellow solution is cooled to –15 °C (internal temperature) over 20 min by means of an ice/salt water bath. The crude mixed anhydride is then added to the benzotriazole solution via cannula over 12 min, ensuring that the internal temperature does not exceed –5 °C (Note 9). Upon completion of addition, the reaction mixture is cooled to –15 °C (internal temperature) by means of a Dewar filled with isopropanol and an immersion cooler (Neslab CC 100) (Note 3) and vigorously stirred until completion by TLC analysis (2 h) (Note 10), at which point the reaction mixture is thick and cream colored.

The reaction mixture is transferred to a single-necked, 1-L round-bottomed flask and concentrated to dryness by rotatory evaporation (40 °C bath temperature, 20 mmHg). Chloroform (50 mL) (Notes 11 and 12) is added to the flask and the solution is concentrated again (40 °C bath temperature, 20 mmHg). The dilution and concentration procedure is repeated 3 times, yielding a white solid. The solid is dried at room temperature, under 0.75 mmHg for 12 h. The white solid is ground with a glass rod and dried at room temperature, under 0.75 mmHg for a further 48 h to afford 29.3 g (99.7%) of N-formylbenzotriazole (Notes 13, 14, and 15).

2. Notes

1. All glass equipment was either flame dried (checkers) or oven dried (submitters) and then maintained under a positive pressure of argon during the course of the reaction. Thermometers and rubber septa were used without any form of drying.

2. Both the submitters and checkers used acetic anhydride (99%) and formic acid (99%) obtained from Acros Organics (used as received).

3. The submitters report the use of an ice/salt bath is sufficient for temperature control at this stage.

4. The submitters recommend the use of a syringe for transfer of the formic acid. An excess of formic acid is necessary to achieve maximum conversion of acetic anhydride into acetic formic anhydride.

5. The reaction was monitored by ^{1}H-NMR analysis of crude reaction aliquots in CDCl$_3$ until the amount of acetic anhydride was minimal. The checkers observed that the reaction was essentially complete in 1.5 h (reached equilibrium with no significant further change in composition).

6. ^{1}H NMR integrations were measured for the formyl H's and equated to the total amount of formyl group supplied. The spectra were obtained within 15 minutes of removal from the reaction mixture.

δ	Compound	Integral	Amount
9.1	Acetic formic anhydride	1.00	(1.00/2.72)*560 mmol = 206 mmol
8.75	Formic anhydride	0.31	0.5*(0.31/2.72)*560 mmol = 32 mmol (64 mmol formyl H)
8.0	Formic acid	1.41	(1.41/2.72)*560 mmol = 290 mmol
	Total	**2.72**	**528 mmol (560 mmol formyl H)**

7. Acetic formic anhydride was maintained under a positive pressure of argon to prevent hydrolysis. See *Org. Synth*, **1970**, *50*, 1 for an alternative synthetic route to acetic formic anhydride and for purification details.

8. Both the submitters and checkers obtained benzotriazole (99%, B11400-100G) from Sigma-Aldrich Inc. and used it as received. The submitters obtained tetrahydrofuran from Sigma Aldrich (CHROMASOLV® Plus, for HPLC, ≥99.9%, inhibitor-free, 34865) and purified though a Pure Solv 400-5MD solvent purification system (Innovative Technology, Inc.). The checkers obtained tetrahydrofuran from J.T. Baker (Cycletaner^{TM}) and purified through a Phoenix SDS Large Capacity solvent system (JC Myer Solvent Systems).

9. Slow addition led to a large amount of the acetyl product.

10. The reaction was monitored by TLC analysis on aluminium sheets pre-coated with silica (Merck Silica Gel 60 F$_{254}$) using 30% ethyl acetate in (40-60) petroleum ether. Spot A = benzotriazole (R$_f$ = 0.2), spot B = *N*-formylbenzotriazole (R$_f$ = 0.5). Middle lane was taken during the course of the reaction.

11. The submitters purchased chloroform from Fisher Chemical (C/4960/17) and the checkers obtained chloroform from Sigma Aldrich (CHROMASOLV® Plus, for HPLC, ≥99.9%, contains 0.5-1% ethanol as stabilizer, 650471)- used as obtained.

12. Removal of the volatile reaction components is best achieved by azeotroping the crude reaction product mixture with chloroform before the product is dried on a vacuum line.

13. *N*-Formylbenzotriazole was obtained in a 300:1 ratio compared to the *N*-acetylbenzotriazole side product as determined by ^{1}H NMR analysis.

1.000/H   0.003/H
Relative Integration

14. The product exhibits the following physical and spectroscopic properties: mp 93–94 °C. IR: v_{max}/cm^{-1}; 3105, 1727, 1605, 1594. ^{1}H NMR (400 MHz, CDCl$_3$) δ: 7.57 (1 H, ddd, J = 8.3, 7.2, 1.0 Hz), 7.70 (1 H, ddd, J = 8.2, 7.2, 1.0 Hz), 8.15 (1 H, dt, J = 8.3, 0.9 Hz), 8.24 (1 H, d, J = 8.2 Hz), 9.86 (1H, s). ^{13}C NMR (100 MHz, CDCl$_3$) δ: 113.6, 120.4, 127.0, 129.9, 130.8, 146.6, 159.8. HRMS [M+H]$^+$calculated for C$_7$H$_5$NO: 148.0506, found 148.0505. This characterization matches the data reported in the literature.[2] Anal. Calcd. for C$_7$H$_5$N$_3$O (147.13): C, 57.14; H, 3.43; N, 28.56. Found: C, 57.06; H, 3.48; N, 28.75.

15. The submitters report that *N*-formylbenzotriazole stored in the freezer showed no decomposition by ^{1}H NMR analysis over a 6-month period. The submitters reported the formation of 59.5 g (98.5%) of product when the reaction is performed at double the reported scale.

Handling and Disposal of Hazardous Chemicals

The procedures in this article are intended for use only by persons with prior training in experimental organic chemistry. All hazardous materials should be handled using the standard procedures for work with chemicals described in references such as "Prudent Practices in the Laboratory" (The National Academies Press, Washington, D.C., 2011 www.nap.edu). All chemical waste should be disposed of in accordance with local regulations. For general guidelines for the management of chemical waste, see Chapter 8 of Prudent Practices.

These procedures must be conducted at one's own risk. *Organic Syntheses, Inc.*, its Editors, and its Board of Directors do not warrant or guarantee the safety of individuals using these procedures and hereby disclaim any liability for any injuries or damages claimed to have resulted from or related in any way to the procedures herein.

Discussion

The importance of formylation to organic chemists is reflected in the myriad of approaches and reagents designed to achieve it.[2] However, this in itself is indicative of the issues often experienced when working with unstable and impure formylating agents.

Formic halides and formic anhydrides are some of the most common formylating agents, however, they tend to suffer from stability problems and degrade easily upon storage.[3]

Formyl fluoride is the most widely used formylated halide, and has found wide-spread use in the formylation of alcohols, phenols and thiols in the presence of base.[3] As an electrophilic aromatic formylating agent, formyl fluoride requires activation *in situ* at low temperatures by catalytic amounts of Lewis acid.[3] Unfortunately, formyl fluoride decomposes at room temperature to carbon monoxide and hydrofluoric acid.[4]

Formic anhydride, on the other hand, has been used for the *O*-formylation of *p*-nitrophenol. Formylation of aromatic rings and other more

362

demanding substrates has proven highly problematic due to its instability above – 40 °C and its high chemical incompatibility.[3] Acetic formic anhydride, on the other hand, is a more stable formylating agent, but is still hydrolysed relatively easily.[5]

Coupling agents (i.e. DCC and EDCi) have also been used in conjunction with formic acid to achieve *N*- and *O*- formylation; however, the yields can be highly variable and the removal of the coupling agents' side products is often labour intensive.[6]

Cyanomethyl formate is a very useful, but difficult to prepare, formylating agent that has been used to formylate alcohols and nitroanilines in the presence of imidazole. Cyanomethyl formate can be synthesised from potassium formate and requires extensive purification to achieve the desired level of purity.[3] Isopropenyl formate is also a very fast and efficient formylating agent which unfortunately requires a lengthy synthesis involving the use of complex ruthenium catalysts.[4]

N-Formylbenzotriazole was initially developed by Katrizky as a stable and convenient alternative to achieve *N*- and *O*- formylation quickly and efficiently.[2] *N*-formylbenzotriazole is often considered to be the reagent of choice as it offers mild and selective formylation of alcohols, amines, and amides.[8]

Katrizky's synthesis begins with benzotriazole which is coupled to formic acid using *N,N'*-dicyclohexylcarbodiimide. The coupling itself proceeds well; however, the separation of the desired *N*-formylbenzotriazole and the urea side product is non-trivial and requires repeated inefficient recrystallisation which severely reduces the yield of the reaction. Furthermore, even after repeated recrystallisation and trituration, the *N*-formylbenzotriazole obtained is often contaminated with urea byproducts making the yields highly variable and irreproducible. Unfortunately, the use of alternative coupling agents, such as, *N,N'*-diisopropylcarbodiimide (DIC) and 1-ethyl-3-(3-dimethylaminopropyl)carbodiimide (EDCi) have been unsuccessful at yielding pure *N*-formylbenzotriazole in significant amounts.

In conclusion, a fast and efficient procedure for the synthesis of *N*-formylbenzotriazole has been developed. This procedure takes advantage of the inherent reactivity of acetic formic anhydride to formylate benzotriazole without the need of coupling reagents. Further to this, we exploit the volatility of the reagents and side products to allow the generation of highly pure *N*-formylbenzotriazole in excellent yield without the need for purification. As a whole, this procedure is a great improvement compared to other methods available currently for the synthesis of *N*-formylbenzotriazole in terms of cost, yield and overall efficiency.[9]

1. Ian Sword Reader of Organic Chemistry and EPSRC Leadership Fellow. WestCHEM, School of Chemistry, University of Glasgow, Joseph Black Building, University Avenue, G12 8QQ. rudi.marquez@glasgow.ac.uk. We thank the Lord Kelvin-Smith programme, Dr. Ian Sword, and the University of Glasgow for postgraduate support (M.M., A.E.P., and A.L.S.). We also thank Dr. Ian Sword and the EPSRC for funding.
2. Katrizky, A. R.; Chang, H. X.; Yang, B. *Synthesis* **1995**, 503–505.
3. Olag, G. A.; Ohannesian, L.; Arvanaghi, M. *Chem. Rev.* **1987**, *87*, 671–686.
4. Fischer, G.; Buchanan, A. S. *T. Faraday Soc.* **1964**, 60, 378–385.
5. Olah, G. A.; Kui, S. J. *J. Am. Chem. Soc.* **1960**, 82, 2380–2382.
6. van Melick, J. E. W.; Wolters, E. T. M. *Synth. Comm.* **1972**, *2*, 83–86.
7. Neveux, M.; Bruneau, C.; Dixneuf, P. H. *J. Chem. Soc. Perkin Trans. 1* **1991**, 1197–1199.
8. a) Mathieson, J. E.; Crawford, J. J.; Schmidtmann, M.; Marquez, R. *Org. Biomol. Chem.* **2009**, *7*, 2170–2175. b) Levesque, F.; Belanger, G. *Org. Lett.* **2008**, *10*, 4939–4942. c) Larouche-Gauthier, R.; Belanger, G. *Org. Lett.* **2008**, *10*, 4501–4504. d) Hayashi, Y.; Shoji, M.; Ishikawa, H.; Yamaguchi, J.; Tamura, T.; Imai, H.; Nishigaya, Y.; Takabe, K.; Kakeya, H.; Osada, H. *Angew. Chem, Int. Ed.* **2008**, *47*, 6657–6660. e) Sewell, A. L.; Villa, M. V. J.; Matheson, M.; Whittingham, W. G.; Marquez, R. *Org. Lett.* **2011**, *13*, 800–803.
9. Pasqua, A. E.; Matheson, M.; Sewell, A. L.; Marquez, R. *Org. Proc. Res. & Dev.* **2011**, *15*, 467–470.

Appendix
Chemical Abstracts Nomenclature; (Registry Number)

Acetic anhydride; (108-24-7)
Formic acid; (64-18-60
Benzotriazole; (95-14-7)
N-Formylbenzotriazole; (72773-04-7)

Dr. Rudi Marquez obtained his DPhil in 1999 at UCLA with Professor Michael E. Jung. Rudi then worked with Professor Ian Paterson at the University of Cambridge before joining the group of Professor Sir Jack Baldwin at the University of Oxford in 2000. In 2003 he began his independent career at the University of Dundee. In 2006, Rudi became the Ian Sword Lecturer at the University of Glasgow. Dr. Marquez's research interests reside at the interface between chemistry and biology, particularly applied to parasitology, cancer, tissue regeneration and crop protection.

Mhairi Matheson received her M.Sci. in Chemistry with Medicinal Chemistry from the University of Glasgow in 2010. As an undergraduate she completed an industrial placement at Syngenta (Jealott's Hill). Mhairi has been the recipient of numerous awards including a Nuffield Foundation summer grant and a Lord Kelvin-Smith Scholarship. Mhairi is currently undertaking doctoral studies under the supervision of Dr. Rudi Marquez in the rational design and synthesis of angiogenesis promoters through the use of small molecules to mimic protein-protein interactions.

Adele Elisa Pasqua was born in Cosenza, Italy and obtained her M.Sc. at University of Calabria in Pharmaceutical Chemistry and Technology. In 2007 Elisa obtained a scholarship at the Bracco Imaging Spa (Trieste, Italy) where she remained for two years before moving to the University of Trieste for a summer scholarship in Prof. Paoletti's group. In 2009, Elisa began her Ph.D. studies under the supervision of Dr. Marquez at the University of Glasgow. Elisa's research is in the area of synthetic methodology and natural product synthesis is supported by the Dr. Ian Sword Scholarship.

Alan L. Sewell is a native of Glasgow, where he obtained a M. Sci. in Chemistry with Drug Discovery at the University of Strathclyde. During his degree he worked with Novartis for 12 months as a medicinal chemist based in the gastrointestinal disease unit at their Horsham research centre. As part of his M.Sci., Alan carried out research on a palladium-catalysed cyclisation to give fluorinated carbocycles. In 2010, Alan began his Ph.D. studies with Dr Marquez at the University of Glasgow working on novel enamide containing small molecules with potential broad-spectrum antiparasitic activity.

Diptarka Hait is currently pursing his undergraduate studies at MIT, and expects to receive his SB degree in Chemistry and Physics in 2016. He joined Professor Movassaghi's lab in 2012, where he is working on the development of new methods for organic synthesis and their application towards the total syntheses of alkaloid natural products.

Jonathan V. Truong pursued his undergraduate studies at Brown University, where he received his Sc.B. degree in chemistry in 2012. As an undergraduate he worked in the laboratory of Professor Jason Sello. In 2012, he joined Professor Mohammad Movassaghi's research group at MIT for his graduate studies where his research has focused on the development of new methodologies for organic synthesis and their application towards the total syntheses of alkaloid natural products.

AUTHOR INDEX FOR VOLUME 90

CUMULATIVE SUBJECT INDEX FOR VOLUME 90

This index comprises subject matter for Volume **90**. For subjects in previous volumes, see either the indices in Collective Volumes I through XI or the single volume entitled *Organic Syntheses, Collective Volumes I-VIII, Cumulative Indices,* edited by J. P. Freeman.

The index lists the names of compounds in two forms. The first is the name used commonly in procedures. The second is the systematic name according to Chemical Abstracts nomenclature, accompanied by its registry number in parentheses. Also included are general terms for classes of compounds and types of reactions.

Most chemicals used in the procedure will appear in the index as written in the text. There generally will be entries for all starting materials, reagents, important by-products, and products, which are indicated by the use of italics.

EDC·HCl: 1,3-Propanediamine, N^3-(ethylcarbonimidoyl)-N^1,N^1-dimethyl-, hydrochloride
    (1:1); (25952-53-8) **90**, 306
ESTERS AND LACTONES **90**, 121, 200
Ethane, 1,2-dibromo-; (106-93-4) **90**, 327
Ethanethiol; (75-08-1) **90**, 10
Ethyl 4-bromobutanoate; (2969-81-5) **90**, 200
Ethyl 4-(4-(4-methylphenylsulfonamido)phenyl)butanoate; (1138239-43-6) **90**, 200
(Ethylthio)methanol; (15909-30-5) **90**, 10
1-Ethynylcyclohexanol; (78-27-3) **90**, 87

Ferrocene; (102-54-5) **90**, 316
α-Fluorobis(phenylsulfonyl)methane (FBSM): Benzene, 1,1'-
    [(fluoromethylene)bis(sulfonyl)]bis-; (910650-82-7) **90**, 130
Fluoromethyl phenyl sulfide: Benzene, [(fluoromethyl)thio]-; (60839-94-3) **90**, 130
Fluoromethyl phenyl sulfone: Benzene, [(fluoromethyl)sulfonyl]-; (20808-12-2) **90**, 130
4-Fluorophenyl boronic acid: Boronic acid, B-(4-fluorophenyl)-, (1765-93-1) **90**, 261
Formic acid; (64-18-60) **90**, 356
N-Formylbenzotriazole **90**, 356

Geranial; (141-27-5) **90**, 215
Geraniol; (106-24-1) **90**, 271

HALOGENATED COMPOUNDS **90**, 1, 10, 130, 229, 251
HETEROCYCLES **90**, 74, 96, 164, 251, 287, 327, 350, 356
1,1,1,3,3,3-Hexafluoroisopropanol: hexafluoroisopropanol; (920-66-1) **90**, 190
Hexamethyldisilazane: Silanamine, 1,1,1-trimethyl-*N*-(trimethylsilyl)-; (999-97-3) **90**,
    130
Hydrazine hydrate; (10217-52-4) **90**, 287
E-1-(1-Hydroxycyclohexyl)ethanone oxime; (62114-93-6) **90**, 87
Hydroxylamine; (7803-49-8) **90**, 87
Hydroxylamine hydrochloride: (5470-11-1) **90**, 62
1-[Hydroxy(tosyloxy)iodo]-3-trifluoromethylbenzene; (440365-98-0) **90**, 1

IMINES **90**, 121, 338
Indoline; (486-15-1) **90**, 251
INSERTION **90**, 41, 74, 164
Iodine; (7553-56-2) **90**, 153
3-Iodobenzotrifluoride; Benzene, 1-iodo-3-(trifluoromethyl)-; (401-81-0) **90**, 1
2-Iodobenzoyl chloride; (2042672) **90**, 164
2-Iodo-N-phenylbenzamide; (15310-01-7) **90**, 164
Iron(II) acetate: Acetic acid, iron(2+) salt (2:1); (3094-87-9) **90**, 62

Lithium aluminum hydride (LiAlH₄); (16853-85-3) **90**, 240
Lithium hexamethyldisilazide; (4039-32-1) **90**, 271

Magnesium; (7439-95-4) **90**, 153
Magnesium, bromo(phenylmethyl)-; (1589-82-8) **90**, 327
METAL-CATALYZED REACTIONS **90**, 41, 74, 96, 182, 200, 287, 327, 338
METALLATION **90**, 316
Methyl 4-aminobenzoate: Benzoic acid, 4-amino-, methyl ester (619-45-4) **90**, 74
4-Methylbenzenesulfonothioic acid, S-[[[(1,1-dimethylethyl)dimethylsilyl]-oxy]methyl]
ester; (1277170-42-9) **90**, 10
4-Methylbenzenesulfonothioic acid, sodium salt (1:1); (3753-27-3) **90**, 10
Methyl 1-(1-(benzyloxycarbonyl)piperidin-4-yl)-2-oxoindoline-5-carboxylate: 1H-Indole-
5-carboxylic acid, 2,3-dihydro-2-oxo-1-[1-[(phenylmethoxy)carbonyl]-4-
piperidinyl]-, methyl ester (1037834-34-6) **90**, 74
Methyl 4-hydroxyphenylacetate; (14199-15-6) **90**, 190
N-Methylimidazole (NMI); (616-47-7) **90**, 240
5-Methyl-1-phenyl-6-hepten-3-ol; (221366-16-1) **90**, 105
3-Methyl-1-phenyl-4-penten-1-ol; (205883-12-1) **90**, 105
(*R*S)-2-Methylpropane-2-sulfinamide: [*S*(*R*)]-2-methyl-2-propanesulfinamide; (196929-
78-9) **90**, 338
2-Methyltetrahydrofuran; (96-47-9) **90**, 174, 287

1-Naphthaleneboronic acid; (13922-41-3) **90**, 153
1-Naphthol; (90-15-3) **90**, 153
Nickel(II) acetylacetonate: Bis(2,4-pentanedionate), nickel(II); (3264-82-2) **90**, 105
Nickel(II) iodide hydrate; (7790-34-3) **90**, 200
p-Nitrobenzaldehyde: Benzaldehyde, 4-nitro-; (555-16-8) **90**, 52
Nitromethane: Methane, nitro-; (75-52-5) **90**, 52
(1S)-1-(4-Nitrophenyl)-2-nitroethane-1-ol; (454217-09-5 **90**, 52

Oxalic acid dihydrate; (6153-56-6) **90**, 251
OXIDATION **90**, 1, 10, 130, 153, 190, 215, 240, 287, 350
OXIMES **90**, 62, 87
*3-Oxocyclohex-1-enecarbonitrile; (25017-78-1) **90**,* 229
Oxone; (70693-62-8) **90**, 130, 153, 350
Oxyma Pure: Acetic acid, 2-cyano-2-(hydroxyimino)-, ethyl ester (3849-21-6) **90**, 306

Palladium acetate: Acetic acid, palladium(2+) salt (2:1) (3375-31-3) **90**, 74
Paraformaldehyde; (30525-89-4) **90**, 10
Pentamethylcyclopentadienylrhodium(III) chloride dimer; (12354-85-7) **90**, 41
Phenanthridin-6(5H)-one; (2413-02-7) **90**, 164
N¹-Phenylacetamidine 4-bromobenzoate; (1207066-26-9) **90**, 174
Phenylacetylene: Ethynyl benzene; Acetylene benzene; (536-74-3) **90**, 96
*4-Phenyl-1,2-butadiene: 2,3-Butadien-1-ylbenzene; (40339-20-6) **90**,* 327
(R)-1-Phenylethanamine: (R)-α-methyl-benzenemethanamine; (3886-69-9) **90**, 338
*(R,RS)-N-(1-Phenylethyl)-2-methylpropane-2-sulfinamide; (247236-71-1) **90**,* 338
*(RS)-N-[1-Phenylethylidene]-2-methylpropane-2-sulfinamide; (220263-59-2) **90**,* 338
Z-*L*-Phenylglycine-OH: Benzeneacetic acid, α-[[(phenylmethoxy)carbonyl]amino]-, (α*S*)-
; (53990-33-3) **90**, 306

Z-L-Phenylglycine-Valine-OMe: L-Valine, (2S)-2-phenyl-N-
    [(phenylmethoxy)carbonyl]glycyl-, methyl ester; (159487-14-6) **90**, 306
Potassium bis(trimethylsilyl)amide: Silanamine, 1,1,1-trimethyl-*N*-(trimethylsilyl)-,
    potassium salt (1:1); (40949-94-8) **90**, 130
Potassium *tert*-butoxide: 2-Propanol, 2-methyl-, potassium salt (1:1); (865-47-4) **90**, 316
Potassium carbonate; (584-08-7) **90**, 287
Potassium fluoride (KF); (7789-23-3) **90**, 130
Potassium hydride (KH); (7693-26-7) **90**, 130
Potassium hydrogen fluoride; (7789-29-9) **90**, 153
Potassium 1-Naphthyltrifluoroborate: Borate(1-), trifluoro-1-naphthalenyl-, potassium
    (1:1), (*T*-4)-; (166328-07-0) **90**, 153
1-Propyne, 3-bromo-; (106-96-7) **90**, 327
Pyridine; (110-86-1) **90**, 200
Pyridine, 2,4-dichloro; (26452-80-2) **90**, 287
4-Pyridinecarboxaldehyde; (872-85-5) **90**, 52

REARRANGEMENTS **90**, 271, 327
REDUCTION **90**, 52, 62, 74, 240, 338
Rhodium, tetrakis[μ-(acetato-κO:κO')]di-, (Rh-Rh); (15956-28-2) **90**, 327
[RuCl₂(*p*-cymene)]₂: Di-μ-chlorodichlorobis[(1,2,3,4,5,6-μ)-1-methyl-4-(1-
    methylethyl)benzene]di-Ruthenium; (52462-29-0) **90**, 338

Silver Hexafluoroantimonate(V); (26042-64-8) **90**, 41
Sodium borohydride: Borate(1-), tetrahydro-, sodium (1:1); (16940-66-2) **90**, 52
Sodium iodide; (7681-82-5) **90**, 200
Sodium *p*-toluenesulfinate hydrate (TolSO₂Na·H₂O); (207801-20-5) **90**, 10
Sodium sulfite; (7757-83-7) **90**, 287
Sodium triacetoxyborohydride: Borate(1-), tris(acetato-κO)hydro-, sodium (56553-60-7)
    90, 74
Styrene; (100-42-5) **90**, 41
SUBSTITUTION **90**, 25, 130, 145, 174, 182, 229, 251
Sulfur; (7704-34-9) **90**, 10
Sulfuryl chloride; (7791-21-5) **90**, 10

Tetrabutylammonium (4-Fluorophenyl)trifluoroborate): 1-Butanaminium, N,N,N-
    tributyl-, (T-4)-trifluoro(4-fluorophenyl)borate(1-) (1:1) , (1291068-40-0) **90**, 261
Tetrabutylammonium bifluoride: 1-Butanaminium, *N,N,N*-tributyl-, (hydrogen difluoride)
    (1:1) (23868-34-0) **90**, 261
Tetrafluoroboric acid diethyl ether complex: Borate(1-), tetrafluoro-, hydrogen (1:1)
    Ethane, 1,1'-oxybis-; (67969-82-8) **90**, 316
2,2,6,6-Tetramethylpiperidine-*N*-oxyl (TEMPO); (2564-83-2) **90**, 240
Thionyl chloride; (7719-09-7) **90**, 251
Titanium(IV) ethoxide: tetraethyl titanate; (3087-36-3) **90**, 338
1,2,4-Triazolo[4,3-a]pyridine, 7-chloro-3-(2-chlorophenyl)-; (1019918-88-7) **90**, 287
Trichloroacetyl isocyanate; (3019-71-4) **90**, 271
2,2,2-Trichloroethanol; (115-20-8) **90**, 174
2,2,2-Trichloroethyl acetimidate hydrochloride; (16507-47-4) **90**, 174

Triethylamine; (121-44-8) **90**, 121, 271
Triethylborane (1.0 M solution in hexanes): Borane, triethyl; (97-94-9) **90**, 105
Triethyl borate; (150-46-9) **90**, 153
Triethyl Phosphite; (122-52-1) **90**, 145
Trifluoroacetic acid; (76-05-1) **90**, 190
2,2,2-Trifluoroethanol; Ethanol, 2,2,2-trifluoro-; (75-89-8) **90**, 1
α,α,α-Trifluorotoluene: Benzene, (trifluoromethyl)-; (98-08-8) **90**, 130
p-Toluenesulfonic acid monohydrate; Benzenesulfonic acid, 4-methyl-; (104-15-4) **90**, 1
p-Toluenesulfonyl chloride: 4-Methylbenzene-1-sulfonyl chloride; (98-59-9) **90**, 200
Triphenylphosphine; (603-35-0) **90**, 271

H-Valine-OMe·HCl: L-Valine, methyl ester, hydrochloride (1:1); (6306-52-1) **90**, 306

Zinc; (7440-66-6) **90**, 200
Zinc Iodide; (10139-47-6) **90**, 145